VIII International Scientific and Practical Conference "Information-Measuring Equipment and Technologies" (IME&T 2017)

MATEC Web of Conferences Volume 155 (2018)

Tomsk, Russia
22-25 November 2017

Editors:

E. Siemens
A. D. Mehtiyev
V. I. Syryamkin
A. V. Yurchenko

ISBN: 978-1-5108-5911-1

Printed from e-media with permission by:

Curran Associates, Inc.
57 Morehouse Lane
Red Hook, NY 12571

Some format issues inherent in the e-media version may also appear in this print version.

Printed by Curran Associates, Inc. (2018)

For additional information, please contact EDP Sciences – Web of Conferences at the address below.

EDP Sciences – Web of Conferences
17, Avenue du Hoggar
Parc d'Activité de Courtabœuf
BP 112
F-91944 Les Ulis Cedex A
France

Phone: +33 (0) 1 69 18 75 75
Fax: +33 (0) 1 69 28 84 91

contact-edps@webofconferences.org

Additional copies of this publication are available from:

Curran Associates, Inc.
57 Morehouse Lane
Red Hook, NY 12571 USA
Phone: 845-758-0400
Fax: 845-758-2633
Email: curran@proceedings.com
Web: www.proceedings.com

TABLE OF CONTENTS

edp sciences

MATEC Web of Conferences

Statement of Peer review

In submitting conference proceedings to MATEC Web of Conferences, I certify to the Publisher that I adhere to the **Policy on Publishing Integrity** of the journal in order to safeguard good scientific practice in publishing.

1. All articles have been subjected to peer review administered by the proceedings editors.
2. Reviews have been conducted by expert referees, who have been requested to provide unbiased and constructive comments aimed, whenever possible, at improving the work.
3. Proceedings editors have taken all reasonable steps to ensure the quality of the materials they publish and their decision to accept or reject a paper for publication has been based only on the merits of the work and the relevance to the journal.

Title, date and place of the conference

VIII International Scientific and Practical Conference "Information and Measuring Equipment and Technologies"

Conference dates: 22 – 25 November 2017
CONFERENCE VENUE Tomsk, Tomsk State University (Faculty of Innovative Technologies)

Proceedings editor(s):

Eduard Siemens- professor of Anhalt University of Applied Sciences (Germany). eduard.siemens@hs-anhalt.de
Ali D. Mehtiyev - Ph.D. in Eng.Sc., Head of the Department of Technologies and Communication Systems of Karaganda State Technical University. barton.kz@mail.ru
Vladimir I. Syryamkin, professor, Director of the Excellence Center "Intellectual technical systems", Head of the Department of Quality Control of the Faculty of Innovative Technologies of National Research Tomsk State University. novorostomsk@gmail.com.
Alexey V. Yurchenko, chairman of the organizing committee - D.Eng.Sc., Senior Researcher of the international laboratory of Vision Systems of Tomsk State University, Professor of Tomsk Polytechnic University. niipp@inbox.ru

Date and editor's signature

EXPERIMENTAL INVESTIGATION OF METAMATERIAL FROM RINGER CONDUCTORS

Aleksandr Mironchev, *Aleksandr* Gorst*

National Research Tomsk State University, 634050, Tomsk, Russia

Abstract. In this work creation of structure from ring conductors having negative index of refraction is shown. Experimental confirmation and finding of index of refraction, namely its real and imaginary part taking into account the error, for single-layer, two-layer, three-layered and four-layer structure.

1 Introduction

At present, metamaterials attract the great interest of the scientific world with their unique properties, unattainable with the use of familiar materials. For antenna technology, metamaterials with its extraordinary properties provide additional possibilities: smaller dimensions, higher antenna gain and wider bandwidth.

In the last decade of the 20th century, intensive research was conducted in many countries on a wide range of issues related to the physics of complex electromagnetic media and artificial materials that are intended for use in modern radio and optical devices and devices. Impressive results in the field of theory were achieved, which made it possible to better understand the radiophysical principles of creating new electromagnetic materials, as well as to reveal potential possibilities and the range of achievable values of material parameters. Therefore, it is not by chance that the term "metamaterial" appeared on the verge of millennia.

2 Experimental study

Earlier authors of the real work presented results of numerical and computer modeling of the flat closed ring conductor and the possibility of creation metamaterial for radio frequency range [1].

For check and confirmation of the stated results it was decided to conduct pilot study.

Creations of the experimental model consists in selection of ring conductors, used at computer modeling. To measure distances between elements and layers of structure.

In quality swore for implementation of elemental annuluses, frothed polystyrene was used. Elemental annuluses have the sizes: Outer diameter is 8 mm, inside diameter is 4 mm, thickness is 1 mm. Elements were located rather each other at distance with 1 cm. Total

* Corresponding author: gorst93@gmail.com

quantity of elements in one layer made 900 pieces. For the experimental model 4 layers with total quantity of elements of 3600 pieces were used (Figure1).

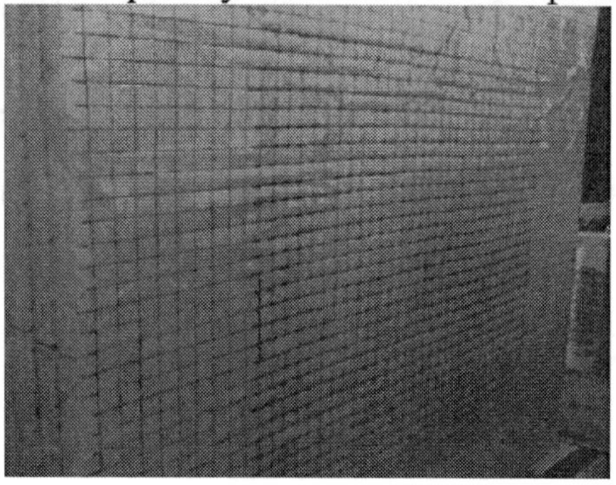

Fig.1. Experimental samples.

Fig. 2. Experimental installation:

1 – SShP the antenna, 2 – the complex plane analysis of chains, 3 – the studied sample.

Experiment carried out by the quasioptical method, based on measurement of parameters in free space.

For implementation of the method the installation presented in the Figure 2 is made. Installation includes the analyzer of chains PNA-L NetworkAnalyzer of the Agilent Technologies company and two transceiving antennas.

As transceiving antennas the combined "Snail" antennas representing constructive association of the electric and magnetic irradiator were used.

During the experiment five principal series of measurements were carried out:

1. The composite from ring conductors one layer, thickness is 1 cm.
2. The composite from ring conductors two layers, thickness is 2 cm.
3. Composite from ring conductors three layers, thickness of 3 cm.
4. The composite from ring conductors four layers, thickness is 4 cm.
5. Organic glass, thickness is 1.8 cm.

It is necessary to make the normalization of the signal which passed through the studied sample on the signal received in its absence for finding of complex index of refraction. The regularization method according to by Tikhonov was used [2].

$$S(f) = \frac{S_2(f)S_1^*(f)}{|S_1(f)| + \alpha}, \quad (1)$$

where S_1 – the signal in lack of the sample, S_2 – the signal which passed through the studied sample α – regularization coefficient.

As is well-known the Real part of index of refraction is determined by slope angle of the phase curve [2]. For this purpose on the chosen frequency interval it is carried out linear regression, presented in the form $y(x) = a + bx$, a and b reflection coefficients of the evaluated line. Using the ratio $\mathrm{Re}(n) = \frac{b \cdot c}{2\pi \cdot d}$, where c – the light speed, d – thickness of the studied sample. Dependence of the phase and amplitude on frequency for the sample of organic glass 1.8 cm thick is given in the Figure 3.

The third straight line represented linear regression of the phase at frequencies from 2.5 GHz to 16 GHz. The schedule is constructed in half-logarithmic scale. On this interval of Re(n)=1.53 that corresponds to tabular values.

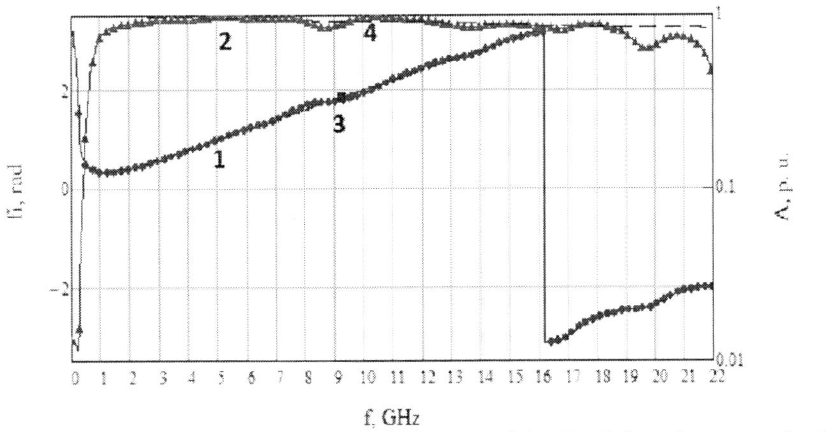

Fig. 3. Dependence of the phase (1) and amplitude (2) past of the signal from frequency, for the layer of organic glass 1.8 cm thick, linear regression of the phase (3), linear regression of amplitude (4).

Let's evaluate the imaginary part of complex index of refraction. As amplitude of last signal has the appearance: $A = \exp(-kn''d)$, where k – wave number, n'' – imaginary part of index of refraction, d – thickness of the studied sample, $n'' = \frac{\ln(1/A) \cdot c}{2\pi d}$.

The regression straight line shows the imaginary part of index of refraction, $\mathrm{Im}(n) = 0.019$. By means of linear regression, perhaps to calculate the error of the received results.

The error of measurements was found with use of standard methods. Thus, we receive complex index of refraction for organic glass at frequencies from 2.5 GHz to 16 GHz of n=1.530±0.001+i (0.019±0.001).

Further with use of the same approach experimental measurements of samples of grids from elemental annuluses were taken. In the Figure 4 schedules of the experiment with the single-layer grid from elemental annuluses are submitted. Dependences of the phase (1) and amplitude (2) past of the signal are given in frequencies from 0.05 GHz to 22 GHz. Lines (3) and (4) represented linear regression for the phase (3) and amplitude (4) respectively. Points noted mean values and confidential intervals.

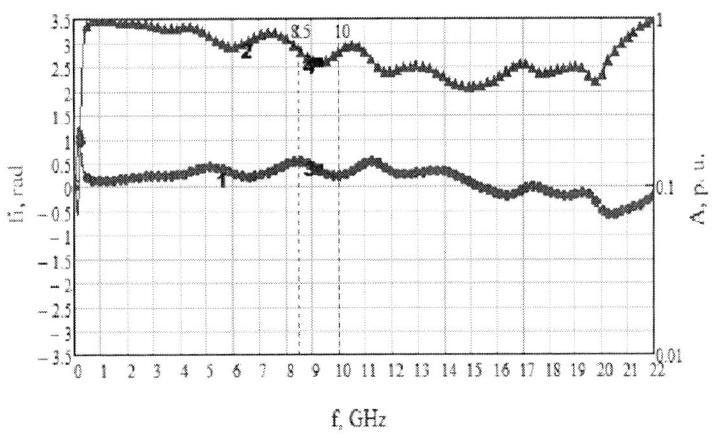

Fig. 4. Dependences amplitude (2) and phase (1) of last signal from frequency for one layer of rings 1 cm thick, linear regression of the phase (3), linear regression of amplitude (4).

It is noticeable that on the interval from 8.5 GHz to 10 GHz at the phase reverse motion is observed.

At frequencies from 8.5 GHz to 10 GHz, using the formulas given above, received complex index of refraction of $n = -0.21 \pm 0.02 + i\ (0.57 \pm 0.04)$. The turned-out result says that even the single-layer grid from the closed elemental annuluses is capable to give negative phase progression.

Let's consider results of the experiment made with two layers the Figure 5. It is possible to notice that, as well as in the previous case with one layer the phase of coefficient of passing has not linear character. Sites with the negative inclination of the phase curve which are characteristic of metamaterial are observed.

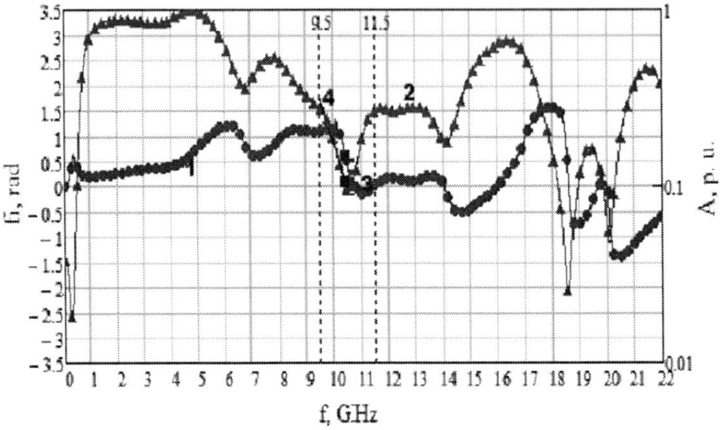

Fig. 5. Dependences of amplitude (2) and the phase (1) of last signal from frequency for two layers of rings 2 cm thick, linear regression of the phase (3), linear regression of amplitude (4).

In the Figure 5 dependences of amplitude and the phase of coefficient of passing are given in the interval from 9.5 GHz to 11.5 GHz. Apparently from the drawing, at frequencies from 9.95-10.9 GHz phase progression reduction takes place. Complex index of refraction of $n = -3.843 \pm 0.13 + i\ (1.8 \pm 0.3)$. Such values of the real part of index of refraction testify to properties of metamaterial. But the imaginary part demonstrates rather strong attenuation at these frequencies. Such result can say that only one of the component the dielectric or magnetic component accepted negative value.

Further are given result of the experiment for three layers of rings 3 cm thick in the Figure 6. The behavior of curves is similar to the schedules provided for two layers with a general thickness of 2 cm. Also frequency intervals on which phase attack has the negative course are observed.

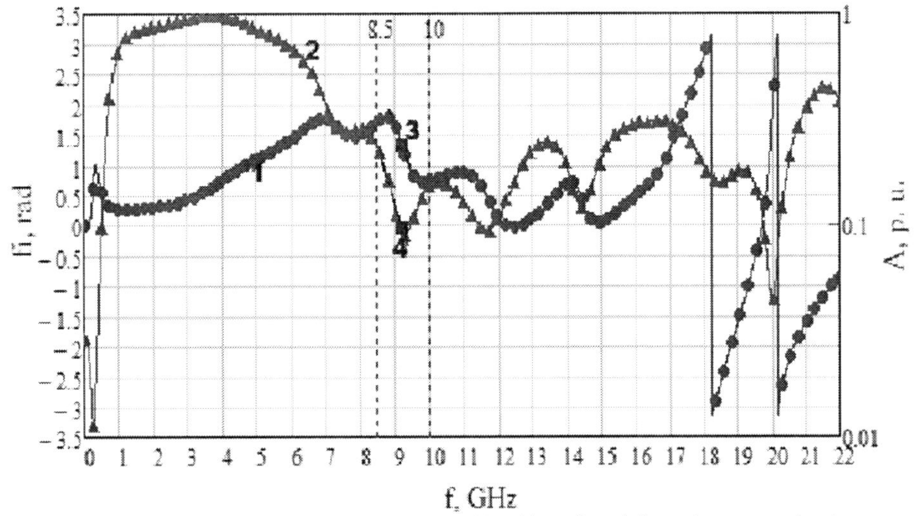

Fig. 6. Dependences of amplitude (2) and the phase (1) of last signal from frequency for three layers of rings 3 cm thick, linear regression of the phase (3), linear regression of amplitude (4).

In frequency band from 8.5 GHz to 10 GHz strong, but smooth recession of signal amplitude of the past through the three-layered grid is observed. At the same time it is possible to notice also decrease of phase attack with growth of frequency.

These observations speak about presence of properties of metamaterial at the studied sample. Complex index of refraction of $n = -1.432\pm0.05+i\ (1.89\pm0.06)$. It is worth noticing that having increased quantity of layers, real part of index of refraction decreased on absolute value, and the imaginary part at the same time remained equal to value for two layers ($n = -3.84\pm0.13+i\ (1.8\pm0.3)$).

Further the case with use of four layers of rings is considered.

The schedules showing dependence of amplitude and the phase of the signal which passed through the four-layer grid 4 cm thick are provided on the Figure 7. Dependence has similar character in relation to the previous cases. Not the linear nature of phase change of last signal is observed.

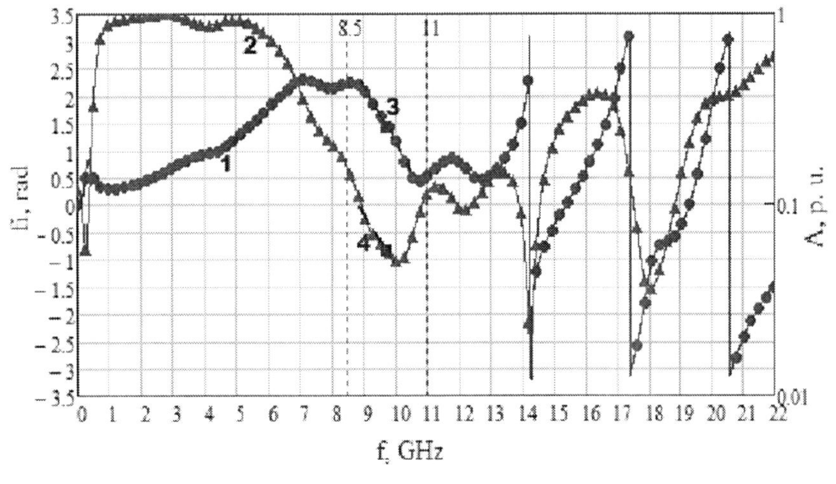

Fig. 7. Dependence of amplitude (2) and the phase (1) of last signal from frequency for four layers of rings 4 cm thick, linear regression of the phase (3), linear regression of amplitude (4).

On the interval of frequencies from 8.5 GHz to 11 GHz there is the decrease, both phase attack, and amplitude. Amplitude of coefficient of passing has low values, and reaches the minimum at 0:05 relative units at the frequency of 10 GHz, at the same time phase attack continues to decrease to the frequency of equal 10.7 GHz. Complex index of refraction at these frequencies of $n = -0.27 \pm 0.01 + i(0.71 \pm 0.01)$.

3 Conclusions

By results of the experiment, it is possible to draw conclusions that grids from the elemental annuluses oriented definitely have properties of metamaterial, namely negative index of refraction. At increase in quantity of layers the real part of index of refraction decreases, at the same time its imaginary part increases. The minimum value of the real part $\mathrm{Re}(n) = -3.84 \pm 0.13$ at three layers. Further increase in quantity of layers, does not lead to reduction of the real part of complex index of refraction, and on the contrary increases it to values, single-layer space grid. It is worth noticing that reduction of the real part of index of refraction is connected with increase in its imaginary part. The submitted schedules show that reduction of phase attack happens only in the presence of attenuation in the sample. It occurs at the expense of the strong interferential fields arising in the near zone of the space grid. As energy of incident wave is spent for emergence of near-fields around each element. There are composite blocked interfering fields which interact with each other. Finally induced near-field does not give increase of the phase, and even compensates the phase attack acquired to the layer.

References

[1] A. Mironchev, A. Gorst, MATEC Web of Conferences **79,** 1 (2016) doi: 10.1051/matecconf/20167901064

[2] G.A. Ponomarev, V.N. Ponomareva, V.P. Yakubov, *Statistical Methods in Radiophysics: A Practical Course Using Dialog-Computing Complexes (*Tomsk University Press, Tomsk, 1989)

RECOGNITION AND IDENTIFICATION BY TIMBRE OF A LIVING CREATURE'S VOICE BASED ON TRAINABLE NEURAL NETWORKS IN REAL-TIME MODE AND THEIR IMPLEMENTATION IN THE INTELLIGENT SYSTEM «NEUROCYBER»

Anton Gafurov, *Oleg* Gafurov[a]

National Research Tomsk State University, 634050, Tomsk, Russia

Abstract. The article considers physical basis for formation and propagation of sound and its psychophysical perception by living creatures. The authors suggest an algorithm of living creatures' identification by the voice timbre and the algorithm's practical realization by artificial neural networks, by initialization of a conversation by the robot in order to reduce ambiguity of responses and to improve the quality of recognition in real-time mode.

1 Introduction

The use of artificial neural networks allows solving the problem of non- linear object control by creating an adaptive control system with trainable neuro-emulator. The problems of identification, synthesis of control systems, and their analysis and hardware implementation are successfully solved because of properties of generalizations and universal approximation of artificial neural networks, which are common to various nonlinear dynamic objects. Training is a process of development of desired reaction to external signals in a control system through multiple system effect and external adjustment. A "teacher" who knowns the desired reaction of a control system to certain effects carries out external adjustment. Thus, during the process of training «teacher» provides the system with additional information whether its reaction is true or false.

The method of recognition and identification by timbre of a living creature's voice based on trainable neural networks in real-time mode and its implementation in the intelligent system "NeuroCyber'' relates to the field of speech analysis technology. In particular, the method relates to the system of formation of robots' adaptive behavior in real-time mode, to system of formation of unauthorized access guarding of material or information resources based on biometric information about the speaker. The technical result is an increase of reliability of recognition and identification by timbre of a living

[a] Corresponding author: author@email.org

creature's voice; improvement of the system of formation of robots' adaptive behavior in real time mode when computational and operating systems must perceive signal of external environment and operate on their own without help of "a person-translator"; and improvement of the systems of high noise immunity recognition with noisy observations. Use of combinations of very different by distinctive features acoustic signal attributes as a parametric description of timbre of a living creature's voice allows achieving technical result. In addition, use of the mechanism of automated training, additional training, retraining of neural networks in real time mode, formation of a new object identifier based on words-commands or recognition of familiar subject in changing external conditions also contributes to achieving technical result [1, 2].

2 Physical basics of sound waves and oscillatory processes

Transfer of semantic and emotional information between two subjects (in the usual sense, living subjects) occurs based on acoustic waves caused by oscillatory modulated processes of throat ligaments (alveolus) of human beings and animals.

Oscillation means some periodic or approximately periodic process, in which the value of a physical quantity is repeated exactly or approximately at equal or approximately equal time intervals.

Robots and robotic systems can perform oscillatory processes in a very wide range including radio waves and sound waves; throughout the range, sound waves occupy a very narrow region. Information transmission between two robots can take place in any range, either in the range of sound unheard for ear of a human being or animal, or in the range of ultrashort radio waves.

Basic concept of oscillatory processes and sound waves propagation in atmospheric air under normal atmospheric pressure and average temperature.

Wave equation:

$$\frac{d^2s}{dx^2}+\frac{d^2s}{dy^2}+\frac{d^2s}{dz^2}=\frac{1}{v^2}\frac{d^2s}{dt^2} \tag{1}$$

v –wave propagation velocity, or

$$\Delta S-\frac{1}{v^2}\frac{d^2s}{dt^2}=0 \tag{2}$$

where $\Delta=\frac{d^2s}{dx^2}+\frac{d^2s}{dy^2}+\frac{d^2s}{dz^2}$ Laplace operator

In atmosphere (ideal gas), wave frequency is equal to frequency of wave oscillation source. Different people and living creatures have individual vocal apparatus; this fact enables us to distinguish them by voice and interpret correctly.

A person with normal hearing ability can perceive as sound only elastic waves with frequency of not less than 16 Hz and not more than 20 000 Hz. The sensitivity of the human ear is not the same to the waves of different frequencies; it is maximal for waves with frequencies of 1.5 – 3 kHz. These is because of structural features of the human organ of hearing. The upper and lower frequency limits of different animals vary: $\gamma = 38$ kHz for dogs, γ max> = 100 kHz for bats and whales. Waves with frequencies γ <16 kHz are called infrasound, with frequencies γ> 20 kHz - ultrasonic, and with γ> 10 to the ninth power kHz – hypersonic.

In this paper, the term "sound" is used to describe the feeling, which a sound wave makes for the human organs of hearing, i.e. sound is a physical process of elastic waves propagation in the medium (physical acoustics), this paper describes psychophysiological processes caused by this physical process (physiological acoustics).

The robot receives and distributes sound in accordance with laws of physical acoustics, but human beings and animals - in accordance with structure of their external and internal ear and other psychophysiological constraints and processes.

Simulation of these processes is that the robot and robotic systems must correctly interpret psychophysical information transmitted to them by man or animal, moreover they are trained to recognize the information on their own or under the guidance of a dominant Figure – a master (Mowgli principle).

Physical acoustics uses numerical characteristics – (γ) frequency of sound wave (frequency spectrum), intensity of sound.

Intensity or sound force is physical quantity **I** equal to modulus of the average value of vector of the sound wave energy flux density (Umov's vector):**I** = | **U** |, $I = \omega * U$, where **U** –group velocity of the wave, **ω** – the average value of the volumetric energy density

$$\omega = \frac{1}{T}\int_0^T \omega * dt \qquad (3)$$

where **T** – is time required to complete one oscillation

In case of sinusoidal wave, velocity **U** coincides with the phase velocity **V**, and

$$I = A^2\omega^2$$

$$\omega = \frac{1}{2}\delta$$

$$I = \frac{1}{2}\delta V^{A^2\omega^2}$$

The SI unit of sound intensity is the watt per square meter (W/m^2).

Physiological acoustics uses concepts of height, timbre and volume to characterize sound sensations.

Sound height is a quality of periodic or almost periodic sound depending on the sound frequency and judged by ear.

For humans, we introduce the concept of sound volume depending on the root-mean-square sound pressure **P** and on the ear sensitivity, which is not the same for sounds of various intensity and frequency.

The hearing threshold depends on the sound frequency, reaches minimal value of about 2×10 to the negative fifth power N/m^2 at frequencies γ = 500 – 1000 Hz and makes about 200 N/m^2. Auditory-sensation area limited by two threshold curves on Figure 1 are projected along the coordinate axes in logarithmic frequency scale and root-mean-square sound pressure **P**.

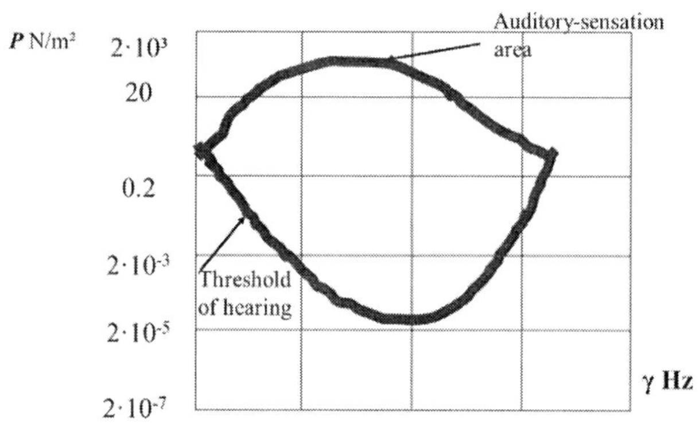

Fig.1. Auditory-sensation area.

Sound pressure is the product of the medium wave resistance ρυ and velocity of oscillations of its particles ds/dt.

Root-mean-square sound pressure $P_0 = \sqrt{\rho v I}$

Values of intensity and root-mean-square pressure for different sounds in air under normal pressure and **20° C**

(ρ v = 418 kg/m²sec) (Figure 1 and Table 1).

Table 1. Values of intensity

Sound	Distance from the source in m^2	Level of sound pressure in dB	Sound intensity in W/m^2	Root-mean-square pressure in N/m^2
Threshold of hearing when*v=1000 Hz*	-	0	10^{-12}	2×10^{-5}
Low sound	1	40	10^{-8}	2×10^{-3}
Loud sound	1	70–80	10^{-5}–10^{-4}	0.06–0.02
Fortissimo of symphony orchestra	10	100	10^{-12}	2
Airplane engine noise	5	120	1	20

Calculations in the «NeuroCyber» are made on the distance of 1 meter from the source to the robot, when a person stands in front of the microphone under normal atmospheric pressure and external temperature **20° C** (***ρ v = 418 kg/m²sec).***

3 Neural networks: basic concepts

Neural network methods are methods based on the use of different types of neural networks (NN). NN consists of elements called formal neurons, which imitate the work of neurons in the cerebral cortex. Each neuron transforms a set of signals coming to its input into an output signal. Figure 2 shows an example of formal neuron work.

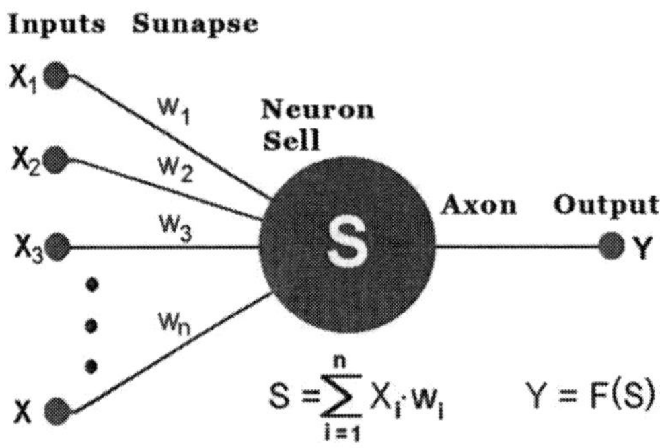

Fig. 2. Formal neuron.

Figure 3 shows neural network structure.

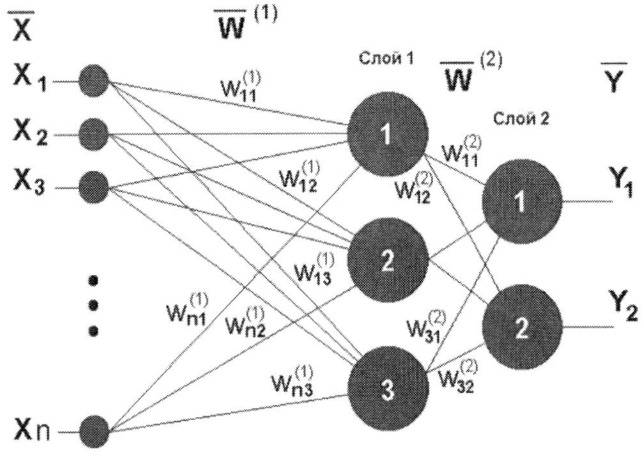

Fig. 3. Neural network.

Connection can exist between individual neurons. Connections between neurons encoded by weight coefficients play a key role in function of the nervous system. On the Figure, connections between neurons are signed by Latin letter W. Index on the top means belonging of a weight coefficient to a layer. One of the advantages of the nervous system is the possibility of parallel operation of all the elements, which considerably increases the efficiency of problem solving in general. This feature of the nervous system is successfully used in image recognition systems. The neural network has inputs X and outputs Y. It represents a system, which generates output state depending on the input state. Presence of weight coefficients that can be determined algorithmically, allows the nervous system to have the most important property – the ability to learn.

During the process of image recognition, a person unconsciously attracts a huge stock of contextual knowledge accumulated throughout the life. Application of neural network methods based on trainable neural networks implemented in the intelligent neuro-informational system "NeuroCyber" potentially can provide the possibility to simulate processes occurring in the human brain during images recognition. In the first approximation, the human brain can be represented in the form of an extremely complex neural network [3, 4].

4 Known methods of speaker identification and recognition

There is a method of automatic identification of a person by means of specific features of pronunciation of a passphrase by this person. Patent of the Russian Federation №2161826, IPC G 10 L 17/00, published on 10.01.2001, consists in the following: the speech signal is divided into voiced zones, time intervals are set in voiced zones - in the area of maximal intensity of the speech signal, and at the beginning of the first and at the end of the last voiced zones. Parameters of the speech signal are determined for the set time intervals and are compared with the etalons formed with mathematical expectations and admissible parameter spread. For this purpose, time intervals are set at the end of the first voiced zone, at the beginning of the last voiced zone and at the beginning and at the end of the other voiced zones. Duration of time intervals is set as multiple of a period of a pitch of speech signal; estimations of correlation coefficients of speech signal parameters are determined and included in the number of positions to be compared with etalons; correlation coefficients of speech signal parameters are additionally considered when forming etalons. Decision on identification of personality is made based on obtained speech signal parameters and corresponding statistical characteristics.

The disadvantage of this method of identification of personality is low noise immunity of the method. Implementation of the method requires separation of exact position of the boundaries of the main voice tone in the input speech signal, which is almost impossible in conditions of acoustic noise (noise from the street, office, etc.).

There is a method of identification of a speaker based on the US patent №6389392, IPC G 10 L 17/00, published 14.05.2002. The method consists in comparison of the input speech signal of an unknown speaker with etalons representing speeches of speakers known in advance, and at least one speaker is presented by at least two etalons. Successive segments of the input signal are compared with etalon successive segments; thus, proximity measure of the compared segments of the input speech signal and etalon signal is obtained.

Composite result of comparison of etalon signal with input speech signal is formed for each etalon of a speaker known in advance, who has at least two etalons. Composite result is formed based on selection of the closest segment of compared etalon for each segment of the input speech signal by the used proximity measure. Further identification of an unknown speaker is made based on composite results of comparison of the input speech signal with etalons. This speaker recognition method is limitedly applied in practice, as it is difficult to realize in real conditions a requirement for availability of at least two etalons of a known in advance speaker. Besides, this method does not provide a high level of reliability of speaker recognition when working in conditions of acoustic noise from offices, streets and vehicles. This is because of the fact, that purely per-segment parametric description of speech signals is heavily influenced by additive acoustic noise and the natural speech variability. Moreover, low reliability of the method in conditions of noise is connected with the fact that the closest by proximity measure segment of compared etalon is searched for each segment of the input speech signal. This leads to the fact that among closest segments there is a great number of presence of the found next segments of a large number of similar pure noise segments corresponding to segments of speech pauses both in etalon and in the input speech signal.

5 Recognition and identification by timbre of a living creature's voice based on trainable neural networks in real-time mode

Principle of sound representation involves digitization – dividing into separate intervals (indications), inside which the amplitude is assumed constant.

Digitization step determines the frequency range of the converted sound; digitization frequency **2 F** is enough to transmit signal with treble frequency **F**.

People and animals hear the same sounds completely different because of the physiology of structure of their hearing and sound-reproducing apparatuses. The robot hears and understands sounds in the most correct way; moreover, the range of its hearing is almost unlimited. With the correct setting, the robot can be a great assistant to humans, it can hear and understand the language of animals, insects, dangerous directed radiation, can register and identify bugs, and can serve as a self-emitting constant source of information.

The proposed method provides a steady recognition of voice of a human being or other living creature after appropriate acquaintance and its identification with the help of a teacher.

The goal is achieved by providing continuous scanning of the surrounding space and continuous filling of the first ring buffer with digitized signal; receiving signal spectrum in quasi-logarithmic scale by applying recursive filter comb to the signal and filling of the second ring buffer with the signal spectrum; determination in the second ring buffer availability and borders of a speech fragment based on adaptive estimation of the noise environment; translation of the spectral components of the speech fragment in a linear analysis buffer.

In order to adapt to the noise environment, determination of availability and borders of a speech fragment in the second ring buffer is made by analysis of spectral components exceeding over adaptive thresholds of noise environment and correspondence of energy and duration limits to etalon records of database. If during the analysis there is a result «not a living creature's voice», thresholds of noise environment are updated by means of a low frequency filter of the first order.

In the case of determination of availability of a living creature's voice in the digitalized record of the scanned window, maximum is defined in a window range containing the voice; and fragment of final predetermined length is cut in order to form a sample, which is supplied to neuro emulator.

Parameterization of the sample is provided by calculation of the fast Fourier transformation (FFT), calculation of instantaneous frequency, signal phases, energy and averaged amplitudes. This multi-dimensional sampling is supplied to neuro emulator.

Recognition is made according to proximity degree or the sample identification by neuro emulator; in case of misrecognition, signal "Do not know" or "Foe" is generated.

Further, if there is a need to identify the voice under the same circumstances, a teacher (human) gives the word command «Get acquainted» or enters the identifier forcibly. Thus, etalon formation occurs, i.e. voice and its identifier are recorded to the database; this sample along with the previous one is supplied to neuro emulator and retraining occurs.

In the case of misrecognition of already familiar voice because of changes in external noise or emotional state of a human being, re-recording of the sample is made forcibly by mans of teacher's commands with the appropriate additions "Alexander is angry", "Dog is in pain". Retraining of neuro emulator is made the similar way.

As regards the device, in order to solve the task of recognition of a speaker, the device contains the source of speech signal; calculating unit of Fourier transformation; unit for determination of parametric description of a speech signal in the form of extractor of the beginning/end of a speech signal; segmentator of a speech signal into a sequence of segments; calculator of a signal power spectrum in the segment and a driver of parametric descriptions of the input speech signal, which are connected in series; neuro emulator unit; unit for making decision on recognizable voice; and a unit for etalons memorization.

5.1 The results of application of different methods of human voice identification

The authors have tried the following groups of methods during the process of development of a mechanism of human voice recognition in real-time mode:

1) Neural network methods.

2) Difference methods for analysis of spectral and cepstral signal coefficients.

Table 2 shows the test results of these methods during analysis of records of various speakers' voices in WAV format. Records of three speakers were analyzed: Artem Soldatenko (a1.wav, a2.wav, a3.wav, a4.wav, a5.wav), Oleg Gafurov (g1.wav, g2.wav, g3.wav, g4.wav, g5 .wav), Alexander Melkozerov (mlkz1.wav, mlkz2.wav).

Table 2. Difference methods. Method of average error. Method of cross-correlation.

Speakers name	File.wav	Neural network	Analysis of spectral coefficients		Analysis of cepstral coefficients	
			Method of average error	Method of cross correlation	Method of average error	Method of cross correlation
Artem Soldatenko	A1.wav	1	0	1	0	1
	A2.wav	1	0.06	0.02	0.005	0.02
	A3.wav	0.3	0.55	0.03	0.006	0.03
	A4.wav	1	0.7	0.04	0.011	0.04
	A5.wav	0.3	0.42	0.01	0.01	0.01
Oleg Gafurov	G1.wav	1	0.6	0	0.03	0
	G2.wav	1	0.7	0.01	0.04	0.01
	G3.wav	1	0.88	0.02	0.03	0.017
	G4.wav	1	0.95	0.13	0.04	0,13
	G5.wav	0.5	0.62	0.01	0.01	0.01
Alexander Melkozerov	Mlkz1.wav	1	0.76	0.02	0.007	0.02
	Mlkz2.wav	1	0.4	0	0.05	0

Neural network methods

The training sample consisted of files a1.wav, g1.wav and mlkz1.wav. Amplitude window with size of 1024 samples was cut out of each file; these windows were inputs of the network.

Correct answers were obtained in 80% of experiments.

Difference methods. Method of average error. Method of cross-correlation.

The same persons were tested as in the previous experiment, and the sample consisted of files a1.wav, g1.wav and mlkz1.wav. Amplitude window with size of 1024 samples was cut out of each file. Then, 1000-band spectrogram in the range of 4000 – 8000 Hz was built on the window using fast Fourier transformation, and 1000 cepstral coefficients were built using the discrete cosine transformation.

Recognition coefficient based on spectral characteristics was unstable and made up $\approx 20\%$

Recognition coefficient based on cepstral coefficients made up less than 10%.

Thus, it is obvious that difference methods of characteristics analysis are not suitable for solving tasks of voice recognition, so the neural network method of amplitudes analysis was selected.

5.2. Results of subjects' recognition and identification in real time mode

The developed technology of subjects' adaptive recognition and identification has been implemented in the system "NeuroCyber" in the subsystem "NeuroAudio" as a component for control of neuro robots. These neuro robots in images of Kuzya, Liza and Bear cub and have been repeatedly demonstrated and participated in appropriate exhibitions in the real exposition mode.

For program testing, 150 subjects were selected and voices of the living creatures (cats, dogs, cows, etc.) were recorded. Test technique was as follows. In the situation of acquaintance, neuro robot Lisa (in the image of a woman secretary), has reacted to appearance of the subject in her field of vision and initiated the conversation on her own without human intervention.

- Hello, what is your name?

From the follow-up response (human voice), the initial segment of the audio file with parameterization was singled out and transmitted to neuro emulator of voice recognition. The whole series of audio files was recorded as the identifier to the database. If it is recognized and identified as one of the known names (Irina, Oleg, etc.), then later, name recorded by the speaker is used in the dialogue, i.e. as the neuro robot's own voice. In case of non- recognition, the recorded file is repeated. In case of a secondary approach, a neural network is used, which has been created in the recent period of time. Or the neuro robot uses additional information by calling all subjects under the common identifier (Alexanders, Olgas, etc.) and creates a neural network trained by parameterized characteristics of voices and individuals. Moreover, if necessary, the robot asks the question, which requires correct and clear answer:

- Are we familiar with you?

Depending on the word command "Yes" or "No", the process of additional training or network starts or the process of recognition by the program "NeuroCyber". Permanently, in case of re-addressing, correction of audio files recording takes place. There is formation of separate files from amplitudes and parameters for different emotional state of the subject under the common identifier, but with additions like "Father is angry", "Mouse is in pain". Respectively, henceforth, the neuro robot comments their state by voice. The test results have shown almost 90% of correct answers.

6 Conclusion

This design project resulted in development of programs of human-robot voice interface with recognition and identification of speech and emotional state by timbre of a speaker's voice based on training of the neural network component.

At the present time, «Neurorobot» company has been the first in the world to create prototypes of neuro robots able to recognize words and complete expression, to determine the emotional component of living beings, as well as to identify individual subjects by the timbre (frequency). Parameterization of digital images of words, phonemic images of alphabet of a particular language and emotional components of the living beings' voices is formed in the training sample for the most typical representatives of human beings and animals; neural network is trained, and the further recognition is made based on it.

Potential partners who might be interested in these technologies include military industrial complex, oil and natural gas industry, factories of computer technology production, aerospace industry, and public health services.

Acknowledgements

The research presented in this paper was supported by the TSU Academic D.I. Mendeleev Fund Program (project No. 8.2.31.2015) and the Russian Foundation for Basic Research (grant No 16-29-04388/17). The authors are grateful to Tatiana B. Rumyantseva from Tomsk State University for her assistance in preparing the paper.

Reference

[1] A.O. Gafurov, The 58th Republican scientific conference of young scientists and graduate students dedicated to the 70th anniversary of the Kazakh National University, 122 (2004)
[2] O.M. Gafurov, A.O. Gafurov, All-Russia scientific and technical conference «Science. Defense. Industry», 34 (2004)
[3] G.A. Kuharev, *Biometric Systems: Methods and Tools for Identification of a Human Person* (Polytechnics, SPb., 2001)
[4] O.M. Gafurov, V.I. Syryamkin, A.O. Gafurov, S.S. Stolyarova, Telecommunications and radio engineering **71**, 1565 (2012)

DUAL-POLARIZED ANTENNA SUB-ARRAY OF 2×2 ELEMENTS FOR MEASUREMENTS OF SHORT ELECTROMAGNETIC PULSES

Evgeny Balzovsky, *Yury* Buyanov, *Vladimir* Koshelev[a], *Eduard* Nekrasov

Institute of High Current Electronics SB RAS, 634055, Tomsk, Russia

Abstract. Results of the investigations of a dual-polarized antenna module of a 2×2 receiving antenna array designed for recording the waveform of short electromagnetic pulses are presented. Each element of the antenna array is an active dipole that is loaded on the single-stage FET-based amplifier. Results of the measurements of the patterns, effective length, and waveforms of the recorded pulses are presented.

1 Introduction

In the tasks of the detection of people behind optically opaque obstacles using ultrawideband (UWB) radiation [1, 2], as well as in studying the electric field structure of short electromagnetic pulses, it is important that the recorded waveforms of the pulses would not have significant distortions. Decrease of the spatial length of the radiation pulses results in the increase of the resolution of the sounded objects. At the same time, a high frequencies in the spectrum of short pulses are intensively absorbed in the obstacle material or in the lossy medium under study. The range of the optimal length of bipolar voltage pulses exciting the transmitting antenna is 1–4 ns. It corresponds to the frequency range of 0.15–2 GHz. To register UWB pulses with a spectrum in this range, receive antennas and antenna arrays are required with the waveform of the voltage at the output of which is proportional to the electric field strength.

The use of UWB pulses allows not only providing high spatial resolution, but also solving the problems of recognition of sounded objects by changing the waveform of the reflected electromagnetic field [3]. In this case, a significant amount of information is contained in the polarization structure of pulses reflected from objects. The elements of the dual-polarized receiving array should register two orthogonal components of the pulsed field. In [4], active antennas consisting of two perpendicular dipoles with aligned centers are suggested. These dipoles allow recording UWB pulses with a spectrum in the frequency range of 0.4–3.5 GHz. However the proposed in [4] antennas have a balancing outputs. This is a disadvantage when combining such dipoles into the antenna array, since it is necessary to use a complex beam-forming circuit and an external UWB balancing-to-unbalancing network (balun). In [5], a lower-frequency active dipole antenna with a bandwidth of 0.15–2 GHz is proposed. Due to the use of a miniature built-in UWB balun,

[a] Corresponding author: koshelev@lhfe.hcei.tsc.ru

this antenna has an asymmetrical output. Since it is difficult to align the centers of such dipoles for recording the orthogonal field components, so the dipoles of vertical and horizontal polarization are shifted by half step of the array relative to each other.

One of the possible schemes of UWB radar can be radiation of objects from one source and electronic steering by a receiving array. To control the array pattern, the signals from each antenna element are summing together with the necessary time shifts. With an increase in the number of elements, the signal summation circuit becomes the key element of the antenna array. The fidelity of the waveform at the array output depends on the frequency characteristics of the summation circuit. It is suggested to use a binary scheme for signal summing which can be realized using multistage ring-type power combiners.

To reveal the features of the construction of the antenna array with a large number of elements, a dual-polarized module consisting of eight dipoles was developed and investigated presenting two sub-arrays of 2×2 dipoles recording two orthogonal components of the short-pulse electric field. The design of sub-array module and the summation unit, as well as the results of research of their characteristics, are described below.

2 Antenna module design

Figure 1 presents the antenna array module physical configuration. The active dipoles 1 form a 2×2 sub-array of the vertical channel, the dipoles 2 form a 2×2 sub-array of the horizontal channel. The feeder lines are covered up with the bulk absorbent material 3 to reduce the effect of reflections from the summation unit 4 located on the rear side of array. The 2×2×2 antenna module has outputs 5 for connection to two channels of the oscilloscope. For dipoles of one channel, the distance between the centers is 67 mm, and the distance between the nearest ends of the dipole arms is 6.7 mm.

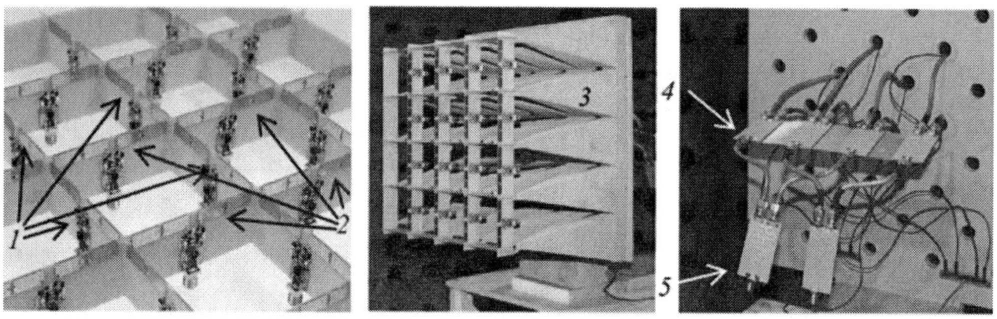

a) *b)* *c)*

Fig. 1. Module of a 2×2×2 element in the antenna array: a) layout of elements; b) side view; c) back view. 1 – vertical channel dipoles; 2 – horizontal channel dipoles; 3 – volume absorber material; 4 – the stage of summation circuit; 5 – output connectors of vertical and horizontal channels.

The construction of the dipole is analogous to that given in [5]. The difference is the implementation of printed-circuit board on double-sided material with metallized through holes. The active dipole (see Figure 2) is made on a 1-mm thick FR4 plate with size of 60×27 mm. The arms 1 and 2 of the dipole are connected to the balancing active FET-based element 3 which is connected to the output connector 5 via the ferrite broadband balun 4.

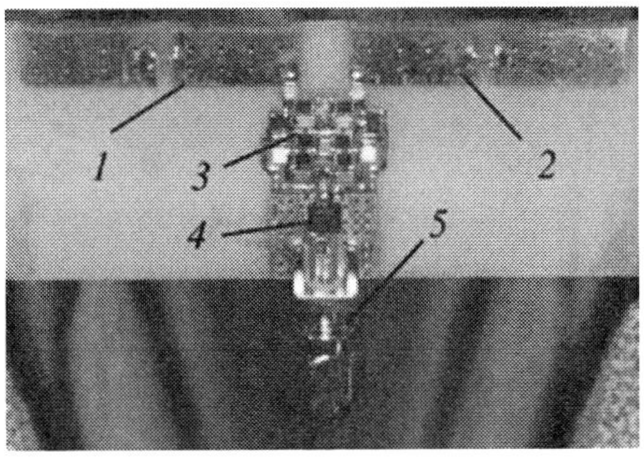

Fig. 2. Appearance of the antenna element. 1 and 2 – dipole arms; 3 – active element; 4 – balun; 5 – output connector.

3 Ultrawideband summation unit

3.1 Design of the summation unit

To reduce the signal attenuation and to increase the decoupling between channels, a circuit based on the ring power combiners was developed and implemented. To extend the bandwidth, the summation unit consists of several stages. It is known that with increasing the number of stage the wave impedance of the near-to-output stage should increase. However, the implementation of a stripline with high wave impedance is related to the limitations of the printing circuit board technology, which does not allow making too narrow conductors. Taking into account the minimum possible width of the conductor, a four-stage summation unit made on a 1-mm thick foil-clad dielectric FLAN having a relative permittivity of 10 was chosen.

Calculations of the summation unit were performed in the approximation of cascade connection of the four-terminal devices. Figure 3 presents the physical configuration of the summation unit, where connectors 1 and 2 are the inputs, and connector 3 is the output. The wave impedances of the line segments L1–L4 are 56, 64, 76 and 85.5 Ohms, respectively. The width of the L4 line is 0.2 mm. The choice of nominal values of the surface-mount resistors is limited by the standard E24 series. The nominal values of the resistors R1–R4 of the standard size 0603 are 470, 300, 160 and 100 Ohms, respectively. The size of the dielectric plate is 46.5×25.7 mm.

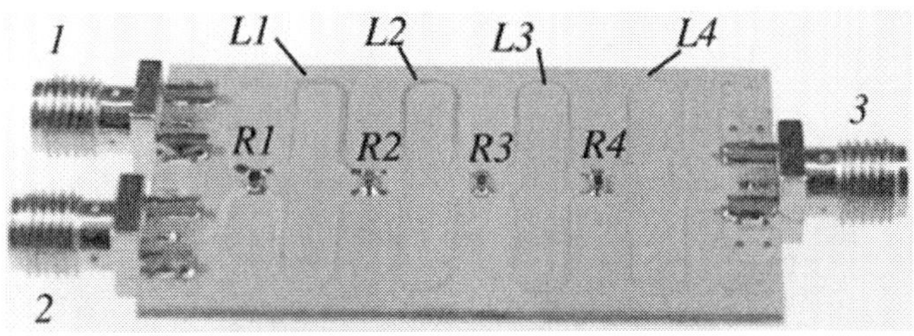

Fig. 3. Physical configuration of the summation unit; 1 and 2 – inputs; 3 – output.

3.2 Characteristics of the summation unit

Figure 4*a* illustrates the voltage standing-wave ratio (VSWR) of the summation unit calculated analytically (curve 1), obtained as a result of full-wave simulation (curve 2) and measurements (curve 3). The VSWR of the inputs is no higher than 1.5 in the range of 0.1–2.2 GHz. Figure 4*b* shows the frequency dependence of transmission coefficient module $|S_{31}|$ of the summing unit. At the central frequency, measured $|S_{31}| = 3.3$ dB. The deviation of the values of the transmission coefficient module is no higher than 0.4 dB in the frequency range of 0.1–2.2 GHz. In this range, the phase-frequency characteristic differs from the linear dependence by no more than 10 degrees. The measured isolation between the inputs 1 and 2 in the frequency range of 0.17–2 GHz exceeds 20 dB, and in the frequency ranges of 0.1–0.17 GHz and 2–2.2 GHz it exceeds 15 dB. The developed UWB summation unit is suitable for synchronous addition of short pulses with a spectrum in the frequency band of 0.1–2.2 GHz with low losses and distortions.

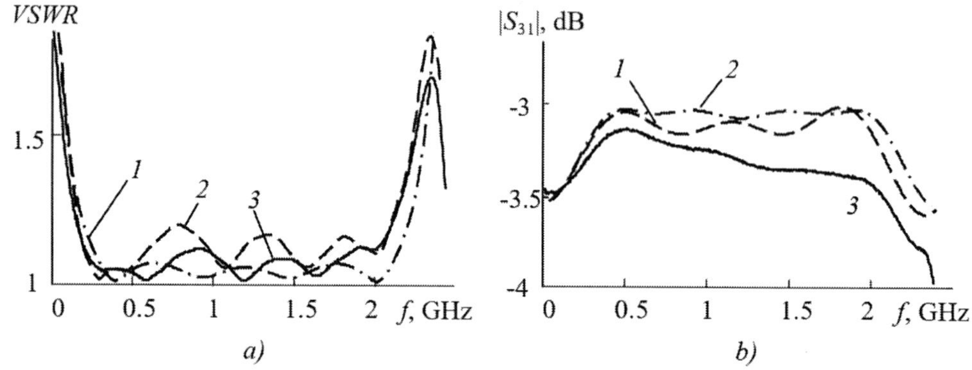

Fig. 4. VSWR (*a*) and transmission coefficient module (*b*) of the summation unit. 1 – analytical calculation, 2 – full wave simulation, 3 – results of measurements.

4 Antenna module characteristics

In the time domain the characteristics of the antenna array module are measured by using a set of generators of short UWB pulses and a set of combined antennas [3], optimized for the effective radiation of pulses with different spectra. Figure 5 shows the waveforms of the voltage pulses $V(t)$ at the output of the receiving antenna array module (curves 1). Curves 2 correspond to the pulses at the output of the TEM-antenna. The spectra (by the level of 0.1 from the maximum value) of the pulses radiated by the antenna occupy the frequency bands of 10 MHz–1 GHz, 50 MHz–1.3 GHz, 80 MHz–2.2 GHz, and 250 MHz–4.3 GHz.

A TEM-antenna with a ground plate size of 120×50 cm and an aperture height of 8 cm was used as a reference antenna for measuring the waveform of UWB pulses. The voltage at the output of the TEM-antenna is proportional to the electric field strength of the incident field, and the effective length is equal to half the height of the aperture in a wide range and has no dependence on the frequency. To quantify the distortion of the waveform of the pulses, the root-mean-square (RMS) deviation σ of the voltage pulse waveform $V(t)$ at the output of the antenna array and the voltage $U(t)$ at the output of the reference antenna were calculated according to the following expression: $\sigma = \sqrt{\int_T [u(t-t_0) - v(t)]^2 dt \Big/ \int_T u^2(t) dt}$, where $u(t) = U(t) \Big/ \sqrt{\int_T U^2(t') dt'}$ and $v(t) = V(t) \Big/ \sqrt{\int_T V^2(t') dt'}$ are the normalized functions, T is the time window, and t_0 is the time shift of $U(t)$ with respect to $V(t)$, at which σ takes the minimum value. The RMS deviation of the waveform of the pulses at the output of the array module from the pulses at the output of the TEM-antenna is shown in Table 1. The effective length of the antenna array calculated as the product of the effective length of TEM-antenna (it is equal to 4 cm) by the ratio of the peak voltage at the antenna array module output to the peak voltage at the output of the TEM-antenna is presented in Table 1 as well.

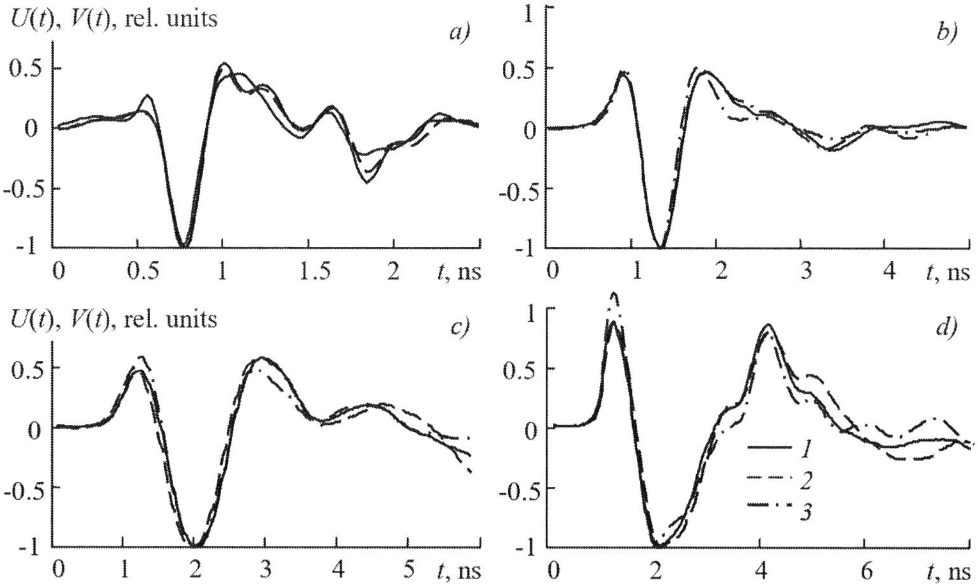

Fig. 5. Pulse waveforms at the antenna module output at the excitation of transmitting antennas by bipolar pulses of the length 0.5 ns (a); 1 ns (b), 2 ns (c), and 3 ns (d). 1 – horizontal channel of the antenna array module; 2 – vertical channel of the antenna array module; 3 – TEM-antenna.

Table 1. Characteristics of the antenna module channels

Duration of pulse of generator, ns	σ		Efective length, cm	
	Vertical channel	Horizontal channel	Vertical channel	Horizontal channel
0.5	0.24	0.23	3.7	3.9
1	0.19	0.21	4.5	4.5
2	0.2	0.14	5.5	5.6
3	0.24	0.2	5.7	6.0

Figure 6 presents the measured patterns of the antenna module in two planes. Here, the pattern is the angular dependence of the value of voltage maximum at the output of the antenna array during the pulse.

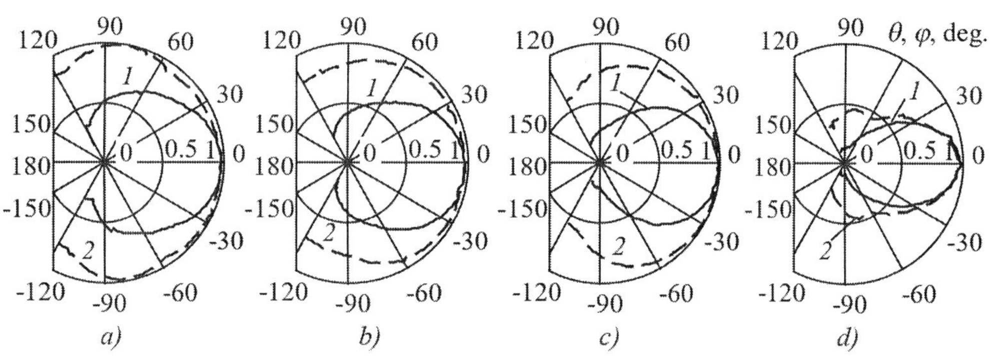

Fig. 6. Patterns of the antenna module at the excitation of the transmitting antennas by bipolar pulses of the duration of 3 ns (a), 2 ns (b), 1 ns (c), and 0.5 ns (d). Curves 1 correspond to the E-plane pattern, and curves 2 correspond to the H-plane pattern.

5 Conclusion

The dual-polarized antenna module of a 2×2 elements of receiving antenna array designed for recording the temporal waveform of short electromagnetic pulses with a spectrum in the frequency range of 0.15–2 GHz was developed and investigated. In the circuit of signal adding from each antenna element, it is provided to connect the controlled delay lines both at the output of each element for steering in a solid angle, and at the outputs of columns or the array lines for steering in the plane. The antenna module was tested by means of short-length (1–4 ns) UWB pulses. The temporal waveform of the voltage pulses at the output of the developed antenna array and at the output of the reference TEM-antenna differ by no more than 20% by the root mean square criterion. When exposed to subnanosecond pulses, in which the upper frequency in the spectrum exceeds 4 GHz, the distortion of the pulse waveform is no higher than 25% as compared to the TEM-antenna. The aim of the further work is to develop an antenna array with a large number of elements.

Acknowledgements

The work was supported by the Russian Science Foundation, project No. 16-19-10081.

References

[1] E. Balzovsky, Yu. Buyanov, MATEC Web of Conferences **79**, 01036 (2016) doi: 10.1051/matecconf/20167901036
[2] S. Shipilov, R. Satarov, I. Fedyanin, E. Balzovsky, V. Yakubov, MATEC Web of Conferences **79**, 01079 (2016) doi: 10.1051/matecconf/20167901079
[3] V. Koshelev, Yu. Buyanov, V. Belichenko, *Ultrawideband Short-Pulse Radio Systems* (Artech House, Norwood, MA, 2017)
[4] E. Balzovskii, Yu. Buyanov, V. Koshelev, J. Commun. Technol. Electron. **55**, 172 (2010) doi: 10.1134/S1064226910020087
[5] E. Balzovsky, Yu. Buyanov, V. Koshelev, E. Nekrasov. IOP J. Phys: Conf. Series **881**, 012001 (2017) doi: 10.1088/1742-6596/881/1/012001

PERSPECTIVES OF THE TOOTH RESTORATION TECHNOLOGY BASED ON THE COMPUTED TOMOGRAPHY DATA

Maxim Putrik*, *Vladimir* Ivanov, *Igor* Antsygin

Federal State Autonomous Educational Institution of Higher Education "Ural Federal University named after the first President of Russia B.N. Yeltsin", 620000, Ekaterinburg, Russia

Abstract. The aim of the study is to create an image processing system, which allows dentists to create tooth surface automatically. X-rays images of jaws from the cone beam tomography or the spiral computed tomography and images from the micro-computed tomography are the initial data for processing. Example of using the combination of optical and x-ray images in the dentistry is described. Also, it was shown how to use the microcomputed tomography method to plan the placement of orthopedic structures on the lost teeth sites.

1 Introduction

Computed tomography is a method of examination, which is widely used in medicine, in particular, in dentistry at present. This method allows you to provide the doctor the exhaustive information which is necessary for diseases diagnosing and treatment planning.

Modern equipment for CT scanning is computer tomographs, among which there are two types:

- spiral computer tomograph;
- cone beam computer tomograph (only for examinations of the maxillofacial area).

Appearance of the above types of apparatus is shown in Figure 1.

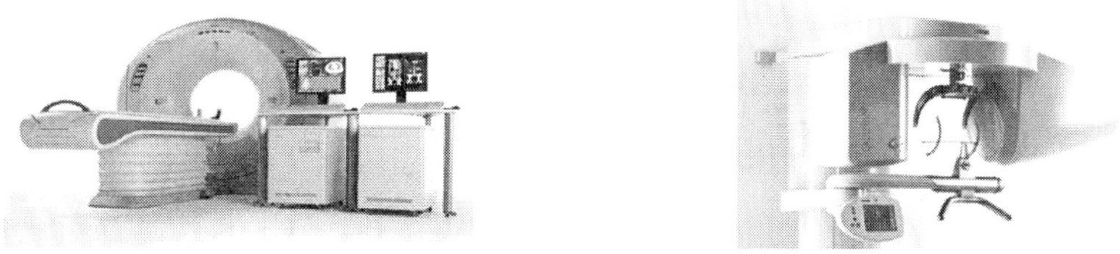

a *b*

Fig. 1. Types of computer tomographs: a – spiral computer tomograph Aquilion (Toshiba, Japan); b – cone beam computer tomograph (Sirona GmbH, Germany).

* Corresponding author: pmb-88@mail.ru

The physical principle of slice formation, which is used in a spiral computer tomograph, is following: the detectors of ionizing radiation are located on the ring, the emitter (X-ray tube) moves inside this ring and generates a pulsed X-ray beam. The patient is located inside this ring on a special bed, which moves inside this ring. As a result of simultaneous rotation of the ring and motion of the bed, a set of digital signals is formed on linear detectors (after passing an X-ray beam through the patient's body), which correspond to the X-ray images of the body section in various projections. Then the inverse Radon transform is applied, the essence of which is to determine the function of two variables (the x-ray image of the body section) over the set of linear integrals (the set of projections) from this function. For the subsequent formation of a three-dimensional image, a set of planar images is used. The determination of the density of human body tissues on the cross-sectional image obtained on a spiral computer tomograph is carried out by means of the use of the Hounsfield scale, based on the character of the interaction of X-ray radiation with matter. The attenuation of X-ray quanta in matter is described by an exponential law.

The section image is formed in gradations of gray color, in accordance with the value of HU (Hounsfield units). For example, areas on the cut with a high HU (bone image) will appear brighter on the monitor screen than soft tissue areas that have a low HU.

When forming a section image on a cone beam computer tomograph, a different physical principle is used. The emitter (X-ray tube) generates a conical shaped pulse beam that passes through the skull of the patient and is fixed to a planar detector that is much larger than the linear detectors of a spiral computer tomograph. Thus, almost all the maxillofacial area is covered in one scan. The volumetric image is formed after scanning a series of planar images at different positions of the emitter and the detector. However, in the case of cone beam tomographs, relative units are used instead of Hounsfield units, in accordance with which for air the luminance value is 0, and the tissue with the maximum density in the special calibration phantom has a luminance value 4095, for example.

The DICOM (Digital Imaging and Communications in Medicine) standard is used to record digital medicine data. The file of this format contains not only a matrix with an image, but important information of a diagnostic nature. The file itself is divided into so-called "Information Elements" (Date Element), which are formed into separate groups. Thus, access to the "atom" of information is carried out by two numbers: the group number and the number of the element within the group. This standard is designed to make medical information independent of device manufacturers. Attached information about the survey and a snapshot make it possible to adequately and simply present the data to the doctor. In the event that the program can not recognize any attribute (element) of the file, it should skip it - for example, not all devices can read and, accordingly, write patient's coordinates to a file, there may also be problems with recognizing Cyrillic characters.

Modern X-ray computer tomographs make it possible to obtain images with a pixel size around 300 μm and software designed to analyse survey results is used to build various sections, synthesize three-dimensional models, virtual operations planning, etc.

Nevertheless, the pixel size of images around 300 μm is not sufficient to solve the modern dentistry problems. That's why the method of microcomputed tomography [1] gradually spreads. It is used for the analysis of techniques for tooth caries removing, assessment of bone tissue density, and studying the structure of the root canals. Images from a microcomputer tomograph have a pixel size from 1 μm and are comparable with optical images in quality [2]. The disadvantages of the microcomputed tomography method are following: a long scan time (several hours), the possibility of studying small samples in vitro only.

Consider the perspectives of using this method for in vivo research. The images which are necessary for the analysis of the internal teeth structure and jaw bones will be obtained after the examination of the maxillofacial area. Also, 3D tooth surfaces synthesized on the basis of these images can be used to model temporary crowns and inlays.

2 Combination optical and x-ray images

The use of the microcomputed tomography method will allow planning the placement of orthopedic structures on the lost teeth sites. An example of such technology [3] (crown restoration of a lost tooth) is realized in the line of CEREC devices (manufactured by Sirona, GmbH): the patient's oral cavity can be scanned by an intraoral camera or dental models made from special gypsum can be scanned too [3]. The essence of this technology lies in the construction of three-dimensional surfaces of jaws (upper and lower). The surface of the lower jaw is scanned to visualize the site of the lost tooth and then to model the dental crown at this site. The surface of the upper jaw is scanned in order to match the surfaces of the modelled crown and the crown of the antagonist tooth: the newly created tooth crown should be modelled not to interfere with the existing teeth. After the crown is modelled, it's image is transferred to a special machine for subsequent production from ceramic blocks.

This technology has the following drawbacks. Modern intraoral cameras allow you to scan without using a special powder, which is sprayed in the mouth and unevenly settles on the teeth. However, the dental row is unevenly covered with saliva when scanning.

Scanning of the dental model made of special gypsum can be performed both by a special scanner and an intraoral camera. It should be noted that the tooth model is modelled manually on the basis of the impression of the patient's jaw. Thus, the quality of the performed manual manipulations has a great influence on the final result.

Planning an implant operation is one of the stages in the process of restoring a lost tooth. Currently, implantation operations are performed only after the virtual planning is carried out, which involves placing the implant model inside the jawbone image: the future implant must be sufficiently immersed in the jawbone, and its axis must pass through the centre of the planned orthopedic structure. To visualize this process, Sirona company implements the possibility of importing optical images (tooth crown) into the result of scanning the maxillofacial area performed on a computer tomograph [3].

Let's describe the virtual planning of the implantation operation [3]. Since the results of the examinations of the maxillofacial area performed on a cone beam computer tomograph GALILEOS are used in this study, the example of virtual planning will be considered for the SIDEXIS software with the additional module GALILEOS Implant (supplied with the GALILEOS computer tomograph).

The SIDEXIS software (see the Figure 2), in addition to the usual viewing of survey results in all planes, allows performing additional constructions, namely: drawing a neural canal and performing a virtual implantation operation. To draw a body of a nerve channel, a dentist or a radiologist on several transverse sections marks the location of the nerve channel in the bone, after the program draws the entire channel in all the pictures. For the virtual implantation operation, first of all, it is necessary to analyze the condition of the bone tissue in place of the missing tooth and choose the number of the tooth. After that, the program will suggest choosing the appropriate implant from the database. The parameters of the implant are its diameter and height, but the implant has a rather complex shape resembling a paraboloid and a ribbed outer surface for secure anchoring in the bone. In the event that the required implant is not in the database, it is possible to create a three-dimensional model, having previously indicated the diameter and height. The simulated implant will take the form of a regular cylinder.

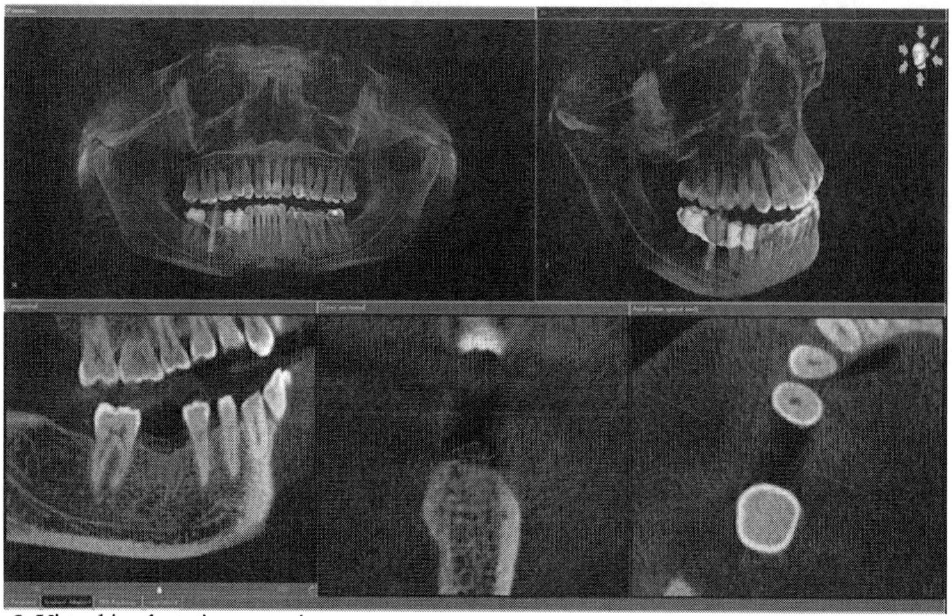

Fig. 2. Virtual implantation operation.

After selecting the implant, its image will appear on the current sections (Figure 2). To move the implant, it is enough to pull it with the mouse, holding the left key, and at the position of the mouse pointer near the implant axis, it is possible to rotate around one of the bases, thus planning the location of the real implant in the jawbone. Moving and rotating the implant are also possible in any of the sections, and changing the coordinates of the implant causes all images to be updated. When several implants are staged, the active implant is displayed in red, and the rest - in white.

3 Use of the micro-computed tomography data

For patients who have dentist's recommendation to remove of one of the teeth, the author suggest a different way of modelling the surface of the crown in place of tooth to be removed. It is necessary to make a survey of the patient's maxillofacial area on a computer tomograph in order to confirm the indications for tooth extraction. If such indications are confirmed, tooth extraction is performed, and the removed tooth is subjected to scanning on a microcomputer tomograph [4] whose detector resolution is significantly higher than the resolution of a CT scanner with a cone beam (CBCT). The removed tooth was scanned at the SkyScan (BRUKER-MICROCT, Belgium) with following parameters: voltage 50 kV, 23 μm pixel size, 11 Mp x-ray detector resolution.

To determine the tooth boundaries (on the images from microcomputer tomograph) using Gaussian filter the following method is used [5]. To apply this method, it is necessary to set the brightness value of pixels corresponding to the image of healthy tooth. It was found experimentally that this brightness value is no less than about 400 HU. Let's consider the mentioned method in detail. The Gaussian filter is applied on the original image at first (see Figure 3*a*) to increase signal-to-noise ratio (see the result at Figure 3*b*). Then pixel brightness values are read from the left, then from the right side and upper side of the image. When the pixel brightness value of the health tooth (400 HU or more) is determined, the border point of the tooth is assigned. Points found in this way are highlighted in white in Figure 3*c*, where the thick boundary is depicted.

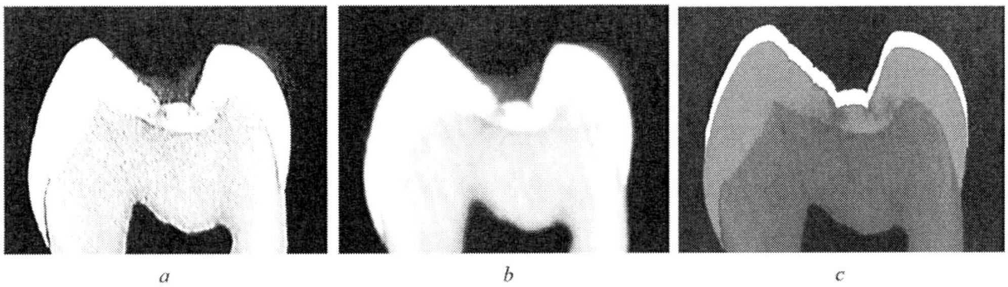

Fig. 3. High resolution tooth cuts: a – original image; b – original image after Gaussian filter application; c – tooth boundary is depicted in white colour.

The next step is to unite all boundary lines to create an entire tooth surface. The surface of the removed tooth obtained in this way is comparable in quality to the image obtained with the use of a three-dimensional intraoral optical camera. Further, the obtained surface (Figure 4) of the removed tooth is used to model and manufacture an artificial tooth that will be placed in site of the removed tooth [2].

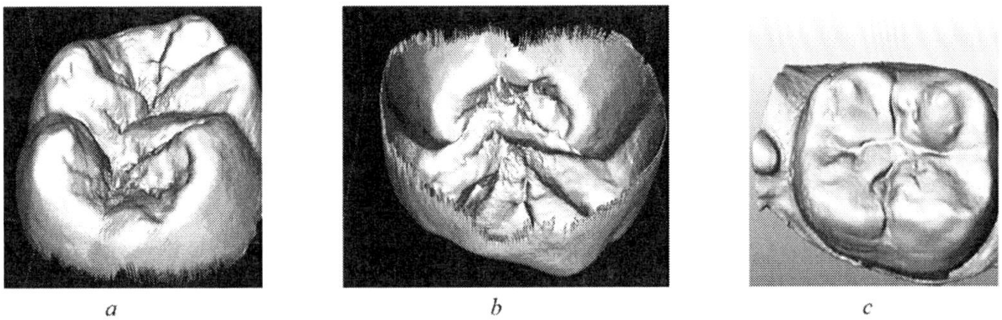

Fig. 4. High resolution tooth surface images [2]: a – tooth surface impression, obtained on the basis of microcomputed tomography data; b – tooth surface impression, obtained on the basis of microcomputed tomography data (inside view); c – optical image.

It is assumed that it is possible to import the image obtained on a microcomputer tomograph in the result of examination of the maxillofacial region.

The presence of the patient is not required, "manual" work is excluded, and therefore manufacturing errors are eliminated, the labour costs of the doctor are significantly reduced, and the process of modelling the dental crown is performed automatically with this approach (the doctor can correct the resulting surface if necessary).

The tooth surface at Figure 4 was used for manufacturing by 3D printer from biocompatible material (A3 colour at Vita scale), see the Figure 5.

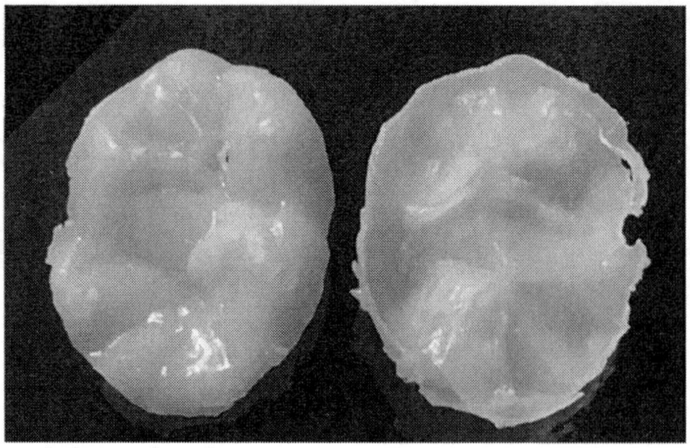

Fig. 5. Printed high resolution tooth surface.

4 Conclusion

Predicting further progress in the technology of the production of computer tomographs in terms of increasing the resolution of images, we can conclude that the surgical guides [3], based on high resolution images, will allow doctors to make accurate and safe implantation operations.

Acknowledgements

The research presented in this paper was supported by the Russian Foundation for Basic Research (grant No 16-29-04388/17).

References

[1] M. Naitoh, A. Katsumata, S. Mitsuya, The Int. J. of Oral & Maxillofacial Impl. **19**, 239 (2004)

[2] M.B. Putrik, J.E. Lavrentyeva, V.YU. Ivanov, News of High. Educ. Instit. Phys. (in Russian) **57,** 70 (2014)

[3] A.N. Ryakhovsky, *Digital dentistry (in Russian)* (LLC "Avantis ", Moscow, 2010)

[4] Skyscan 1272: desk–top high–resolution micro–CT, [Online] URL: http://bruker-microct.com/products/1272.htm.

[5] M.B. Putrik, Y.E. Lavrentyeva, I.N. Antsygin, Biomed. Eng. **48**, 135 (2014) doi: 10.1007/s10527-014-9437-y

THE FLUXGATE MAGNETOMETER SIMULATION IN COMSOL MULTIPHYSICS

Andrey Kolomeytsev[a], *Pavel* Baranov and *Ivan* Zatonov

National Research Tomsk Polytechnic University, 634050, Tomsk, Russia

Abstract. This article describes the fluxgate magnetometer simulation in Comsol Multiphysics software package. The simulation results coincide with the experiment described earlier. Decomposition of the output signal by the Fourier coefficients shows a frequency doubling.

1 Introduction

Weak magnetic field measurement (from 10^{-11} to 10^{-4} T) is a crucial objective in terms of space and geophysical research, attitude control [1–6], magnetic tomography, defectoscopy, non-destructive testing [7–10] etc.

A fluxgate magnetometer is the most commonly used to measure weak magnetic field. However, despite the high level of achieved characteristics, measurement accuracy improving is possible.

Electrostatic, electromagnetic, thermal and gas-dynamic interactions influence on fluxgate magnetometer characteristics. It is extremely difficult to solve this type of problems analytically. Fluxgate magnetometer successful development requires physical and simulation modelling more reliable predicting of the device behaviour. The solution to this problem is a finite-element simulation. It significantly reduces efforts to an experimental study and gives an understanding of the subject.

Modern systems of the finite-element simulation are able to implement a wide variety of fluxgate magnetometer design and evaluate their parameters. The need for such simulation is undeniable in the context of the search for new solutions. Nevertheless, there is no described technique for fluxgate magnetometer simulation with finite-element methods.

In this paper, we describe the main stages fluxgate magnetometer simulation in Comsol Multiphysics software package. The simulation input data is get from [11].

2. Model definition

Today, differential fluxgate magnetometers are the most common. Such a fluxgate magnetometer consists of two identical permalloy cores of rectangular cross-section that are parallel to each other. Around each core is a coil. They form excitation circuit. A secondary coil is wound on both primary coils. In a couple with indicator device, it forms a measuring circuit. Primary coil current i induces magnetic fields in cores that is opposite in direction

[a] Corresponding author:aak65@tpu.ru

to each other. Provided that the cores are identical and are in the external field H_0, magnetic flux density can be written as:

$$\begin{cases} B' = B(H_0 - H_1) \\ B'' = B(H_0 + H_1) \end{cases}, \quad (1)$$

According to the Faraday's law of induction, output voltage in the secondary coil is:

$$e = -sw_2 \frac{d}{dt}(B' + B''), \quad (2)$$

We approximate $B = f(H)$ by a polynomial with degree 3:

$$B = aH + bH^3, \quad (3)$$

where a and b – is the approximate coefficients. They depend on the material and shape of the core.

Assuming that $H_1 = H_m \cdot sin(\omega t)$ and $H_0 = const$, on substitution (3) into (2), we obtain:

$$e = 6\omega b s w_2 H_0 H_m^2 \sin 2\omega t, \quad (4)$$

where ω – is excitation current frequency.

It is seen from (4) that the output voltage has a doubled frequency and is proportional to the magnitude of the measured external field. The principles of the differential fluxgate magnetometer theory are described in the presented equations.

The model created by us consists of two cores, two primary coils and a measuring coil (Figure 1). Each primary coil is wound on a core. The measuring coil encircles both primary coils.

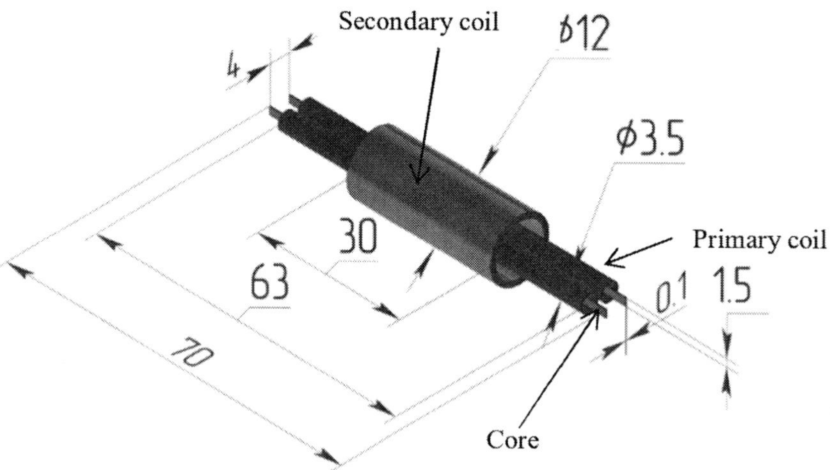

Fig. 1. Model illustration of a fluxgate magnetometer.

2.1 Electric circuit

The model assumes that the primary and secondary coils are made of thin wire and consist of many turns. The primary coils have their own active resistance R_p and are connected to an alternating current source. The primary and secondary coils are inductively coupled, as shown in Figure 2. The secondary coil also has its own active resistance R_s.

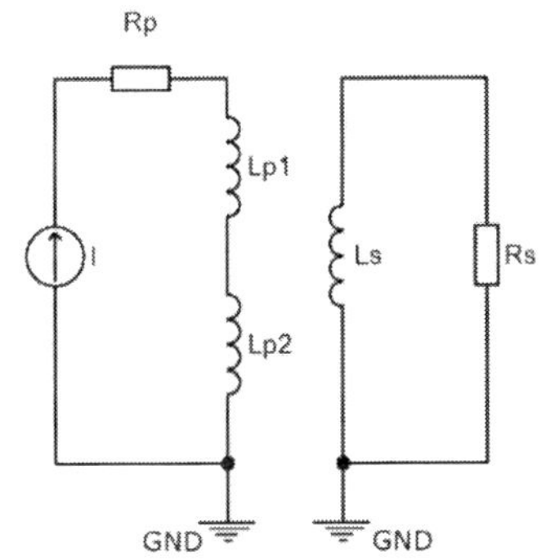

Fig. 2. Current source connected with a primary coil.

The current source generates sine signal with a frequency of 1.5 kHz and amplitude of 20 mA. The primary coil consists of 315 turns. Each primary coil has an active resistance of 4.6 ohms. The secondary coil consists of 2450 turns and its active resistance is 530 ohms.

The measured magnetic field is directed along the fluxgate magnetometer sensitivity axis and has a value of 10^{-4} T.

2.2 Materials

The air cylinder is around the fluxgate magnetometer. Copper from Material Library is selected for the coils. Nickel Steel Square 50 is chosen as the core material. Jiles-Athertron model describes non-linear material characteristics. Due to this, it is possible to take into account the hysteresis in the core. The parameters of the Jiles-Atherton model are determined experimentally. The core material parameters are summarized in Table 1.

Table 1. Core material parameters

Property	Value	Unit
Electrical conductivity	2.088	MS/m
Relative permittivity	1	1
Maximum magnetization parameter	1.6	MA/m
Langevin slope parameter	560	A/m
Pinning parameter	1200	A/m
Reversibility parameter	0.1	1
Interdomain coupling parameter	0.0007	1

2.3 Mesh

Correct adjustment of a mesh is great importance in simulation. When studying fluxgate magnetometer, it is crucial that the cores must interact with the field equally. In this paper, we set up the mesh each domain manually. Thereby, the mesh becomes uniform and domains are more symmetrical than with the physics-controlled mesh. The resulting mesh is shown in Figure 3.

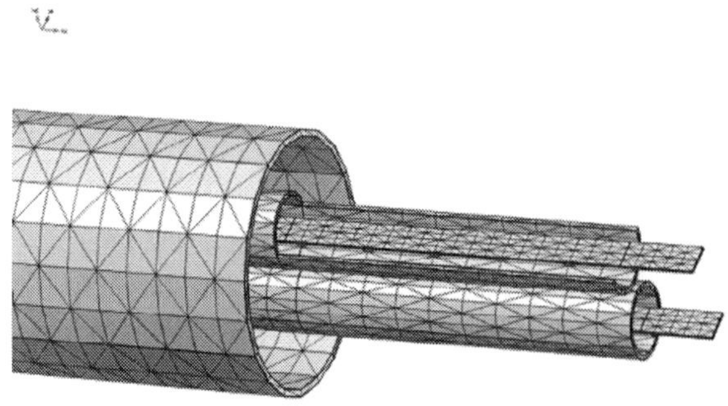

Fig. 3. Finalized mesh.

3. Results and discussion

The Figure 4 shows the distribution of magnetic flux density in the cores and the currents direction at $t = 0$, $t = \frac{T}{4}$, $t = \frac{T}{2}$ and $t = \frac{3T}{4}$.

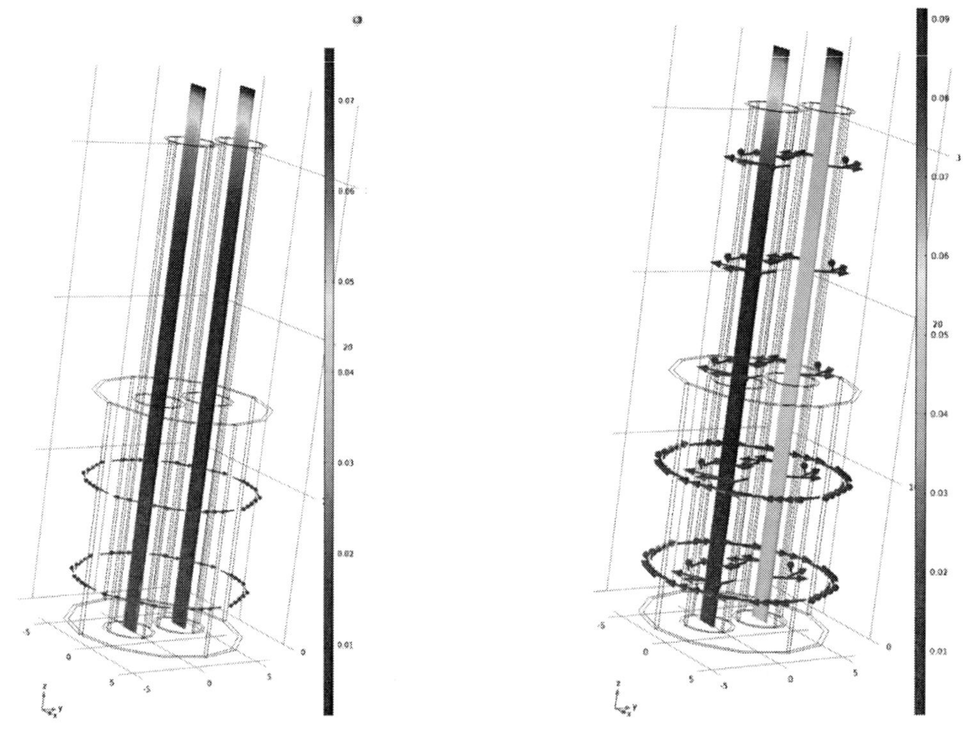

(a) $t = 0$ (b) $t = \frac{T}{4}$

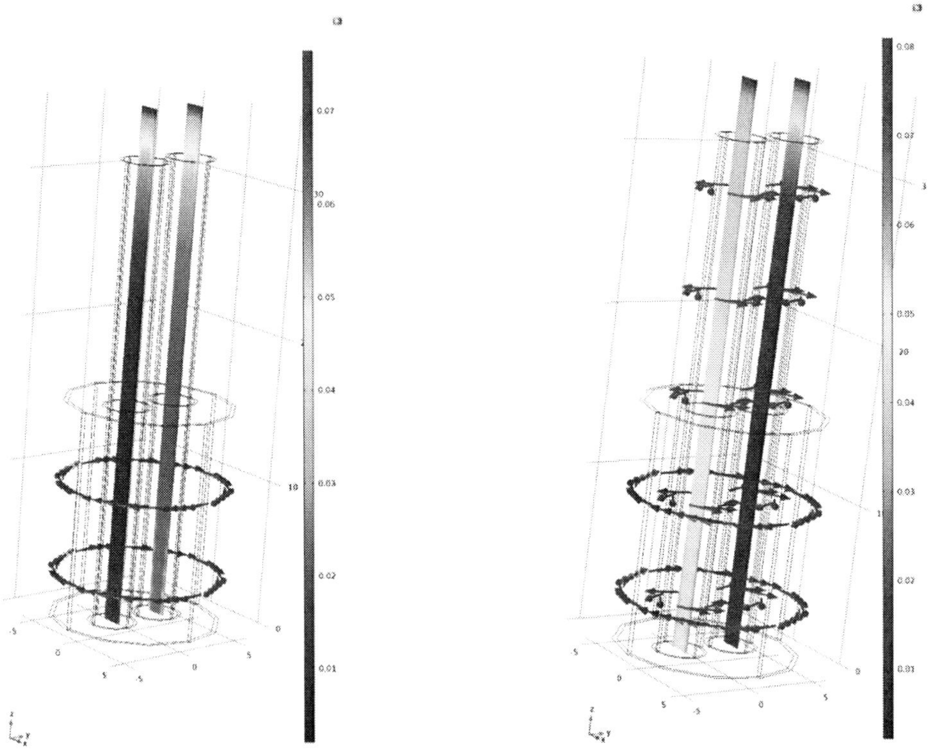

(c) $t = \frac{T}{2}$ (d) $t = \frac{3T}{4}$

Fig. 4. Magnetic flux density in cores (scaled) and current direction (red arrows).

The Figure 5 shows the current against time in the primary (a) and secondary coils (b).

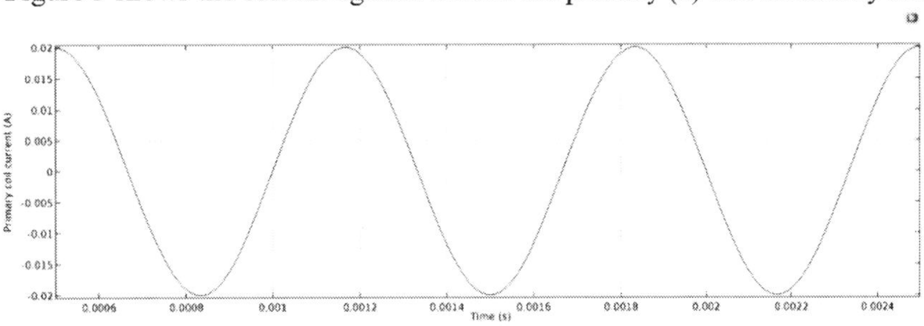

(a)

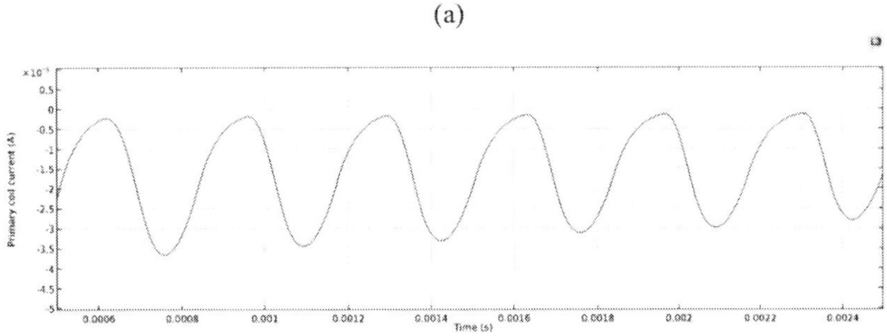

(b)

Fig. 5. Primary and secondary coil current timing diagram.

The waveform coincides with the results of the experiment described in [11]. The output voltage has an error of 15-20%. This can be explained by the difference between the core material used in the simulation and the experimental studies.

Decomposition into Fourier coefficients confirms the presence of a pronounced second harmonic in the output signal (Figure 6).

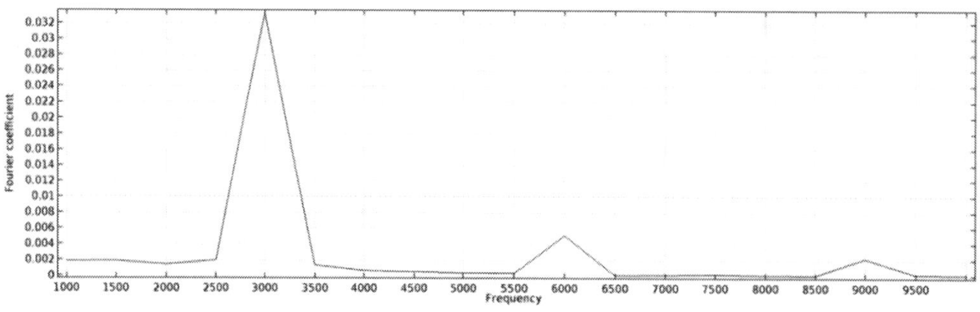

Figure 6. Factorization Fourier coefficients.

4. Conclusion

The paper describes the simulation main stages of a fluxgate magnetometer in Comsol Multiphysics software package. The result confirms the possibility of simulating similar measuring device by finite element methods. The shape and amplitude of the output signal in the measuring coil coincide with sufficient accuracy with the results of experimental studies performed by other authors. In the future, we plan to carry out a construction thermal analysis for the purpose of calculation the heat release, depending on the shape of the excitation signal.

Acknowledgments

The research is funded from Russian Science Foundation (RSF), Grant Number 17-79-10083.

References

[1] P. Ripka, *Magnetic Sensors and Magnetometers* (Artech House, Boston, 2001)
[2] F. Kaluza, A. Gruger , H. Gruger, Sens. Actuators A, Phys. **106**,48(2003)
[3] P. Ripka, Sens. Actuators A, Phys. **33** (3), 129 (1992)
[4] P. Ripka, IEEE Sens. J. **10**(6), 1108 (2010)
[5] B. Andta, S. Baglioa, A. R. Bulsarab, C. Trigonaa Sens.&Actuat. **151**,145 (2009)
[6] S. Uchaikin, A. Likhachev, F. Cioata, et al., LT26 (2011)
[7] M. Janosek, P. Ripka, Sens.&Actuat. **151**,141(2009)
[8] W. Magnes, M. Díaz-Michelena, IEEE Trans. on Magnet. **45**(10), 17 (2009)
[9] M. Vopàlensky, P. Ripka, A. Platil, Sens. Actuators. **106**, 38(2003)
[10] V. Baranova, et al., XXI IMEKO, 4 (2015)
[11] U. Afanasiev, *Ferrozondovie pribory* (Energoatomizdat, Leningrad,1986)

EVOLUTION OF THE BRAIN COMPUTING INTERFACE (BCI) AND PROPOSED ELECTROENCEPHALOGRAPHY (EEG) SIGNALS BASED AUTHENTICATION MODEL

Qaseem Ramzan[1, 2*], *Stanislav* Shidlovskiy[1]

[1]National Research Tomsk State University. 634050, Tomsk, Russia
[2]toggleTechs (a creativity platform), Islamabad, Pakistan

Abstract. With current advancements in the field of Brain Computer interface it is required to study how it will affect the other technologies currently in use. In this paper, the authors motivate the need of Brain Computing Interface in the era of IoT (Internet of Things), and analyze how BCI in the presence of IoT could have serious privacy breach if not protected by new kind of more secure protocols. Security breach and hacking has been around for a long time but now we are sensitive towards data as our lives depend on it. When everything is interconnected through IoT and considering that we control all interconnected things by means of our brain using BCI (Brain Computer Interface), the meaning of security breach becomes much more sensitive than in the past. This paper describes the old security methods being used for authentication and how they can be compromised. Considering the sensitivity of data in the era of IoT, a new form of authentication is required, which should incorporate BCI rather than usual authentication techniques.

1 What is a Brain Computer Interface?

What is a Brain Computer Interface (BCI)? A brain computer interface, sometimes called a mind-machine interface (MMI), direct neural interface (DNI), or brain–machine interface (BMI), is a direct communication pathway between an enhanced or wired brain and an external device. BCIs are often directed at researching, mapping, assisting, augmenting, or repairing human cognitive or sensory-motor functions [1].

1.1 History of BCI Research

In the 1970s, research on BCIs started at the University of California, which led to the emergence of the expression brain–computer interface. The focus of BCI research and development continues to be primarily on neuroprosthetics applications that can help to restore

* Corresponding author: qaseemramzan0@gmail.com

damaged sight, hearing, and movement. The mid-1990s marked the appearance of the first neuroprosthetic devices for humans. BCI does not read the mind accurately, but detects the smallest of changes in the energy radiated by the brain when you think in a certain way. A BCI recognizes specific energy/ frequency patterns in the brain.

John Donoghue and his team of researchers from Brown University formed a public traded company, Cyberkinetics, in 2001. The goal was to design a brain computer interface, the so-called BrainGate, commercially. The company has come up with NeuroPort ™, which was its first commercial product. Researchers from Columbia University Medical Center have successfully monitored and recorded electrical activity in the brain with improved precision. According to researchers, NeuroPort™ Neural Monitoring System enabled them to identify micro-seizure activity prior to epileptic seizures among patients [2].

1.2 Current Standing of BCI

BCIs comprise an active area of research and could start to integrate advances from adjacent fields such as neuroscience, nanomaterials, electronics miniaturization, and machine learning. For example, one neuro-imaging research project is starting to make guesses as to what participants see during brain scans, purporting to be able to distinguish between a cat and a person. Merging this kind of functionality with BCIs might produce new applications. Other experimental BCI projects have been proposed. One is Neocortical Brain-Cloud Interface: autonomous nanorobots that could connect to axons and neuronal synaptic clefts, or embed themselves into the peripheral calvaria and pericranium of the skull. Another project, Brainets, envisions linking multiple organic computing units (brains) to silicon computing networks. The third project is Neural Dust, in which thousands of 10-100 micron-sized free-floating sensor nodes would reside in the brain and provide a computing processing network [3].

1.3 Future of BCI

A number of developments have been taking place in the field. By 2050, it is has been suggested that BCI could become a magic wand, helping men to control objects with their mind. The day isn't far off when man may be able to guide an outside object with their thoughts in order to consistently execute both natural and complex motions of everyday life.

So far BCIs have been conceived primarily as a solution for medical pathologies. However, it is possible to see BCIs more expansively as a platform for cognitive enhancement and human-machine collaboration. The BCI functionality of typing on a keyboard with your mind suggests the possibility of having an always-on brain-Internet connection. Consider what the world might be like if each individual had a live 24/7 brain connection to the Internet. Just as cell phones connected individual people to communications networks, BCIs might similarly connect individual brains to communications networks. We propose a variety of BCI applications and concepts throughout the rest of this paper, all of which are speculative.

In one sense, ubiquitous BCIs are expected. It is contemplated that communication technology, already mobilized to the body via the cell phone, could be "brought on board" even more pervasively. BCIs are merely a next-generation improvement to the current situation of people constantly staring at their phones. In another sense, though, BCIs are not only a "better horse" technology: they are also a "car" in that it is impossible to foresee the full range of future applications that might be enabled from the present moment. BCIs pose a variety of practical, ethical, and philosophical issues. Life itself and the definition of what it is to be human could be quite different in a world where BCIs are widespread. Some of the immediate practical concerns of BCIs could include invasiveness, utility, reversibility, support, maintenance, upgradability (hardware and software), anti-hacking and anti-virus protection, cost, and accessibility. Beyond practical concerns, there are ethical issues regarding privacy and security. For example, neural data privacy rights are an area where standards need to be defined. There could be at least three classes of BCI applications introduced in graduated phases of risk and complexity: biological cure and enhancement; information and entertainment; and self-actualization (realization of individual cognitive and artistic potential). Each of these is for separate discussion [4].

1.4 BCI in view of IoT (Internet of Things)

The IoT (Internet of Things) is a network of physical devices, vehicles, home appliances, and other items embedded with electronics, software, sensors, actuators, and network connectivity, which enable these objects to connect and exchange data. Each thing is uniquely identifiable through its embedded computing system but is able to inter-operate within the existing Internet infrastructure. Experts estimate that the IoT will consist of about 30 billion objects by 2020 [5].

The advancement of consumer electroencephalography (EEG) devices opens the technology up to a completely new world of BCI options. Combining EEG technologies with other IoT technologies like heart rate monitoring, facial emotion recognition, smart homes, home security systems, and personal devices will put new threats of privacy and security breach.

1.5 Operation principles of BCI

The purpose of a BCI is to detect and quantify features of brain signals that indicate the user's intentions and to translate these features in real time into device commands that accomplish the user's intent. To achieve this, a BCI system consists of four sequential components.

1. Signal Acquisition.
2. Feature Extraction.
3. Feature Translation.
4. Device Output.

These four components are controlled by an operating protocol that defines the onset and timing of operation, the details of signal processing, the nature of the device commands, and the oversight of performance. An effective operating protocol allows a BCI system to be flexible and to serve the specific needs of each user.

Electrical signals from brain activity are detected by recording electrodes located on the scalp, on the cortical surface, or within the brain. The brain signals are amplified and digitized. Pertinent signal characteristics are extracted and then translated into commands that control an output device, such as a spelling program, a motorized wheelchair, or a prosthetic limb. Feedback from the device enables the user to modify the brain signals in order to maintain effective device performance. BCI is brain-computer interface; ECoG is electrocorticography; EEG is electroencephalography [6].

1.5.1 Signal Acquisition

Signal acquisition is the measurement of brain signals using a particular sensor modality (e.g., scalp or intracranial electrodes for electrophysiologic activity, fMRI for metabolic activity). The signals are amplified to levels suitable for electronic processing (and they may also be subjected to filtering to remove electrical noise or other undesirable signal characteristics, such as 60-Hz power line interference). The signals are then digitized and transmitted to a computer.

1.5.2 Feature Extraction

Feature extraction is the process of analyzing the digital signals to distinguish pertinent signal characteristics (i.e. signal features related to a person's intent) from extraneous content and representing them in a compact form suitable for translation into output commands. These features should have strong correlations with the user's intent. Because much of the relevant (i.e. most strongly correlated) brain activity is either transient or oscillatory, the most commonly extracted signal features in current BCI systems are time-triggered EEG or ECoG (Electrocorticography) response amplitudes and latencies, power within specific EEG or ECoG frequency bands, or firing rates of individual cortical neurons. Environmental artifacts and physiologic artifacts such as EMG (Electromyographic) signals are avoided or removed to ensure accurate measurement of the brain signal features.

1.5.3 Device Output

The commands from the feature translation algorithm operate the external device, providing functions such as letter selection, cursor control, robotic arm operation, and so forth. The device operation provides feedback to the user, thus closing the control loop.

1.5.4 Feature Translation

The resulting signal features are then passed to the feature translation algorithm, which converts the features into the appropriate commands for the output device (i.e. commands that accomplish a user's intent). For example, a power decrease in a given frequency band could be translated into an upward displacement of a computer cursor, or a P300 potential could be translated into selection of the letter that evoked it. The translation algorithm should be dynamic to accommodate and adapt to spontaneous or learned changes in the signal features and to ensure that the user's possible range of feature values covers the full range of device control.

1.6 Programming Using OpenBCI

OpenBCI is an open source brain-computer interface platform created by Joel Murphy and Conor Russomanno after a successful Kickstarter campaign in late 2013. OpenBCI boards can be used to measure and record electrical activity produced by the brain (EEG), muscles (EMG), and heart (EKG), and is compatible with standard EEG electrodes. The OpenBCI boards can be used with the open source OpenBCI GUI, or they can be integrated with other open-source EEG signal processing tools.

1.6.1 Hardware

The OpenBCI 32-bit Board uses the ADS1299, an IC developed by Texas Instruments for biopotential measurements. The OpenBCI uses a microcontroller for on-board processing — the 8-bit version (now deprecated) uses an Arduino-compatible ATmega328P IC, while the 32-bit board uses a PIC microcontroller — and can write the EEG data to an SD card, or transmit it to software on a computer over a bluetooth link. In 2015, OpenBCI announced the Ganglion board with a 2nd Kickstarter campaign. It costs $100 (1/5 the cost the 32-bit board). It has four input channels for measuring EEG, EMG, and EKG, and is also bluetooth enabled [7]. Table 1 presents price details of OpenBCI Hardware.

Table 1. Price Details of OpenBCI Hardware (as of December 2017)

Index	Hardware	Price ($)
1	Gangolion Board (4-channels)	199.99
2	Cyton Biosensing Board (8-channels)	499.99
3	Cyton + Daisy Biosensing Board (16-channels)	949.99
4	WiFi Sheild	149.99
5	Ultra Cortex Mark "IV" EEG Headset	349.99
6	Gold Cup Electrodes	29.99

1.6.2 Software

OpenBCI has released an open-source application for use with the OpenBCI written with Processing. Display and processing software written in NodeJS and Python are also being developed [8]. Table 2 presents the major open-source repositories of OpenBCI.

Table 2. Major Open-Source Repositories of OpenBCI (as of December 2017)

Index	Repositories	Details
1	OpenBCI_GUI	OpenBCI_GUI is a cross platform application for the OpenBCI Cyton and Ganglion. Tested on Mac, Windows and Ubuntu/Mint Linux. Ganglion requires some extra configuration on Linux/Windows.
2	OpenBCI_WIFI	Firmware library that runs on the Push The World WiFi Shield to stream data from OpenBCI Boards
3	OpenBCI_HUB	An electron application for the ganglion.
4	OpenBCI_WiFi_Sheild	Schematics for the OpenBCI wifi shield
5	OpenBCI_NodeJS_WiFi	NodeJS driver for the Wifi Shield
6	OpenBCI_NodeJS	Node.js SDK for the OpenBCI Biosensor Board written by Push The World
7	OpenBCI_32bit_Library	Repository for OpenBCI 32bit Arduino Libraries
8	OpenBCI_JavaScript_Utilities	Utilities for OpenBCI boards
9	OpenBCI_NodeJS_Cyton	The official NodeJS driver for the Cyton board over Serial.

2 Security Challenges in The Era of IoT

This paper proposes a need and plan of switching to more secure authentication methods for the age of Brain Computer Interface and IoT. This new security method should use BCI as a source of authentication method instead of old authentication methods. In the proposed method, authentication should be checked by the EEG Signals from brain and authentication should be granted by verifying the pattern with previously stored pattern.

2.1 Authentication Methods Being Used Currently

Authentication is an essential element of a typical security model. It is a process aimed to confirm identification of a user (in some cases, a machine) that is trying to access resources or log in. There is a number of different authentication mechanisms, but all serve this same purpose. Some of the security methods being used for Secure Authentication these days are listed below.

1. Passwords.
2. PIN Code.
3. Finger Print.
4. Pattern Recognition Systems.
5. 2-factor Authentication.

Security of all these methods could be compromised one way or the other. There is a proven history of the breach of security with all specified methods.

3 A Look into BCI Authentication Methods

BCI Authentication methods will use the pure EEG or EEG + EMG signals to authenticate a user over computer or mobile devices. This will allow the user to be authenticated by brain signals safely without using any other input methods. Such authentication system will be based on EEG signals.

3.1 Electroencephalogram (EEG) Model for Authentication

Electroencephalography (EEG) is the recording of electrical activity along the scalp through measuring voltage fluctuations accompanying neurotransmission activity within the brain. The electrodes are attached in a cap-like device. It has unique usability advantages over other types of brain signal recording that recommend it for commercial use. It is portable, inexpensive, and easy to use. EEG recording also provides high temporal resolution. However, its signal to noise ratio and spatial resolution represent a limitation compared to other methods [9].

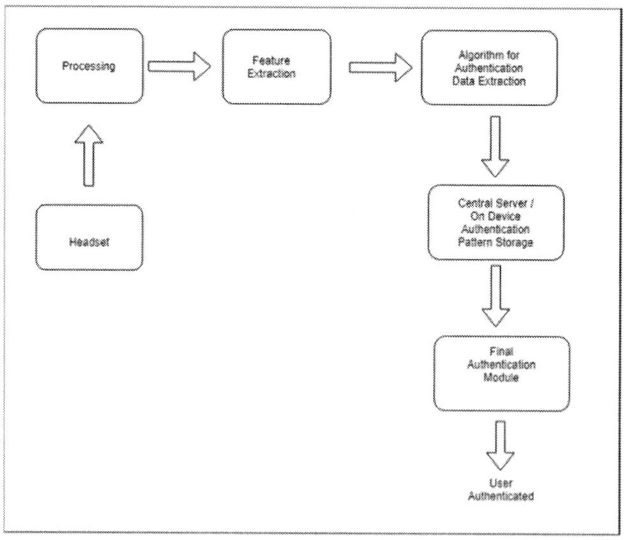

Fig. 1. Proposed Authentication Model Using EEG Signals.

Figure 1 shows an authentication model using EEG signals proposed by the authors. Signals acquired from headset will be processed and then the important features will be extracted from the raw EEG signals. A custom authentication algorithm extracts important authentication related information and makes a package, which could be checked for authentication. This authentication data package is then compared with previously stored data pattern on either a server or local storage in order to verify if that it is the same user leading to authentication or rejection of user.

3.2 Current Challenges

Currently there is a number of challenges causing difficulty for the research process. Once the research is successful, still there will be some limitations to produce this as a locally useable device.

3.2.1 Usability Challenge

The headsets we can find currently in the market are costly due to expensive sensors. These devices cannot be used directly and require to setup the electrodes and put a conducting gel on the scalp of a person for signals reading.

3.2.2 Training Process & ITR (Information Transfer Rate)

Training processes express the limitations facing the user acceptance of BCI technology. They include the issues related to the training process necessary for class discrimination. ITR (Information transfer rate) is one of the system evaluation metrics that combine both performance and acceptance aspects [10–13].

Users training is a time-consuming activity either in guiding the user through the process or in the number of recorded sessions. It takes place either in preliminary phase or in the classifier calibration phase. The user is taught to deal with the system as well as to control his or her brain feedback signals in the preliminary phase, while in the calibration phase the trained subject's signal has been used to learn the used classifier.

ITR (Information Transfer Rate) is a widely used evaluation metric for command BCI systems. It depends on the number of choices, the accuracy of target detection, and the average time for a selection. Thus compared to imagery BCI, selective attention strategies achieve higher ITR as their offered choices are larger.

3.2.3 Non-linearity

The brain is a highly complex nonlinear system, in which chaotic behavior of neural ensembles can be detected. Thus, EEG signals can be better characterized by nonlinear dynamic methods than linear methods. The training sets are relatively small because usability issues influence the training process. Although heavily training sessions are considered time consuming and demanding for the subjects, they provide the user with necessary experience to deal with the system and learn to control his/her neurophysiological signals. Thus, a significant challenge in designing a BCI is to balance the trade-off between the technological complexity of interpreting the user's brain signals and the amount of training needed for successful operation of the interface [14].

4 Conclusion

Brain Computer Interface is a newly emerging technology, which is soon going to surround all other technologies and will be involved in daily life activities. It can be easily understood that BCI is going to integrate with IoT on very large scale and this integration will be a new rise in the technological advancements. At the same time, there is a need to find new authentication methods, which are more secure and do not use conventional authentication techniques but rely totally on EEG & EMG Signals of Brain Computing Interface.

Acknowledgements

The research presented in this paper was supported by the Russian Foundation for Basic Research (grant No 16-29-04388/17). The authors are grateful to Tatiana B. Rumyantseva from Tomsk State University for English language editing.

References

[1] Brain Computer Interface. URL: https://en.wikipedia.org/wiki/Brain–computer_interface
[2] History of Brain Computer Interfaces. URL: http://www.brainvision.co.uk/blog/2014/04/the-brief-history-of-brain-computer-interfaces
[3] Future of Brain Computing Interfaces (Block-chaining Your Way into a Cloud Mind). URL: http://jetpress.org/v26.2/swan.htm
[4] Melanie Swan. Future of Brain Computing Interfaces (Block-chaining Your Way into a Cloud Mind). URL: http://jetpress.org/v26.2/swan.htm
[5] Internet of Things.URL: https://en.wikipedia.org/wiki/Internet_of_things
[6] Jerry J. Shih. Brain-Computer Interfaces in Medicine [Electronic resource] URL: https://www.ncbi.nlm.nih.gov/pmc/articles/PMC3497935/
[7] OpenBCI. URL: https://en.wikipedia.org/wiki/OpenBCI
[8] OpenBCI Repositories. URL: https://github.com/OpenBCI
[9] Sarah N. Abdulkader. Egyptian Informatics Journal. Brain Computer Interface Applications and Challenges. URL: http://www.sciencedirect.com/science/article/pii/S1110866515000237
[10] Sarah N. Abdulkader. Egyptian Informatics Journal. Information Transfer Rate. URL: http://www.sciencedirect.com/science/article/pii/S1110866515000237
[11] V. Syryamkin, V. Shidlovsky. *Correlation-extremal direction-finding systems* (Tomsk State University publishing house, 2010)
[12] D.V. Shashev, S.V. Shidlovskiy, Opt. Inst. D. Proc. **51**, 227 (2015)
[13] D.V. Shashev, MATEC Web Conf. **79**, (2016)
[14] Sarah N. Abdulkader. Egyptian Informatics Journal. Non-Linearity. URL: http://www.sciencedirect.com/science/article/pii/S1110866515000237

THE PROCESS AUTOMATION OF TEMPERATURE MEASUREMENT BY MEASURING FACILITY WHICH IS CONSISTS FROM AGILENT E8363B AND FREEZER/HEATING CHAMBER TNC-80

Olga Dotsenko, *Kirill* Frolov*

National Research Tomsk State University, 36 Lenin Ave., Tomsk, Russia 634050

Abstract. The vector network analyzer Agilent PNA series E8363B and freezer and heating chamber TNC-80 were used to measure electromagnetic parameters of natural and artificial substances under temperature influences. LabView 2016 software was used to the virtual instrument generation. This virtual instrument is needed for efficient electromagnetic parameter and temperature measurement automation. The virtual instrument framework and faceplate of the virtual instrument are shown. The present work involves an experimental study of the magnetic permeability of ferrites with hexagonal structure under temperature influences. It is show that there is a nonlinear influence of temperature on magnetic permeability of W-type ferrites.

1 Introduction

The temperature measurements of electromagnetic parameters of natural and artificial substances are very a laborious process if seen in the context of the major staff time. There are many factors which are affecting on the duration of the process. They are volume temperature gradient, the temperature range that is chosen for research, number of selected parameters of test samples, aggregate state (solid, crystal powder, or liquid) and chemical composition of the test samples, etc. The timing of the experiment shows that about 95% of the staff time is wasted. In that time the temperature increases or decreases to the set value.

There are huge amount freezer or heat measuring facilities which cost in the middle price range. These measuring facilities are not adapted to cooperate with peripheral facilities [1]. In particular there are the measuring facilities of high technological level. For example it is Agilent PNA series E8363B which will be regarded below in this article. These facilities are adapted to cooperate with peripheral facilities and have application software for these purposes. These facilities have the serial interface and cost in the high price range [2]. Standard features for the equivalents measuring facilities have not had special functions for such research. But they have a lot of upgrade opportunities. As a result

* Corresponding author: FrolovKirill.O@yandex.ru

the forward-looking task is to creation of fully automate measuring facilities for research effect of temperature on electromagnetic parameters of substances.

2 Hardware and software

The present work deals with the creation automate measuring facility to research effect of temperature on electromagnetic parameters of different substances. That facility based on vector network analyzer Agilent PNA series E8363B and freezer and heating chamber TNC-80 and personal computer. LabVIEW (National Instruments, 2016) is the software environment capable to creation virtual instrument framework. It is used for creation a virtual automate measuring facility.

To resolve the assigned task both NI LabVIEW 16 and software module NI LabVIEW Real Time 15 SP1, and drivers Agilent PNA series E8363B [3] were used. NI VISA 15.5, NI 4882_15 software drivers are open to access on the Company official website NI. Agilent 82357B USB/GPIB was used for connecting Agilent PNA series E8363B and personal computer. TESTA software and LAN interface were used for connecting freezer and heating chamber TNC-80 and personal computer.

Figure 1 shows a schematic diagram of automate measuring facility to research effect of temperature on electromagnetic parameters of different substances: 1 – Agilent PNA series E8363B; 2 – Agilent 82357B USB/GPIB cable; 3 – personal computer; 4 – Ethernet cable; 5 – freezer and heating chamber TNC-80.

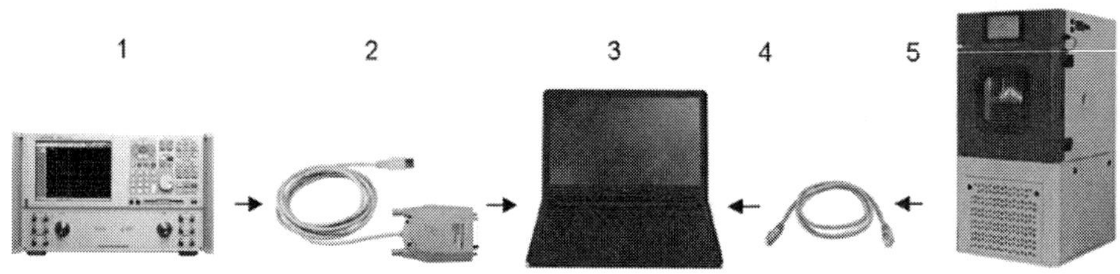

Fig. 1. Schematic diagram of automate measuring facility based on Agilent PNA series E8363B.

Agilent PNA series E8363B vector network analyzer were used to measure S-parameters of measurement cell. It is a main part of automate measuring facility. The Agilent PNA series E8363B network analyzer technical characteristics are following: the authorized frequency bandwidth is from 10 MHz till 110 GHz, the dynamic range is from 110 dB till 0.006 dB.

The volume multimode rectangular cavity was used as measurement cell. The cavity perturbation techniques were used for the evaluation of magnetic permeability of experimental samples at SHF range. In this technique, a cavity was designed with a very small slot at the center of the broad/short wall of rectangular cavity in order to insert the sample material. A coaxial cable was used to connection Agilent PNA series E8363B vector network analyzer and measurement cell.

The freezer and heating chamber TNC-80 is the second part of automate measuring facility. The freezer and heating chamber TNC-80 technical characteristics are following: the range of operating temperatures is from -70 °C to +150 °C with the maximum deviation by ± 0.5 °C [4]. This chamber has two technological gaps to input measurement cell which is connected with Agilent PNA series E8363B vector network analyzer.

The proper system software was installed on a personal computer. Also it was install TESTA software at same computer. The interaction between personal computer (3) and Agilent PNA series E8363B (1) accessed via an Agilent 82357B USB/GPIB cable (2). The interaction between personal computer (3) and freezer and heating chamber TNC-80 (5) accessed via an Ethernet cable (4).

TESTA software was used for gathering temperature data. The current chamber TNC-80 temperature and other current session data write to the file. This file is recorded into the TESTA program directory in real time procession every 5 seconds.

Before starting the measuring procedure it is necessary to perform the actions described below. Start TESTA program. Set the function heating and cooling chamber process program. Start Virtual Instrument program (Figure 2). Download file path TESTA into Virtual Instrument program. Connect to the Agilent 82357B via VISA resource name Agilent. The measurement points set up in the window «Temperature measurement».

a)

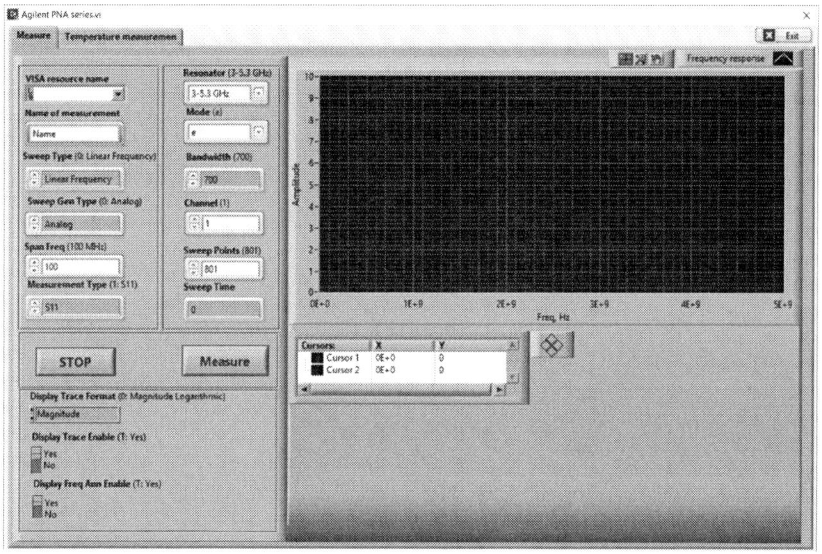

b)

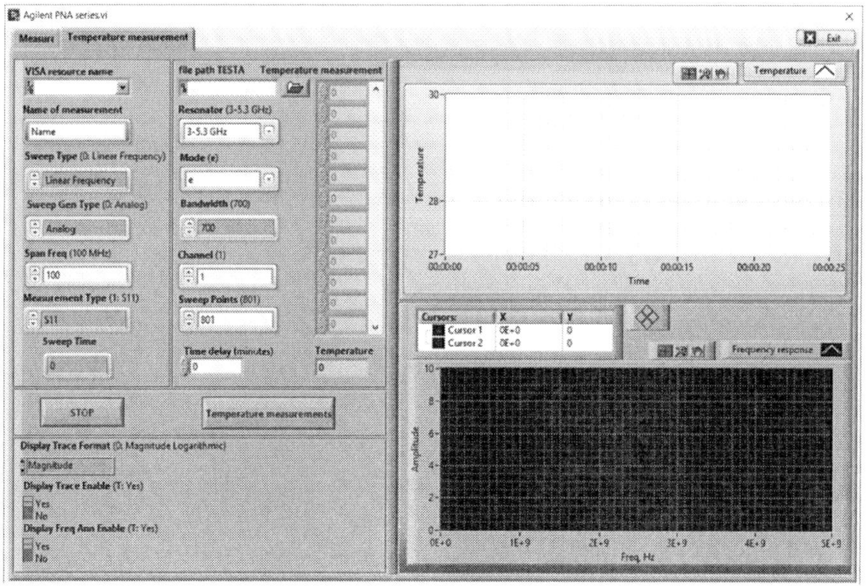

Fig. 2. Agilent PNA series E8363B Virtual Instrument program interface.

The remaining empty fields are given by operator during experiment. They are not mandatory. The contents of empty fields are given by the Agilent PNA series E8363B measurement conditions.

After pushed on button «Measure/Temperature measurements» the measurement system starts operating in the stand-alone mode. It monitors the current temperature in the chamber and records the electromagnetic parameters of measurement cell when the temperature reaches a set value. The results of measurement save into folder which was creation by Virtual Instrument program. This folder and Virtual Instrument program are stored in the same directory.

Since the chamber port interacts with TESTA program, the main difficulty is the direct interaction between Virtual Instrument program and the temperature sensors.

A block diagram of the Virtual Instrument program is demonstrated in Figure 3.

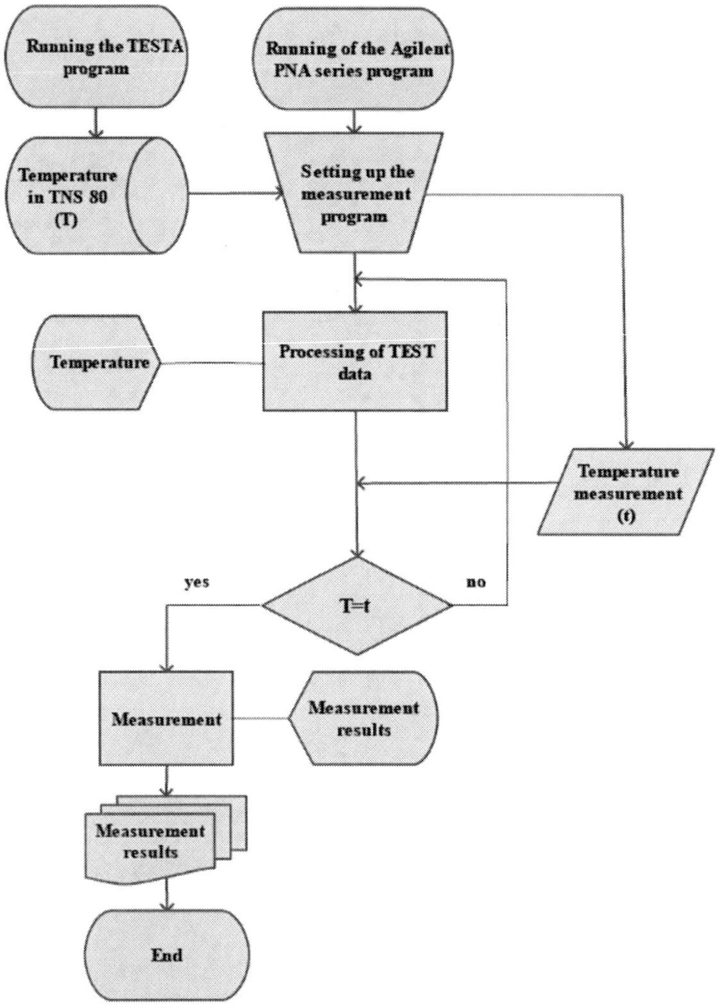

Fig. 3. A block diagram of the Virtual Instrument program for temperature measurements.

A block diagram shows the software algorithm, when the Virtual Instrument program is worked according Figure 2*b*. The essential difference between Figure 2*a* and Figure 2*b* is that a Figure 2*b* is necessary to set temperature measurement points. Running a program according Figure 2*a* (Setting up the measurement program) is an amplitude frequency response measurement. The amplitude frequency response saves to a local folder.

3 Results and discussion

To establish correct functioning of the Virtual Instrument program it is necessary to measure the effect of temperature on the magnetic permeability of W-type hexaferrite powder [5]. The hexaferrite powder was put into quartz tube with inner diameter 2 mm. After that the quartz tube was put into rectangular cavity. After that the rectangular cavity with test sample was put into chamber. The measuring point was equal 7.29 GHz. The measurements were carried out under conditions of temperature from 230 K till 315 K. Temperature correction coefficient of frequency was defined before measure. It was equal 17.45×10^{-6} K^{-1}. It was used to calculate of magnetic permeability of test sample. The measurements were carried out every 10 K.

The complex magnetic permeability was calculated by following formulas:

$$\frac{\mu'}{\mu_0} = 1 - \frac{(l^2 + a^2 p^2) V_c \Delta\omega'}{a^2 p^2 V_s \omega_0}, \quad (1)$$

$$\frac{\mu''}{\mu_0} = \frac{(l^2 + a^2 p^2) V_c \Delta\omega''}{a^2 p^2 V_s \omega_0}, \quad (2)$$

where $\Delta\omega' = \omega - \omega_0$ is the shift of resonance frequency to left side, $\Delta\omega'' = (\delta\omega - \delta\omega_0)/2$ is the change in the half-width of the resonance line when the sample is placed in the cavity, V_c is the volume of the test sample, Vs is the volume of the cavity, p is the number of wave variations along cavity, a is the widthwise of the cavity, l is the lengthwise of the cavity.

Figure 4 shows the calculation results.

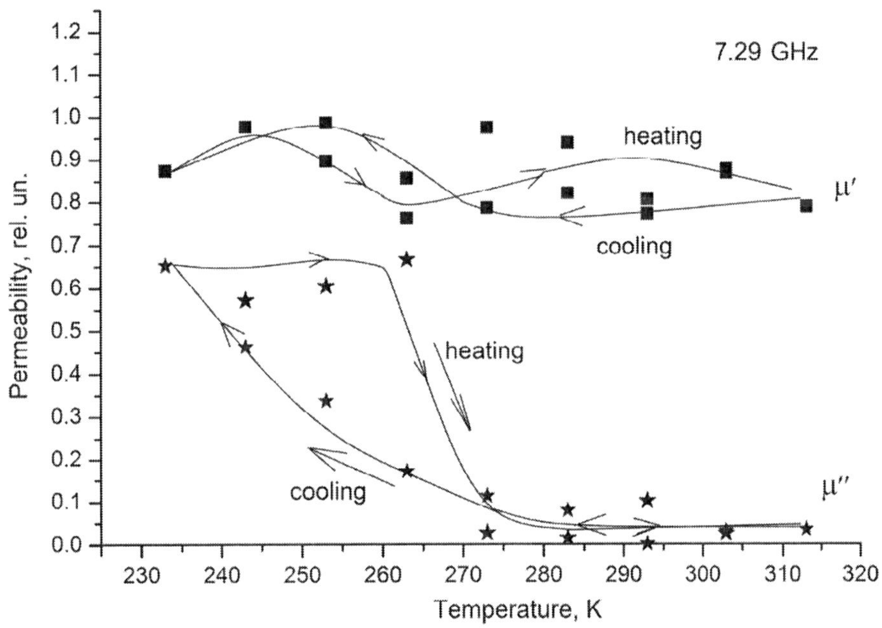

Fig. 4. Complex permeability of W-type hexaferrite powder.

In order to provide an interpretation of the magnetic permeability results, we will use a data from article [5]. This article shows, that W-type hexaferrite bulk sample have nonlinear temperature dependence of magnetic permeability. Note that the similar dependence is observed in our experiment. Since we measured hexaferrite powder we saw that there is a correlation (though quite weak) between the bulk sample and the powder sample. Also we saw that there are the different shapes of the curves both cooling and heating.

4 Conclusions

Virtual Instrument program is very intuitive and easy to use. Even young researcher may operate by virtual automate measuring facility. There is the opportunity to modernize and improve this virtual automate measuring facility in the future. Firstly, there is the opportunity to connect other peripheral facilities. Secondly, there is the opportunity for remote control of measurement process.

The virtual automate measuring facility for research effect of temperature on electromagnetic parameters of substances is used on the Center Radiophysical measurements, diagnostics and research parameters of natural and artificial materials of Tomsk State University. There were plans to adaptation of this program to other measure equipments.

Acknowledgements

The authors would like to thank the Center Radiophysical measurements, diagnostics and research parameters of natural and artificial materials of Tomsk State University for the equipment provided. Authors are thankful to D.V. Aksentev of National Research Tomsk State University for helping in measurements.

This Research is supported by Tomsk State University Competitiveness Improvement Program.

References

[1] Klimaticheskie kamery tepla-kholoda, URL: http://miroborudovaniya.ru /klimaticheskie_kamery_teplo-kholod/
[2] Kamera tepla-holoda KTH-74-65-165-65, URL: http://www.sktb-spu.ru/ katalog/klimaticheskie-kamery/kamery-tepla-holoda/kamera-tepla-holoda-kth-74-65-165-65/
[3] Instrument driver network. Agilent PNA Series Analyzer, URL: http://sine.ni.com/apps/utf8/niid_web_display.download_page?p_id_guid=04755C3E B05960C5E0440003BA7CCD71
[4] Klimaticheskie kamery, URL: http://testa7.ru/testa_equipment_productio n/testa-medium-format-camera/
[5] E.P. Naiden, V.I. Maltsev, G.I. Ryabtsev, Phys. Stat. Sol. **120**, 209 (1990)

METHODS AND APPROACHES FOR AUTOMATIC PROCESSING AND STORAGE OF HIGH-POTENTIAL ELECTROCARDIOGRAM REGISTERED BY HARDWARE AND SOFTWARE COMPLEX ON NANOSENSORS

Sergey Rybalka[1], *Mikhail* Yuzhakov[a1], *Maxim* Ivanov[1], *Nguyen* Dang Kuang[1], *Radik* Kodermjatov[1], *Guo* Wenjia[1], *Ivan* Maksimov[2], *Ilya* Zimin[2]

[1] National Research Tomsk Polytechnic University, 634050, Tomsk, Russia
[2] Institute for Cardiology of the Russian Academy of Medical Sciences, 634012, Tomsk, Russia

Abstract. The problems of designing the system for automatic analysis of cardiograms are considered. The structure of the system for recording and storing electrocardiograms was formed. The procedures that make up the complex of automatic analysis are listed.

1 Introduction

Despite the advances in the field of cardiology, a number of research studies are currently being pursued [1]. A large number of studies are devoted to the investigation and development of means to monitor cardiac functioning without surgical interference. For many years, scientists of Tomsk Polytechnic University together with colleagues from Tomsk Scientific Research Institute of Cardiology have been studying bioelectrical activity of the heart in an extended frequency range of 0–10,000 Hz [2, 3]. A hardware and software complex based on nanosensors capable of recording signals from 0.3 μV with a sampling frequency of 32 kHz and 64 kHz was developed. ECG is currently being studied and analyzed [4, 5]. The preclinical studies carried out in healthy volunteers showed the possibility of recording the cardiac micropotentials from 0.3 μV with duration from 0.3 ms and the relationship between the micropotential parameters and the human heart state [6]. Unlike conventional electrocardiography in a narrow frequency range of 0.05–150 Hz [7], recording of a high-precision ECG requires a large amount of memory, especially in case of regular heart monitoring at home. For this reason, it is necessary to develop new approaches to record and store the obtained data.

The tasks can be divided into several groups:

1. Analysis and classification of the ECG form.
2. Analysis of the cardiac pulse elements.
3. Evaluation of the patient's condition: healthy–pathology.

[a] Corresponding author: libra2000@yandex.ru, libra@tpu.ru

The paper aims to formulate the requirements for the system for automatic analysis of ECGs, to consider some aspects of the cardiac signal recording and to focus on the individual cardiac signal properties in order to identify the system of requirements imposed on the processing and storage system.

A multi-channel system with a high polling frequency and a long-term periodic monitoring of the patient's heart requires the storage of a large amount of data. Modern electronics enables the storage of large data volumes in miniature units. To store very large data volumes, the system provides two separate units, a miniature recorder and a data collection and storage unit (DCSU). Today, the DCSU can be implemented on smartphones or special units of a similar size [8]. The patients' ECGs are gradually copied and stored in a single database, for example, in the clinic (see Figure 1).

During periodic monitoring of the heart, it is important to know what the patient is doing at the moment when the problem arises and prior to that [9]. To record the patient's behavior, the system can be supplemented with portable means to record video and audio signals. That is, a short video fragment showing the external environment can be shot – the room and those present in the room. The image may be of low resolution and should remind the patient of what was happening at a certain moment of time.

During ECG recording, video recorders of the cardiac signal transmit the data to the DCSU and indicate the time of recording. This allows synchronization of the ECG and the external environment.

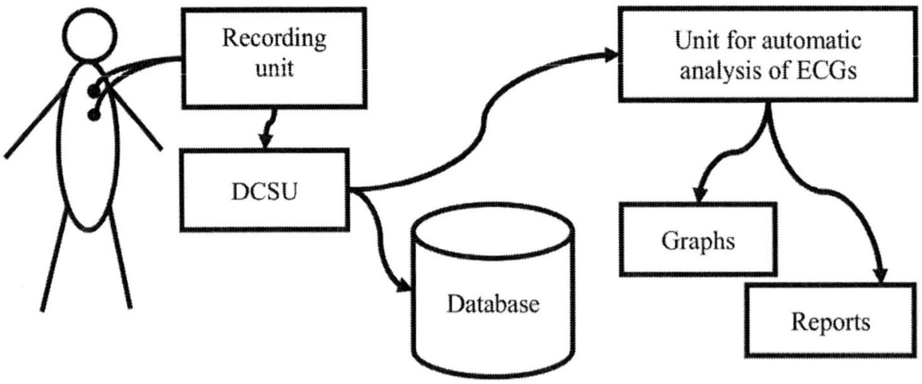

Fig. 1. Structural scheme of the complex for ECG recording and automatic.

The accumulation of a large amount of data requires the development of a robotic system for automatic processing of the ECG signal. This system should automatically process the data and provide signals indicating abnormalities.

2 Problems of the complex for ECG recording and automatic processing

The recorded cardiac signals can be divided into two groups – short and multiply repeated signals with a certain cycle during dynamic monitoring of the heart. Short ECGs are similar to conventional ones, and they are recorded when a patient is at rest and after physical activity. The recording is performed within several minutes. The recording environment (at rest, load type) is registered in the log.

The examination of a large number of patients requires storage of large data volumes. Therefore, the system should be able to maximally compress the data.

This can be achieved by combining several methods:

1. Removal of old data. Since the system is capable of automatic detection of abnormalities or changes in the ECG form, these can be early revealed through monitoring of a healthy person's condition. However, the storage of all data on the normal condition of a patient is meaningless. It is enough to store the data on the current normal state of the heart. Therefore, the ECG itself can be removed or a sample fragment of the normal ECG can be stored. This enables the analysis of the ECG dynamics with regard to the patient's age.

2. Identification of typical areas:

a. Typical cardiac pulses are pulses that are similar to each other and an overwhelming majority of them can be observed on the ECG.

b. Not typical cardiac pulses are pulses which form differs from that of typical ones. They constitute a small percentage of the total signal duration. These cardiac pulses should be divided into several groups in case they significantly differ from each other.

If the signal is partially removed, the corresponding video fragments are also deleted. The video data compression is possible by substituting this data for the data about the patient's condition – rest, load type, and the type of activity the patient is involved at the moment. However, processing should be done manually.

In addition to compressing the stored data, the system should automatically recognize pathologies in typical cardiac pulses and in pulses that are different from typical ones. If there are abnormalities in the ECG or it significantly differs from the previous ECGs of the patient, the system should issue a warning. The doctor analyzes the ECG in an automated mode, prescribes other types of analyses, etc.

The analysis, which can be performed using both short ECG recordings and those repeated with a certain cycle, has its own characteristics. Therefore, it is necessary to create a list of requirements for the system for automatic processing of short recordings separately from the requirements for processing of repeated recordings.

3 Requirements for the unit for automatic analysis of ECGs

The feature of the unit for processing short recordings of cardiac signals is that the obtained recording can be fully loaded into the computer's RAM. In this case, all the analysis tasks that are part of the program can be solved using this recording. The main stages of processing short recordings, the units to implement these stages and the corresponding requirements can be as follows.

Setting the parameters for short signal processing. The parameters of the subsequent automatic processing stages are set at this stage that is performed manually. The processing system is supposed to be automatic. However, the parameters of each of the processing stages should remain unchanged. This increases the processing system flexibility.

Short signal loading. At this stage, files are loaded in the internal format of the device. The high-frequency signal recording requires a high-speed apparatus and simple pre-processing methods. Therefore, the signal in the recording device is typically stored in an analog-to-digital converter (ADC) data format. This allows making the device tiny and reliable and provides a high speed of signal recording.

When setting a short signal loading stage, you need to specify which single file you want to load, or to create a list of files to be processed.

Removal of final samples. At this stage, the initial and final samples of the cardiac signal are removed. The ECG recording is complicated. Therefore, the initial and final cardiac pulses can be significantly distorted. For example, the patient has not come to a calm state yet, or, on the contrary, he began to show activity at the end of recording.

When setting this stage, the duration of the initial and final sections to be removed is specified.

Short signal conversion. The loaded signal has an integer format, and the range of values equals to the ADC range. To analyze and process the ECG, the values should be in the range of ± 300 mV and in a real format. Therefore, the signal is converted from an integer format into a real one and scaled.

This stage does not require any settings. The ECG recording unit will be modified with time. ECG recording will be performed at different frequencies. The conversion unit should receive the data for correct conversion into the real format.

Processing of short duration ECG recordings. It is the main unit of the system that involves a set of independent and interrelated procedures for processing the initial ECG or the recording that has been processed using other procedures. The set of procedures should include:

- procedures for plotting and calculating characteristics of the heart rate; detection of the cardiac rhythm disturbances;
- procedures for automatic filtration of different types;
- procedures for approximation of a noisy cardiac signal;
- procedures for separation of the ECG recording into individual cardiac pulses;
- procedures for extraction of groups of similar cardiac pulses – typical and not typical of this ECG; classification of cardiac pulses;
- procedures for analysis of the cardiac pulse forms; identification of critical abnormalities;
- procedures for formation of the charts and reports on the processing results, the lists of normal ECGs, with obvious or suspected pathology.

During the system development, the list of procedures will be supplemented with other types of processing. The increased capabilities of this unit and the improved quality of signal processing will enable automatic operation. The doctor will only examine ECGs with obvious or suspected pathology.

The unit setting involves the creation of a list of procedures that must be performed while processing the list of files with ECGs. It is also necessary to determine the types of graphs and the composition of reports for each of the procedures.

In case of repeated studies, for example, at home, processing will be carried out according to the procedure for short ECGs, the data that allows evaluation of the patient's heart state over time will be stored in the memory unit. The program that implements Simson's method has been developed [10, 11]. The program is designed to process high-resolution ECGs [12–14]. This software includes the automation elements and allows detection of ventricular late potentials in both the averaged signal and single cardiac pulses.

A procedure chart of the program is shown in Figure 2.

Some units of the program algorithm (Figure 2) include the following automatically performed actions:

In unit 2, after the signal is loaded, the duration of the cardiocycle is determined. The process of calculating the duration of a cardiac cycle is realized by correlation: the two most correlated adjacent cardiocycles are selected from a larger number of similar ones in the entire ECG record and the distance between the peak R is measured. The obtained time characteristic is used in calculating the search step and the search area of the peak R for detecting all cardiocycles in the signal.

In unit 3, the user selects the type of research. Depending on the choice, the program performs the following: for the study of a single cardiocycle appears interface in the form of a vertical line on the graph and the slider with which it moves; to study the averaged signal, the program starts the algorithm for checking the averaging capability.

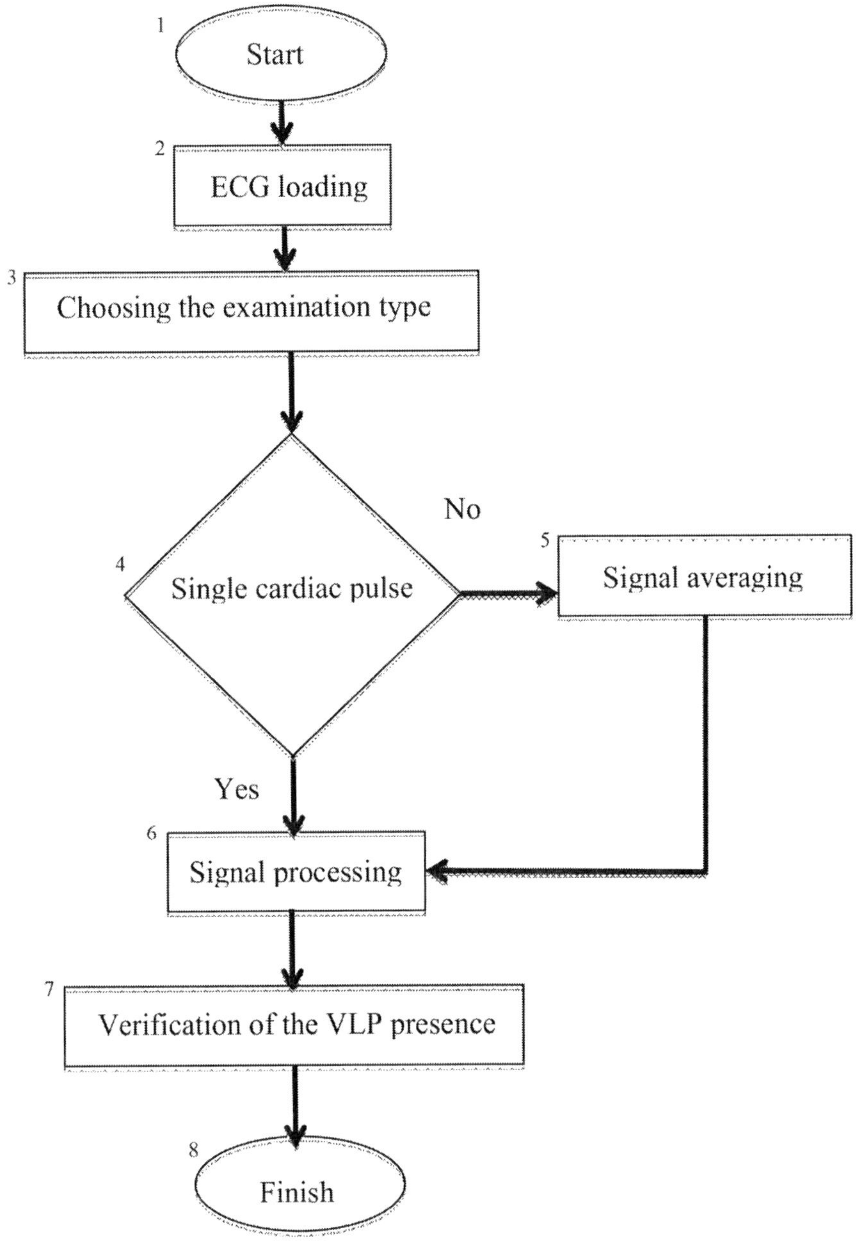

Fig.2. Algorithm for the program.

In unit 5, the signal is averaged. Prior to averaging, the direction and center point of the QRS complex are automatically determined. This procedure is carried out with the help of filtering, which removes excess extremes, and the derivative of the first order. The result of the algorithm for determining the center of the QRS complex can be seen in Figure 3. The summation of the cardiocycles is superimposed on each other depending on the central points of the QRS complex and using the time interval defined in unit 2.

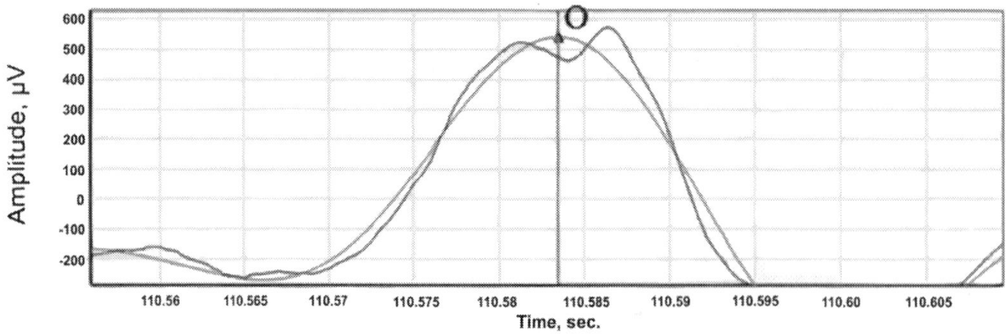

Fig. 3. Centre of the R-wave.

In unit 6, bi-directional bandpass filtering and summing of the signal from the three leads are performed to obtain the resulting Simson's curve using the formula:

$$F(X,Y,Z)=\sqrt{X^2+Y^2+Z^2}\ ,$$

Bi-directional filtration is carried out in two stages: in the first filtration is performed from the beginning of the cardiac cycle to the central point of the QRS complex, in the second from the end of the cardiocycle to the central point of the QRS complex. On the resulting curve, the algorithm automatically calculates the reference points for the diagnosis of the patient.

The actions related to the automatic adjustment of settings or points on the graphs are supported by the possibility of manual adjustment to improve the accuracy of results. Addition of the automation elements to the program allows significant acceleration of the work associated with processing of high-resolution ECG signals.

4 Conclusion

The development of the system for recording and automatic analysis of ECGs requires the design and coordination of operation of several devices. A separate task is to compile a list of procedures for automatic analysis of ECGs and to determine the types of graphs and reports that reflect the analysis results. The paper lists the main requirements for these systems.

References

[1] S. Narayanaswamy, Indian Pacing Electrophysiol J. **2**, 50 (2002)
[2] M.M. Yuzhakov. *Development and investigation of methods and technique of nanovolt and microvolt level for electrophysological researches: dissertation* (Tomsk, 2012)
[3] M.L. Ivanov *Development and investigation of the ECG hardware and software complex based on nanosensors for recording the heart micropotentials in real time without averaging and filtering: dissertation* (Tomsk, 2015)
[4] D.K. Avdeeva, V.Y. Kazakov, N.M. Natalinova, M.L.Ivanov, M.A. Yuzhakova, N.V. Turushev, European Journal of Physical and Health Education **6**, 1 (2015)
[5] M.G. Grigoriev, N.V. Turushev, Applied Mechanics and Materials **770**, 526 (2015)
[6] N.V. Turushev. *Electrocardiograph for non-invasive recording of spontaneous activity of myocardial cells to early detect the signs of sudden cardiac death: dissertation* (Tomsk, 2016)

[7] D. Ricciardi, I. Cavallari, A. Creta, G. Di Giovanni, V. Calabrese, N. Di Belardino, S. Mega, I. Colaiori, L. Ragni, C. Proscia, A. Nenna, G. Di Sciascio, Journal of electrocardiology **5**, 691 (2016)
[8] Guo Shu-Li, Han Li-Na, Liu Hong-Wei, Si Quan-Jin, Kong De-Feng, and Guo Fu-Su, J. Geriatr Cardiol **6**., 528 (2016)
[9] S. Levine, B.J. Coyne, L.C. Colvin, *Clinical Exercise. Electrocardiography* (Jones & Bartlett Publishers, Burlington, 2015)
[10] M.B. Simson, Circulation **64**, 235 (1981)
[11] J.A. Vassallo, D. Cassidy, M.B. Simson et al, The American journal of cardiology **55**, 985 (1985)
[12] I.A. Lezhnina, K.V. Overchuk, A.A. Uvarov, V.A. Perchatkin, A.B. Lvova, Journal of Physics: Conference Series **881,** 012008 (2017)
[13] D.K. Avdeeva, M.L. Ivanov, S.A. Rybalka, I.V. Maksimov, M.V. Balahonova. Biology and Medicine **4**, 1 (2015)
[14] I.A. Lezhnina, A.A. Boyakhchyan, K.V. Overchuk, A.A. Uvarov, Journal of Physics: Conference Series **881,** 012041 (2017)

EXPERT SYSTEMS AS THE BASIS OF DECISIONS IN THE KNOWLEDGE SOCIETY

Svetlana Kvesko[a1], *Anna* Kornienko[2], *Bronislav* Kvesko[3], *Nataliya* Chicherina[2]

[1]National Research Tomsk State University, 634050, Tomsk, Russia
[2]National Research Tomsk Polytechnic University, 634050, Tomsk, Russia
[3]Siberian Federal University, 660041, Krasnoyarsk, Russia

Absract. The paper discusses the intelligent information systems as expert systems to meet the challenges of management in the knowledge society. These allow optimization of decision-making and simulation of the human and society activities. The focus is made on the relation between the information structure and globalization. It is emphasized that the information structure becomes the basis of public life functioning, which is based on complex industry of information processing through information technologies. Modern management approach focuses on the "strategy" construct. The latter allows timely and adequate effect. This adequacy, in turn, makes it possible to determine the long-term development of an object. The strategy appears to be optimal if it is supported by resources. Using innovations solves the problem of prediction of generated knowledge. An innovation system is able to initiate transformations in the knowledge system. The tasks solved by the expert systems involve the tasks of combined type, namely the tasks of learning, forecasting and monitoring that ensures forecast completeness, which employs a parametric dynamic model.

1 Introduction

Cyber science is one of the areas that implement the results of the research in the intelligent information systems. Database processing in the "science of great potential", the forms of representation and types of knowledge models (logic, production and network models), representation and processing of fuzzy knowledge, search techniques in complex spaces, neural network models, methods and design of neural networks are the issues related to intelligent information systems, which make the basis of cyber science.

In the area of cyber science, the knowledge society, which is a service society in its developed form with knowledge-based service, implements its services and functions. The development of intelligent information systems, which has become an obligatory condition for cyber science development, and implementation of research findings on artificial intelligence to advance cyber science are considered to be relevant due to the role played by the "science of large wharehouses" in the knowledge society through its service function.

[a] Corresponding author: svetla_kvesko@mail.ru

2 Expert systems as the basis of administrative decisions in the knowledge society

Each information system is created to perform the following functions: reception of user requests for information and input data, the use of the input data and the data and knowledge stored in the system for generating and retrieving the requested information. Generally, knowledge is divided into two categories: declarative knowledge (affirmative one that describes the problem domain objects, and their properties and relations) and mandatory knowledge (imperative one that describes the steps to implement the required transformations). In information systems, knowledge is divided into two types referred to as operational and factual knowledge:

- *factual knowledge* is the data content (user-entered and stored in the information system) used to satisfy the request for information;
- *operational knowledge* implies interpretation of data (identification of the request-related data content) and retrieval of requested information (knowledge useful for solving the user's problems).

Currently applied special storage structures represent complex spatio-temporal and conceptual relations between the data. The data warehouse is used to perform data pre-processing and storage. When the data warehouse is filled, three interrelated tasks are being solved: data acquisition, data cleaning and data consolidation. Data acquisition is the process of data transmission from external sources to the data warehouse and its replenishment. When the data warehouse is filled with consolidated data, the data access from sources should run in accordance with the metadata. As data is consolidated in terms of a particular science, consolidation rules are to be defined. Data cleaning is data modification in the course of data warehouse filling up: unwanted duplicate cleaning, missing data restoring, data formatting, unwanted symbol removing, data type unifying and integrity checking. Most currently available products are provided with different sets of data cleaning tools and appropriate diagnostic tools.

Finally, the data mart (sometimes referred to as a data market) is a specialized data warehouse that serves one of the areas of science. The online analytical processing (OLAP) technology is based on a multi-dimensional conceptual view, a natural view of the controlled object by management personnel. Multi-dimensional analysis is simultaneous analysis of several measurements. Each measurement includes sets of consolidated data in the sequence of generalization levels, where each upper level corresponds to data aggregated across an appropriate measurement. The operation of disclosure or drilling down corresponds to processing from higher consolidation levels to lower ones, and a convolution operation or rolling up implies processing from lower to higher levels.

OLAP products are built so that the analyst could be able to: carry out an analysis within a common conceptual scheme; set the defined transformations in functionally complete formal language in a natural and easy-to-use user interface; access the data in user's language (as a rule, the user does not know programming languages); to make request freely, which means processing from lower to higher levels.

MOLAP employs a special non-relational storage model. ROLAP is based on application of specific database structures (*star* and *snowflake* schemas, indexing and aggregation storage) that allows the use of relational database management systems. A hybrid version of OLAP is referred to as HOLAP (hybrid OLAP). Using both ROLAP and MOLAP architecture, one can combine the best features of both architectures. For most OLAP products, preliminary calculation of the aggregates is a key strategy to ensure the performance gain. However, data pre-aggregation is associated with significant costs, since the number of aggregates can easily exceed the volume of the detailed data source that causes a sharp increase ("explosion") in the stored data volume. The Oracle8i server is used as a means of storage in relational data warehouses and data marts, and the Express Server is that for storage in multi-dimensional data marts. The central tool, Oracle Warehouse Builder, is built based on the Common Warehouse Metadata architecture, and it enables the design of the target repository structure, creation of the procedure of removing, matching and uploading data from various sources, and metadata generation.

To address high level analytical challenges, a special technology implemented by the Oracle Express family of OLAP-products is used. Express Server provides a powerful set of analytical calculations, which includes more than one hundred built-in functions that can be used to quickly build new estimates for decision-making at different management levels. To improve the effectiveness of the MS SQL Server in decision-making systems, data warehouses are used. Microsoft, together with other companies, has developed the Microsoft Data Warehouse Framework that reduces the cost of the data warehouse ownership and increases the effectiveness of its management and operation.

Modern databases employ a variety of tools and technologies that enhance their intellectual capabilities. This s primarily associated with multi-dimensional organization of data in the data warehouse, natural language interface and implementation of *what-if* scenarios. Data Warehouse is a repository for relevant information extracted from the operational database for rapid data analysis (OLAP-technology implementation).

Special techniques (Data Mining and Knowledge Discovery) are used to extract relevant information from databases, which indicate the service-oriented science, when the process of science institutionalization is studied in light of the potential of information and communication technologies. Modern databases currently used in cyber science are oriented towards mechanisms and technologies enhancing their intellectual potential. All these mechanisms are obtained from artificial intelligence research. Cyber science emerging in the knowledge society is inherently one of the ways to implement the concepts of artificial intelligence.

In 1969, the first International Joint Conference on Computer Science was held in Washington, at which the term "artificial intelligence" was introduced, and artificial intelligence became an extensively explored aspect. After 10 years, the Dictionary of Cybernetics defined artificial intelligence as an artificial system that simulates complex problem solving by the human during his lifetime, i.e., it emphasized the "secondary" meaning of the functioning principles of the devices that implement the principles of "intelligent machines". Intellectual tradition of the artificial intelligence study dates back to the "Physics" of Aristotle, who introduced the distinction between matter and form. This distinction is a philosophical basis for the principles of symbolic calculation or data abstraction. The Aristotle's Logic, which interprets the idea that the study of thought is the basis of knowledge, is close to the ideas of artificial intelligence [1].

The concept that the brain resembles a machine and uses encoded knowledge was formulated in the domain of philosophical knowledge; the thought itself is used to select an effective action. Mathematics has made a great contribution in the development of artificial intelligence. Mathematicians created the means to manipulate logically valid statements (and invalid probability statements). Mathematicians laid the foundation of the theory of algorithms. With the advent of the theory of artificial intelligence, the importance of the domain specific knowledge, the so-called expert knowledge, became evident.

In the monograph "Artificial Intelligence: Strategies and methods for solving complex problems", G.F. Luger interprets the expert knowledge as a combination of theoretical understanding of the problem and a set of heuristics to solve it, the rules, which effectiveness is evident in a particular subject area [2].

Expert's knowledge is codified and takes the form used by the computer to solve the problem. The techniques employed by various expert systems currently used in the field of design, research, education, business and medicine were developed within the framework of DENDRAL, MYCIN, PROSPECTOR, INTERNIST and XCON projects. The viability of these techniques was proved through creation of expert systems as a dominant area for practical application of artificial intelligence [3].

However, analysts caution against overoptimism in assessing the potential of expert systems technology due to the limited knowledge available to the expert system, difficulties in transferring profound knowledge in a particular subject area, lack of flexibility and finally, inability to provide meaningful explanations, difficulties related to expert systems testing and limitations of learning through experience.

G.F. Luger emphasizes the feature of the intelligent system that it is not just the subject knowledge, but also the knowledge that it knows the subject. This knowledge is defined in specific and generalized terms. When the intelligent system describes the knowledge, it learns through interaction with the world. Analysts refer this "awareness of the intelligent system of its knowledge" to meta-knowledge rightly attributing meta-knowledge to a higher level of knowledge [4]. Since 1980, the field of artificial intelligence turned into industry when Digital Equipment Corporation implemented the R1 commercial expert system. In the 80s, major US corporations had groups focused on the use and study of artificial intelligence. In 1981, Japan started a fifth-generation computer project.

The 10-year project was intended to develop the computer using the Prolog language. At the same time, a research consortium was established in the US to ensure the industry competitiveness. The concepts of artificial intelligence made the strategy aimed to promote the chip design and a human-machine interface investigation. Although the ambitious goals were not completely achieved, the field of artificial intelligence was turned into the industry in the 80s of the past century. The emerging knowledge-based society engaged in the development of artificial intelligence is interested in fundamental problems of knowledge representation and search.

The knowledge representation implies the use of the formal language employed in the computer manipulations to gain new knowledge. The search is the problem solution method which includes the problem states, the concept that reflects the stages of the problem solution [5]. The main issues of the analysis of artificial intelligence and its application are the structure of symbols and operations designed to intellectually solve the problem, and the search for the solutions caused by certain symbol structures and operations. The research in subject domain of artificial intelligence is based on the knowledge representation and search. Processing of the knowledge arrays is a dominant paradigm of the intelligent technologies in the knowledge-based society. The systems based on the knowledge or domain models described in an ultra-high-level language close to the natural one are often referred to as intelligent systems [6].

Facing the problems of high complexity (interpretation of natural language, support for difficult solutions, interpretation of visual messages), the human refers to intelligent systems. The intelligent systems are crucial for using the weakly formalized knowledge of the specialists with great and invaluable practical experience. Expert systems are the type of intelligent systems. Expert systems are effective for the fields of vast experience gained by highly qualified specialists in various subject areas, where difficult solutions depend on examination results. These fields acquired the status of "expert fields".

Intelligent systems fall into expert systems to solve different types of tasks. The type of analysis tasks (many solutions can be listed and included in the system) includes data interpretation, diagnostics and solution support. The type of synthesis tasks (the number of solutions cannot be listed and the solution array is derived from sub-problem solutions) includes design, planning and management. The possibilities of the emerging cyber science to solve the above mentioned tasks cannot be overestimated.

It should be emphasized that that the expert system development exhibits the difficulties, namely the complexity of the knowledge extraction and structuring, and the knowledge extraction and formalization studied in the field of knowledge engineering as a direction of artificial intelligence aimed to develop the models, methods and systems of obtaining, structuring and formalizing the knowledge of experts in order to design the knowledge domain.

We accept the opinions of the analysts who believe that knowledge engineering is doubly gnoseological due to a triple interpretation: first, the reality is reflected in the expert mind; secondly, the expert experience is interpreted by the knowledge engineer; and, thirdly, the expert system is interpreted in the field of knowledge, the informal description of the domain concepts identified from the system of the expert knowledge [7].

The knowledge formation as a society dominant resource and the generation of new forms of knowledge in the knowledge-based society require new forms and models of knowledge representation. This is largely due to the fact that in the 70s of the twentieth century, the area related to the expert system creation arose from the field of artificial intelligence. This is caused by the need to create the expert systems that can be applied to solve the problems of the expert analysis. The importance and complexity of the problem is largely determined by the knowledge structure, which in turn is determined by the scope of the knowledge application. This structure consists of the facts of the designated area, relations between these facts and the production rules; in addition, the structure should consider the knowledge used to incorporate the knowledge into the expert system [8].

Cyber science is focused on different schemes of knowledge representation. The latter includes the logic model, frame-based and production systems, and semantic networks. These schemes are the most typical models of the knowledge representation implemented within the cognitive area of artificial intelligence investigation. The analysis of the theoretical schemes for knowledge representation is performed within the logical area. For example, the scheme of knowledge representation was created based on the first-order logic predicates. The other knowledge representation schemes are based on mathematical formalization [9]. The considered schemes correlate with a specific knowledge structure. It is noteworthy that the above mentioned knowledge representation models are being employed at present.

The feature of the knowledge representation systems is informal simulation of the human activity. The knowledge representation models are focused on the information that comes from the experts; the information is often contradictory, but the specific character of its application implies that this kind of information is to be presented as the solution in accordance with the knowledge representation form. The knowledge base as an integral part of the knowledge-based systems includes not only the knowledge description, but also the mechanisms of obtaining the solution [10, 11].

-One of the typical knowledge representation models (within the cognitive area of artificial intelligence investigation) is a model based on the use of frames. A frame is an abstract image to represent the stereotype of the object, concept or situation; a generalized and simplified model or structure. What is the specificity of the knowledge representation in terms of frames? The frames were interpreted as relatively large units of knowledge representation, and their hierarchical structure was revealed with due regard to the abstraction degree. In addition, the possibilities of representing a combination of declarative and procedural knowledge were analyzed.

The interpretation states that a man completely changes his attitude towards the current circumstances and refers to the structure-frame when he gets into a new situation. The frame is a unit of the knowledge representation formed in the past. However, the details of the frame can be changed depending on the current situation [12]. The frame is a network composed of several vertices and relations: the recorded information about the true state of the object and the terminal slots (terminals), which should be filled with specific knowledge and data.

The structure representing these components is referred to as a frame. The hierarchical structure with its classification and generalizing properties arises due to the analogy between the conceptual objects. The structure implies the relation "abstract-concrete". Complex objects are the combination of some frames called a frame-based network. The frame itself is one of the knowledge organizing schemes to create expert systems.

3 Conclusion

It can be argued that the most important problems associated with the epistemological principles of the intelligence organization as a physical system are the problems of representation, the nature of interpretation (the symbol value can be understood only in the interpretation context), the representation uncertainty, as well as the problems of limitations and symbol interpretation in the neural networks. The above stated and similar problems are crucial for further development of the artificial intelligence and are still open for investigation. The invention of the digital computer marked the birth of the cognitive science, otherwise known as the science of intelligent systems. The urgent question today is whether the science of intelligent systems is capable of creation the artificial intelligence.

One of the promising directions of the search for answers to these questions is the autoepistemic logic based research. These systems are capable of automatic generation of the concepts for expert systems that enables the possibility of generalization, the central operation of the learning process. The knowledge-based society has changed the technological basis of the society under the influence of scientific knowledge.

Today we can talk about the functions of information technologies, namely sociogenic, culturegenic and gnoseogenic ones. "Computers" exhibit their potential within the gnoseogenic function, the potential of information networks which form the operation potential to ensure the growth of new knowledge. The parametrical characteristics of the knowledge-based society are free automatic access to the available information and knowledge, accessibility of the information technology, and the availability of infrastructures that make the basis for national information resources.

It is necessary to emphasize the organic relation between the emerging information structure and the globalization process. The information structure is the basis for functioning of a new social life based on complex information processing by means of electronic equipment and new information technologies. The cyber science advances in the knowledge-based society enable information processing, transmission and production via computers, software and telecommunications; knowledge, information and information infrastructures acquire the status of production forces.

References

[1] Dzh. F. Lyuger, *Iskusstvenny intellekt. Strategii i metody resheniya slozhnyh problem* (Izdatel'skii dom "Vil'yams", Moscow-Sankt-Peterburg-Kiev, 2005)
[2] A. P. Chastikov, T. A. Gavrilova, D. L. Belov, *Razrabotka ekspertnyh system* (Izd-vo "BHV-Peterburg", Sankt-Peterburg, 2003)
[3] *Predstavlenie i ispol'zovanie znanii* (Izd-vo "Mir", M., 1989)
[4] S. Osuga, *Obrabotka znanii (*Izd-vo "Mir", M., 1989).
[5] R. Dekart, *Rassuzhdenie o metode* (Izd-vo Akademii nauk SSSR, Moscow, 1953).
[6] *Upravlenie znaniyami* (Izd-vo "Vysshaya shkola menedzhmenta", Sankt-Peterburg, 2010)
[7] R. B. Kvesko, S. B. Kvesko, N. E. Salkova, T. N. Shinn, Symp. on Sc. and Techn., KORUS-2005 **1**, 1089 (2005)

[8] W. N. Press, et. al., *Numerical recipes in C: the art of scientific computing* (Cambridge University Press 1992)
[9] E. Wenger, W. M. Snyder, H. Bus, Rev. January-February **139,** (2000)
[10] S. Rassel, *Norvig Piter. Iskusstvennyi intellekt. Sovremennyi podhod* (Izdatel'skii dom "Vil'yams", Moscow-Sankt-Peterburg-Kiev, 2006)
[11] I. V. Plotnikova. Europ. Proc. of Soc. & Behav. Sc. (EpSBS) **7,** 173 (2016)
[12] V. I. Syryamkin, E. N. Bogomolov, V, S. Kutsov. Adv. Mat. Res. (2014)

PARAMETRIC NONLINEAR LOCATOR

Vladimir Antipov [1,*] ,*Sergey* Shipilov [2]

[1] Siberian Physicotechnical Institute named after the academician V.D. Kuznetsov of Tomsk State University, 634050, Tomsk, Russian Federation
[2] National Research Tomsk State University, the Department of Radiophysics, 634050 Tomsk, Russia

Abstract. The device relates to the field of radar techniques for locating or objects detecting using reflection or reradiation of radio waves. This device can be used to detect and identify the objects containing non-linear electrical contacts. The non-linear radar contains two sources of signals that generate the signals in close frequency rangers, and a signal combining scheme associated with the antenna. The source of the first signal is provided with a low-frequency modulator. A vector reflectometer is included between the source of the second signal and the signal combining circuit. The design is simplified and the detectability of the non-linear radar is improved due to these design features.

1 Justification of the structural scheme

The device is intended for detecting and identifying the objects containing non-linear electrical contacts such as a metal-metal, a semiconductor-metal or a semiconductor-semiconductor, for example, the concealed weapons or the electronic devices for unauthorized retrieval of information. The operation of the most nonlinear radars is based on the reception of harmonics of a powerful probing signal. In contrast, the concept of parametric nonlinear location [1] implies a change in reflection from a nonlinear object under the action of an additional usually electromagnetic effect. A typical example is a two-frequency locator [2] containing two sources of signals that generate the signals with close-range frequencies f_1 and f_2 and the signal combining circuit associated with the antenna. This radar contains a receiver capable to receive the signals besides the harmonics of frequencies f_1 and f_2 at the sum and difference frequencies that do not coincide with the harmonics of frequencies f_1 and f_2. As a result, the design is simplified by reducing the requirements for suppressing harmonics. Nevertheless the additional antennas of the appropriate ranges are needed to receive these signals since the sum and difference signals are far away from the transmitted signals by frequencies. This circumstance leads to a complication in the design. In addition a high level of both sounding signals is required for the formation of combinative (sum and difference) signals, and this requires a doubled number of powerful amplifying cascades. The radar's detectability is low under the limited power of the amplifiers.

* Corresponding author: antipov50@mail.ru

In this paper we solve the problem of combining the advantage of a simpler scheme for selecting of harmonics in a two-frequency locator and the advantage of a smaller number of powerful amplifying cascades in a single-frequency locator. The source of the first signal is provided with a low-frequency modulator for this in the nonlinear radar containing two signal sources that generate signals in close frequency rangers and a signal combining circuit associated with the antenna. And a vector reflectometer is switched between the source of the second signal and the signal combining circuit. The result is an improvement in the detectability of a non-linear radar and a simplification of its design.

In the circuit under study the low-frequency modulator of the first signal provides a time-dependent build-up of the nonlinear characteristic of the object. Wherein its averaged conductivity at different frequencies, including at the frequency of the second signal, is also modulated. The vector reflectometer in the second signal channel provides a formation of the modulated response in accordance with the modulation law of the first signal. It is unnecessary to increase the power of the second signal to the power level of the first signal to obtain a modulated response. Due to this the design of the nonlinear radar is simplified by reducing the number of powerful amplifying cascades. The design is also simplified at the expense of using the only antenna needed for transmitting the first and second signals as well as for receiving the reflected modulated signal at the frequency of f_2.

The improvement of the detectability in comparison with the known scheme is due to the nature of the spectrum of the build-up of the nonlinear volt-ampere characteristic of the object by a powerful signal. The appearance of harmonics or combinative frequencies is associated with the effect of harmonic fluctuations on a nonlinear element. As a rule, semiconductor elements have an exponential volt-ampere characteristic. The harmonics of the current J_n are proportional to the modified Bessel functions of the first type $I_n(Z)$ under the influence of the harmonic signal with a relative amplitude *Z*. The conductivity *G* in the element with the exponential volt-ampere characteristic is proportional to the current, that is it contains harmonics, which are also proportional to the functions $I_n(Z)$. The second harmonic of the current, that responsible for the formation of the received signal in a conventional nonlinear radar, is proportional to $I_2(Z)$. In the scheme under study the average conductivity *G*, that is proportional to $I_0(Z)$ and receives a small increment δ under the influence of the first signal, is responsible for the formation of the signal entering the reflectometer. Considering the expansion of the functions $I_n(Z)$ into series in accordance with the well-known formula

$$I_n(Z)=\sum_{k=0}^{\infty}\frac{\left(\frac{Z}{2}\right)^{2k+n}}{k!\Gamma(k+n+1)} \tag{1}$$

one can see that the current component at the second harmonic is proportional to $\frac{Z^2}{8}$, while the change of the average conductivity is proportional to $\frac{Z^2}{4}$, that is twice more then for the second harmonic. Besides the detection efficiency is enhanced on account of the nature of signal processing in the vector reflectometer. The signal generated in the reflectometer is proportional to $\frac{(1-\delta G)}{(1+\delta G)}\approx 1-2\delta G$ according to the relation between the conductivity of an element and its reflection coefficient. Due to this circumstance the conductivity modulation appears in the reflected signal twice as effectively as in the case of a harmonic response in a conventional nonlinear locator. In other words, the advantage of coherent detection is realized in comparison with quadratic detection. These considerations confirm an improvement in the detectability in the proposed scheme.

2 Experimental results

The model of the nonlinear radar based on laboratory radio measuring instruments was assembled and tested in accordance with the structural diagram (Figure 1).

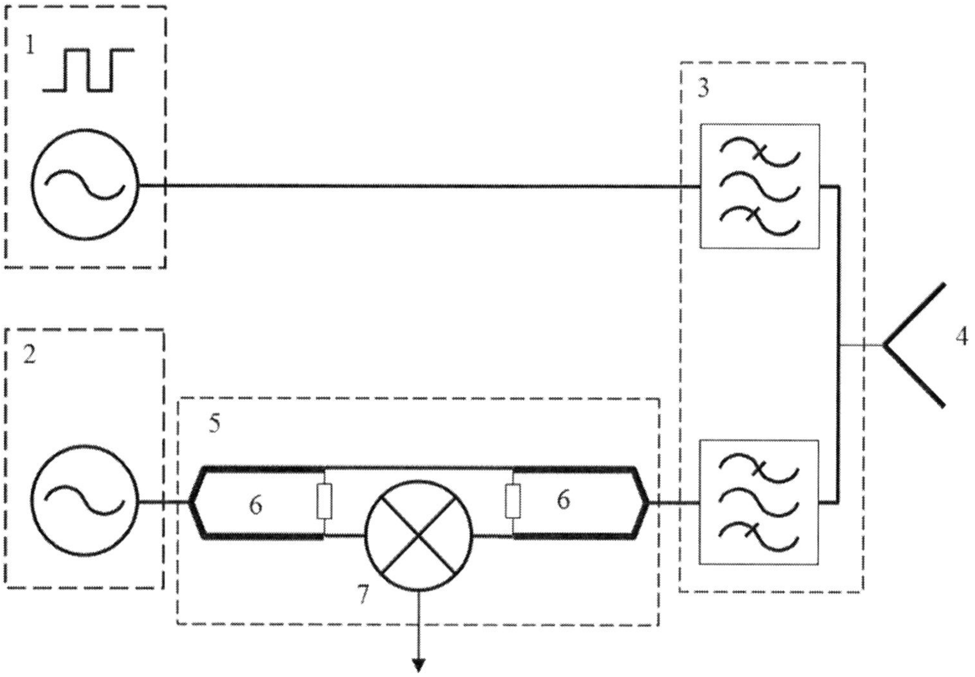

Fig. 1. Structural diagram of the model of the parametric nonlinear locator.

The generators of standard signals are used as the sources of microwave oscillations. The first generator 1 operates at the frequency of 1300 MHz in the mode of internal amplitude modulation of 1 KHz, the second generator 2 operates in continuous generation mode at the frequency of 1600 MHz. The power of the generators is 60-80 mW. The signal combining circuit 3 is made on five-link pintle filters and provides independent relation of the frequency channels with the antenna. The antenna 4 has the form of a vibrator above the plane, equipped with one director to expand band of reconciliation frequencies. The reflectometer 5 is represented by the combination of Wilkinson's three-decibel divisors 6 with the balance mixer 7. The output signal is amplified by a selective microvoltmeter and is observed on an oscilloscope. The antenna is directed to the object possessing the nonlinear volt-ampere characteristic. The reflection coefficient of the object at the frequency of the second source also turns out to be modulated in accordance with the law of modulation of the first generator. As a result, a low frequency signal indicating the presence of a nonlinear object in the antenna coverage zone is generated in the vector reflectometer.

With the indicated power values this scheme ensured the detection of typical test objects at a distance of up to 0.5 m, which is comparable with similar indicators of serially produced non-linear radars with power of about 1 W.

3 Prospects for characteristics improvement

The increase in the energy potential of the nonlinear locator is primarily connected with the increase in the power of the modulating signal of the frequency f_1. The specificity of the parametric method of nonlinear location consists in the presence of two fundamentally different signals: the first signal must be powerful enough to change the conductivity of the nonlinear object, and the second one can have an arbitrary power, that would provide the excess of the noise threshold.

In accordance with the formula (1) the modulation effect of the average conductivity of the nonlinear object is proportional to the square of the amplitude of the field Z, respectively, this effect is inversely proportional to the square of the distance to the object D. We emphasize that for the nonlinear locator the key parameter is the extent of buildup, which is characterized by a certain "threshold" level for a semiconductor element. The Z_{cr} level is the critical for a reliable detection of nonlinearity. The specific gravity of the higher terms of the expansion in the formula (1) begins to increase at this level. The reflection coefficient of the object acquires a noticeable modulated increment if this level is exceeded.

We select in the field of view of sounding antenna the elementary cell possessing a nonlinearity and that has the reflection coefficient K_{neg} at the frequency f_2. The possibility of registering the increments of the reflected signal at the frequency f_2 is limited mainly by its excess over the amplitude and phase noises of the generator. As a rule in the radius of the nonlinear locator action this excess remains big enough and weakening of the reflected sounding signal, that is proportional to the fourth extent of the distance, is easily compensated by amplification increase. Thus, within certain limits the threshold power of the modulating signal can remain proportional to the detection distance in the second extent, and not in the fourth extent or more, as it happens in the locators of traditional structure.

The above arguments are illustrated by the diagram in Figure 2.

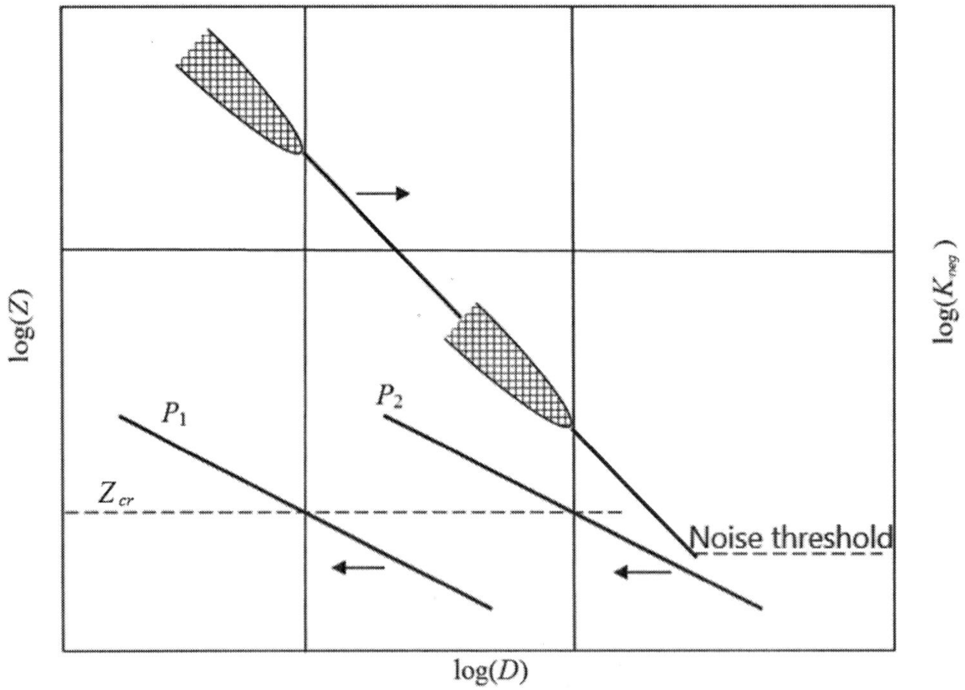

Fig. 2. The diagram of signal levels in the parametric nonlinear locator.

The distances D to the object are plotted along the horizontal axis on a logarithmic scale, and the characteristic levels of signals are plotted along the vertical axes as well on a logarithmic scale. The lines with the signatures of P_1 and P_2 correspond to two different power levels of the modulating signal, and relating to them the vertical axis from the left corresponds to the buildup level Z of the nonlinear element. The slope of the lines takes into account the weakening of the signal in proportion to the square of the distance. The Z_{cr} level corresponds to the threshold of nonlinearity demonstration. The line with the hatching areas, referring to the vertical axis to the right, corresponds to the transmission coefficient of the sounding signal K_{neg}. The hatching areas reflect the variations in the reflection coefficient observed under the action of the modulating signal at the powers P_1 and P_2. The slope of the line corresponds to weakening of the signal in proportion to the fourth extent of the distance. The dotted line conditionally displays the noise threshold of the sounding circuit, determined mainly by the amplitude noises of the frequency generator f_2. The location of the areas in which variations in the reflected signal are manifested under the action of the modulating signal, relative to the noise threshold, evidences that the detection distance which is far from this threshold depends solely on the efficiency of the build-up of the nonlinear element by the modulating signal, that is. This distance is proportional to the ratio of the power of the frequency generator f_2 to the square of the distance to the object.

We use the basic radar equation from [3] to estimate the relation between the signal levels and the noise threshold of the sounding circuit:

$$P_c = \frac{P_i D_a S_{effc} S_{effa}}{(4\pi)^2 D^4}, \tag{2}$$

where P_c is the power of the received signal, D_a is the factor of directed action of the antenna. S_{effc} is the scattering area of the target, S_{effa} is the effective area of the antenna, D is the distance, herewith the factor of directed action and the effective area of the antenna at the wavelength of λ are related by (3):

$$D_a = \frac{4\pi S_{effa}}{\lambda^2}. \tag{3}$$

We substitute in these formulas the values of the wavelength equaled 0.19 m for the frequency $f_2 = 1600$ MHz, the antenna size is of the order of $\frac{\lambda}{2}$, the area of the tested object is of the order of 10^{-4}m^2 (1 square centimeter) and the distance is 1 m. Under these conditions the ratio of the received and emitted signals is -56 dB. Taking into account that the dynamic range of modern vector analyzers reaches 120 dB, it can be judged that the range of measured signals at the distances in units of meters is much higher than the noise threshold, so that the operability of the circuit under investigation is actually determined mainly by the power of the modulating signal. Thus, we can expect an increase in the distance detection of test objects in 4 times, that is up to 2 m if the power of the modulating signal is increased from the above value of 60 mW to 1 W, that is approximately in 16 times.

Improving the operational characteristics of the circuit under study is also associated with eliminating so-called "blind" distances in which the modulation of the reflection from the object is perceived in the mixer as phase modulation and does not produce the desired response. In connection with this the reflectometer in the diagram of Figure 1 should be equipped with a quadrature channel. In this case the "blind" distances will differ by a quarter of the wavelength and the signal will be detected at least in one of the channels.

The integrated microcircuits of generators, amplifiers and mixers as well as a planar antenna and miniature dielectric filters will be applied in the projected pilot sample of the parametric nonlinear locator.

4 Conclusion

The considered scheme of the parametric nonlinear locator with a relatively powerful modulating signal and a relatively low-power sounding signal differs significantly from the analogues of simple design. At the same time it is able not to concede in the detectability to known constructions and even probably to surpass them.

Acknowledgements

The research was supported by the Russian Ministry of Education and Science as a part of the state order No 3.2068.2017/4.6.

References

[1] G.N. Shcherbakov, Special Equipment. **4**, 52 (2000)
[2] System and method of radar detection of non-linear interfaces//US 6765527 B2, 20 Jul. (2004)
[3] G.B. Belotserkovsky, Radiolocation fundamentals and radar devices (Soviet Radio, Moscow, 1975)

SYNTHESIS OF ULTRASOUND FIELD SOURCES BASED ON PHASE SCREEN APPROXIMATION

Dmitry Sukhanov[1], *Ksenia* Zavyalova, *Nadezhda* Erzakova

Tomsk State University, 634050, Tomsk, Russia

Abstract. Here is proposed the method for synthesizing the sources of an acoustic field on the basis of an approximation of the phase screen. The technology of manufacturing ultrasonic phased arrays providing the formation of a field of a given distribution is proposed. An experimental setup has been developed for the formation of a vortex field at a distance of 10 cm.

1 Introduction

The formation of an ultrasonic field of the required distribution is an urgent task in the development of ultrasonic sounding and tomography systems, technologies of acoustic levitation and remote particle control, augmented reality devices for the transmission of tactile information [1]. The main direction of solving similar problems is the use of phased arrays [2, 3] of ultrasonic radiators (Figure 1). For the formation of the field, it is advisable to control the signal of each radiator separately, but such a scheme is rather complicated and expensive. A simpler solution is to use a single signal generator, which is fed through the delay lines to the emitters, thereby ensuring a phase difference between the elements. Since the ultrasonic signal is relatively low-frequency, too long electrical delay lines are required.

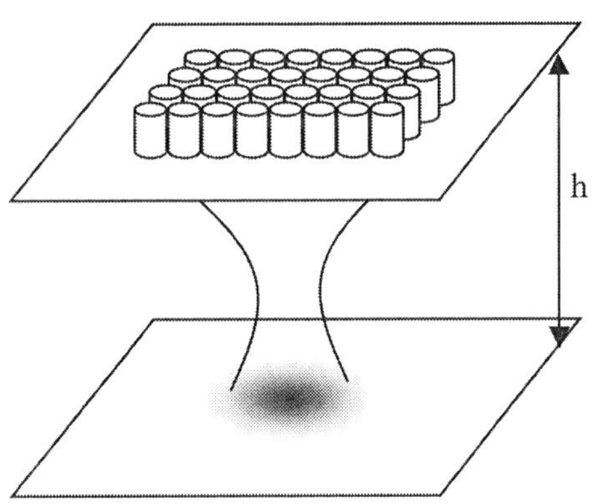

Fig.1. Scheme of field generation.

[1] Corresponding author: sdy@mail.tsu.ru

It is proposed to use acoustic delay lines formed after each radiator. In this case, the length of the delay lines does not exceed the length of the ultrasonic wave, in this case, 8.25 mm at a frequency of 40 kHz for a sound velocity in the air of 330 m/s.

2 Field formation return task

Let us consider the possibility of forming a vortex beam with the help of a grid of emitters placed in a hexagonal grid with a step of 11 mm. In Figure 2 the required distribution of the field is shown, which must be created at a distance of 10 cm. The required field distribution we define by following formula:

$$U(x,y) = \exp(i\varphi - ikr/2)\cdot(r \geq R_1)\cdot(r \leq R_2), \qquad (1)$$

where $r = \sqrt{x^2 + y^2}$, $R_1 = 1$ cm, $R_2 = 5$ cm.

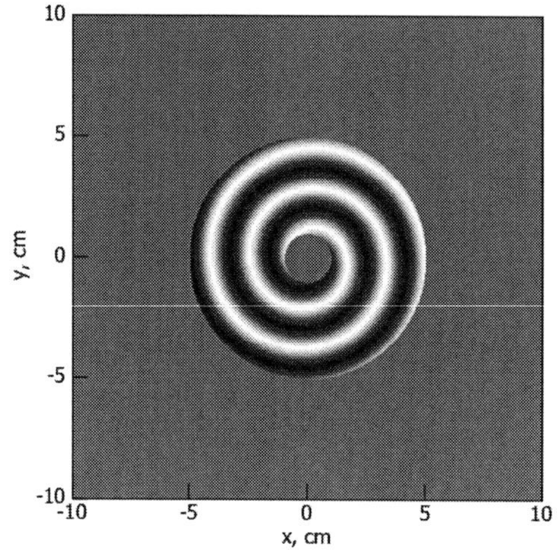

Fig.2. Required field distribution.

To obtain the amplitude-phase distribution of the sources from the target field, we use the wave reversal method [4, 5]. We propose to reverse propagate only homogeneous waves. Let us solve this task by means of spatial plane wave spectrum representation:

$$\tilde{U}(k_x,k_y) = \frac{1}{4\pi^2}\int_{-\infty}^{\infty}\int_{-\infty}^{\infty} U(x,y)\exp(-ixk_x - iyk_y)dxdy. \qquad (2)$$

It is proposed to multiply the spatial spectrum of this field by $\exp(ik_z h)^*$, where * – is a complex conjugation operation, $k_z = \sqrt{k^2 - k_x^2 - k_y^2}$ – is the projection of the wave vector on the axis perpendicular to the plane of the lattice, the wave number, and the h – distance from the array of radiators to the plane of the target field distribution. The plane wave spectrum of field in the plane of sources is:

$$\tilde{P}(k_x,k_y) = \tilde{U}(k_x,k_y)\cdot(\exp(ik_z h))^*, \qquad (3)$$

Reconstruction of field distribution if the source plane is:

$$P(x, y) = \int_{-\infty}^{\infty} \int_{-\infty}^{\infty} \widetilde{P}(k_x, k_y) \exp(ixk_x + iyk_y) dk_x dk_y \,. \qquad (4)$$

That is, using the inversion of waves in the frequency domain, the required amplitude-phase distribution is calculated on a given plane. Formula (3) allows to calculate homogeneous waves and to avoid enhancement of heterogeneous waves. For the case shown on Figure 2, it was calculated required distribution of sources field on 40 kHz (Figure 3).

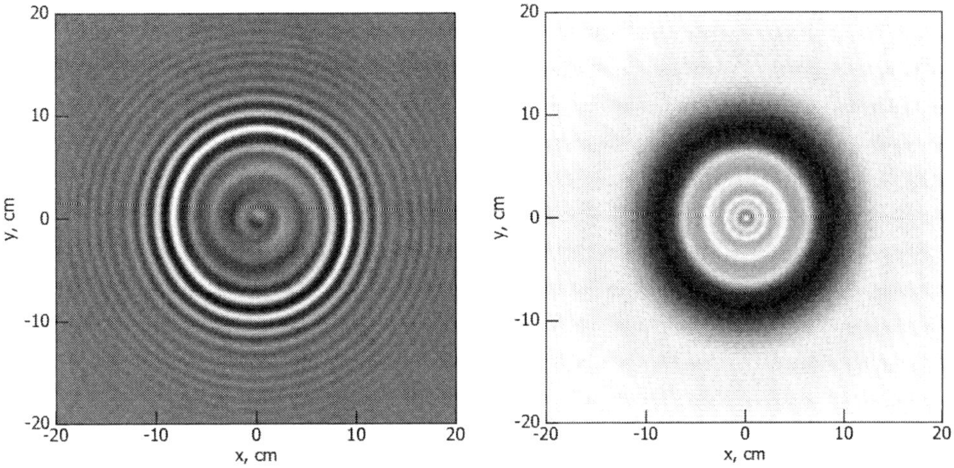

a b

Fig.3. Reconstruction of sources distribution: a – field of sources; b – magnitude of sources field.

3 Sources placement and numerical simulation

We can see that the maximum magnitudes of sources is concentrated in a narrow circle area. That is why it is possible to place sources only in this area to minimize a number of elements. The most dense placement of circular elements on a plane is a cellular hexagonal placement. We propose to approximate the circle by hexagon of transmitters, as shown on Figure 4. On Figure 4*b* the phase of sources is shown on grayscale picture. The magnitude of each source is the same.

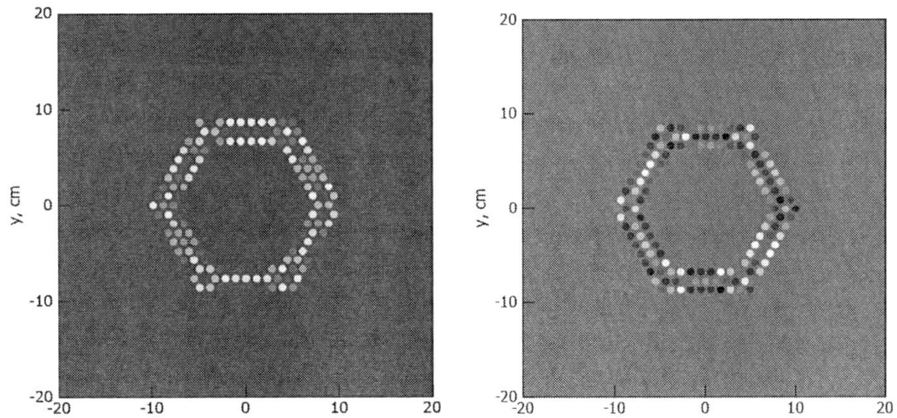

a *b*

Fig.4. Placement of elements (a) and phase distribution of sources (b).

In Figure 4*a* shows the result of the calculation of signals in the emitters, where each color channel corresponds to: red – amplitude; green – cosine quadrature; blue is a sine quadrature.

To calculate the forward field we use the same approach we used for sources reconstruction, but in forward direction:

$$\widetilde{B}(k_x,k_y)=\widetilde{P}(k_x,k_y)\cdot(\exp(ik_z h)) \tag{5}$$

$$B(x,y)=\int_{-\infty}^{\infty}\int_{-\infty}^{\infty}\widetilde{B}(k_x,k_y)\exp(ixk_x+iyk_y)dk_x dk_y \tag{6}$$

After calculating the direct propagation (5-–6) of the field in the spatial spectra, the expected distribution in the target plane was obtained (Figure 5). It can be seen that the field has a vortex character, as expected, but there are side lobes, which is due to the sparseness of the grating.

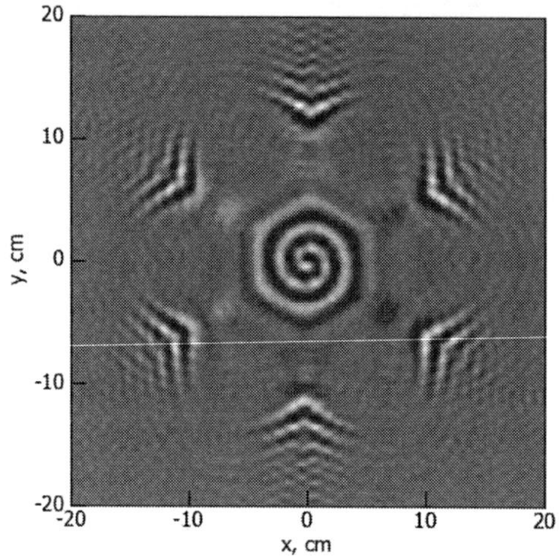

Fig.5. Numerical modeling of field in the focusing plane.

A plurality of in-phase point emitters located in a hexagonal grid with a pitch of 11 mm are set at a distance from the median plane, so that this distance provides the required phase shift, based on the approach of the phase screens.

4 Experimental studies

The experimental setup is made of 144 emitters installed in a plate with holes of the calculated depth (Figure 6). As a result of measuring the plane distribution of the field at a range of 10 cm (Figure 7), it was found that it corresponds with the numerical model qualitatively.

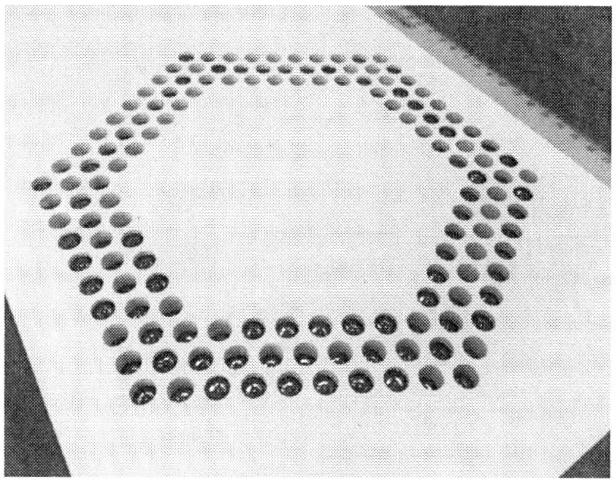

Fig.6. Photography of the phased array.

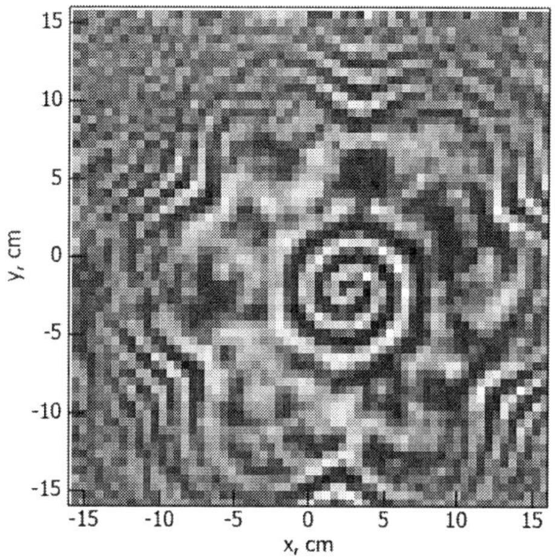

Fig.7. Measured field.

5 Conclusion

A method is proposed for synthesizing ultrasonic field sources in air to form a given field distribution. The possibility of its application has been shown numerically and experimentally.

Acknowledgments

The scientific research was carried out with the support of the grant of the Russian Science Foundation No. 17-79-20051.

References

[1] S. Hashizume, A. Koike, T. Hoshi, Y. Ochiai, ACM SIGGRAPH 2017 Posters, No a57, 129315 (2017)
[2] A.I. Korobov, M.Y. Izosimova, S.A. Toschov, Physics Procedia **3**, 201 (2010)
[3] A. Ganguli, R.X. Gao, K. Liang, J. Jundt, A. Ordonez, IEEE Transactions on Instrumentation and Measurement **59**, 5353739 (2010)
[4] Y. Jing, F.C. Meral, G.T. Clement, Physics in Medicine and Biology **57**, 901 (2012)
[5] H. Nicacio, V.L.S.N. Button, R.G. Dantas, E.T. Costa, Proceedings of the IEEE Ultrasonics Symposium **1**, 559 (2002)

INTELLECTUAL MULTI-LEVEL SYSTEM FOR NEURO-FUZZY AND COGNITIVE ANALYSIS AND FORECAST OF SCIENTIFIC-TECHNOLOGICAL AND INNOVATIVE DEVELOPMENT

Sergey Gorbachev [a]

National Research Tomsk State University, 634050 Tomsk, Russia

Abstract. The article is devoted to description of structure and operation algorithms of a multilevel intellectual system for analysis and forecast of scientific-technological and innovative development of objects and territories based on expert methods, trainable models of fuzzy logic and fuzzy cognitive analysis. The article analyses distinctions and advantages of the system in making forecast evaluation and scenarios of the dynamics of multidimensional feature objects and processes. The authors have built a neutrosophic cognitive map for interaction of technologies based on the results of foresight studies.

1 Introduction

The current state of research in the field of dynamics of scientific and technological progress is closely connected with the methods of monitoring and analysing the emergence of a new technological order. For example, in [1] S.Glaziev gives a characterization of a new way as a certain set of conjugate industries located approximately at the same stage of technical development. In a long-term perspective, scientific and technological progress will be carried out through renovation of the traditional technological order. Therefore, when creating a model of modern technical and economic processes, researchers meet the problem of blurriness and uncertainty of the factors of technological change [2].

Global modelling of these processes began in the 1970s in the works of J. Forrester and D.Meadows [3], in which technological development and resource constraints were taken into account. These studies had great influence on the work of other researchers and public opinion and consequently the ideas of sustainable (self-sustaining) development gained significance.

Based on the emerging issues of the adequacy of theoretical models, optimization and imitation groups were developed to formalize socio-economic and scientific-technological processes.

---

[a] Corresponding author: spp03@sibmail.com

An optimization model is characterized by the non-closure of the equation systems describing them; it means that some of the variables corresponding to the control actions are not specified.

In this case, investigation of objects consists in finding the values of these variables, which optimize a certain functional as an achievement of the specific goal set beforehand.

Among the simulation models, first, it is worth noting the models of Forrester, Meadows and Mesarovic-Pestel. They are a system of nonlinear differential equations. Using this approach, a model is provided with a pre-selected and fixed control mechanism, which in fact is a decision-making model built on the system development background. In this case, the research mechanism is associated with a search of various "policies" and an analysis of their influence on the system behaviour. By "politicians", we mean various combinations of hypotheses about functional relationships, structure of the control mechanism, and numerical values of the parameters. This helps to understand the dynamics of the processes under study and to identify possible critical development trends. At the same time, the fixation of the control mechanism does not presuppose a conscious change of the existing development tendencies by a person and does not allow analysing all the possibilities of the system with admissible control actions.

The purpose of this study is to develop global models and use them in studying various scenarios based on the development of a multilevel intellectual system for analysing and predicting the dynamics of socio-economic, technical-economic and scientific-technological processes. A set of mathematical, intellectual, expert, and cognitive methods with logical interpretation of the research results can form the core of this system. For the first time in the world science, the authors of the article developed and implemented a neuro-fuzzy model of the trajectory of world technical and economic development, which allows measuring the level and rates of world technical and economic development and calculating the parameters of countries relative to the standard [4]. The development of these studies was the development of hybrid methods for training fuzzy models based on the synthesis of trained fuzzy decision trees with the conclusion of diagnostic decision rules that allow solving the problem of multicriteria classification of scientific and technological directions and technologies by the target factor.

In our opinion, foresight methods represent an effective analytical tool for identifying priority areas of science, technology and innovation based on expert assessments of strategies and prospects for innovative development, identifying technological breakthroughs that may have an impact on society and the economy in the medium and long term [5].

2 Structural and functional organization of the intellectual system for analysis and forecasting

Foresight methods allow creating an adequate system for monitoring, analysing and forecasting the development of science and technology. Thus, a multilevel and nonlinear system of scientific priorities and innovative technologies can be presented as a hierarchically constructed system with such specific features as the principles of adaptation to the existing social and economic situation of global and national scale, and evolution over time taking into account the trends of the world technical and technological development.

Based on the results of the study, a structural-functional scheme of an intellectual system for analysis and forecast of scientific and technological development was developed. The system includes three levels linked together by analysis and forecast (Figure 1).

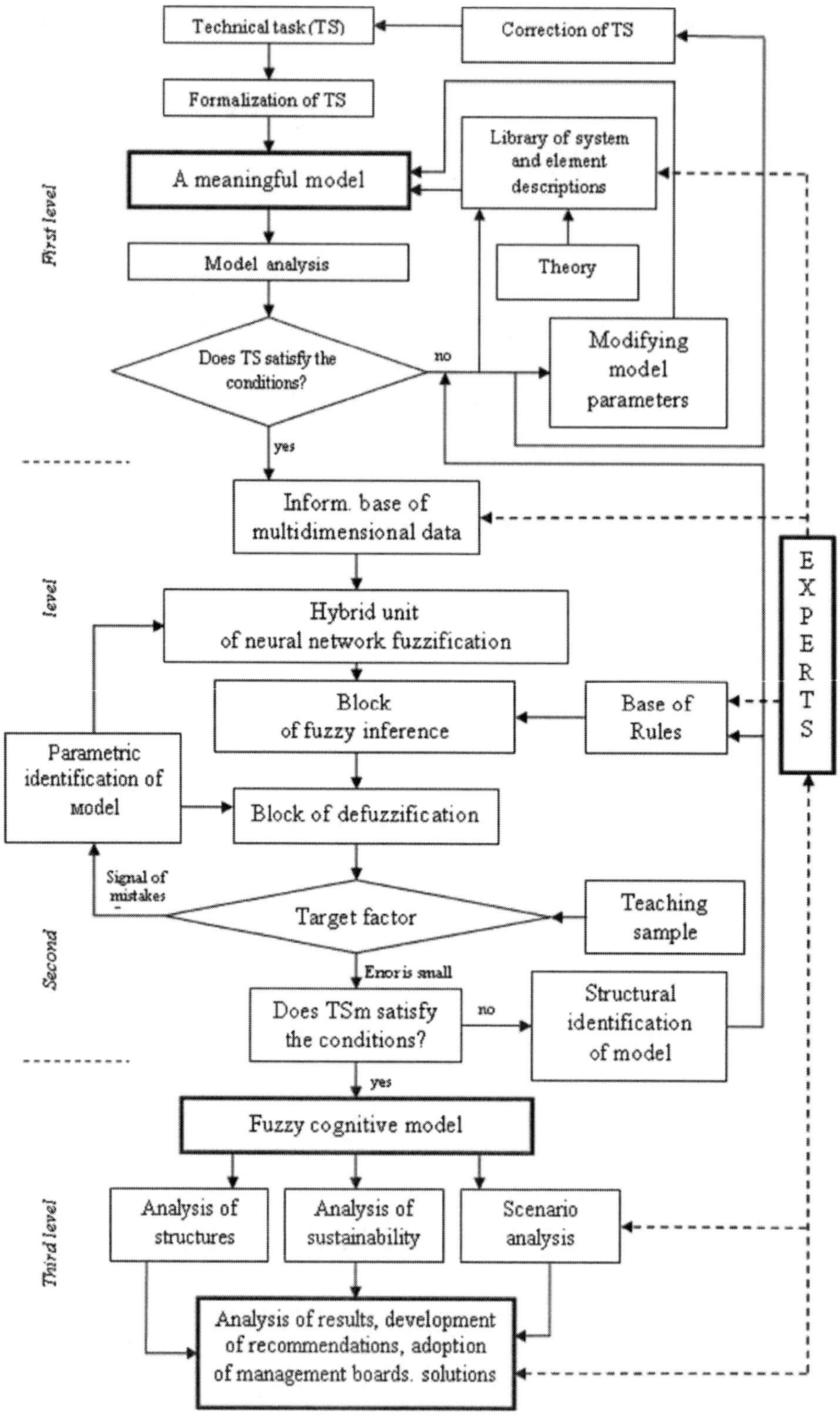

Fig. 1. Structural-functional scheme of intellectual multi-level system for analysis and forecast of scientific, technological and innovative development.

2.1 A meaningful model (level 1)

At the first level, we developed a conceptual methodology for forming a meaningful research model depending on the target factor based on statistical and other information that can be quantitative or qualitative. At the same time, we studied the methodological problems of forming the pool of experts taking into account their professional identity as well as the issues of qualitative representation of experts' knowledge during the surveys [6]. During the expert seminars, we examined the scope, limitations and assumptions, as well as the principles for formation and selection of indexes and indicators for the expert evaluation of the effectiveness of scientific and technological directions and technologies. During the examination process, we considered not only the features and requirements of innovative development of the Russian economy, its technological, production, scientific and personnel potential, but also the trajectory of world technical and economic development.

At the initial stage, the content model is characterized by a set of quantitative and qualitative characteristics that can be identified by experts. In the future, it can be replenished with other indicators, depending on social-economic, political, and other target factors.

2.2 Multidimensional adaptive model of neuro-fuzzy inference (level 2)

The second level is connected with formation of the information base of the characteristics by the results of foresight research, solution of computational analytical problems with the conclusion of fuzzy diagnostic decision rules. The construction of classical adaptive neuro fuzzy inference systems (ANFIS) is associated with formalization of input parameters and target indicators of the effectiveness of scientific (innovative) projects and technologies in the form of a vector of interval values (fuzzy interval); the hit in each interval of this vector is characterized by some degree of uncertainty [7]. This procedure is called "fuzzification". Based on the initial information, experience and intuition, experts often and confidently can make quantitative evaluation of the boundaries (intervals) of possible (permissible) values of parameters and the range of their most possible (preferred) values. For example, experts can characterize the primary data using *triangular fuzzy numbers* with the following membership function $\mu_A(x)$ of the fuzzy set A:

$$\mu_A(x)=\begin{cases}0,\ if\ x\le a_{\min}\\ \dfrac{x-a_{\min}}{c-a_{\min}},if\ a_{\min}<x<c\\ \dfrac{a_{\max}-x}{a_{\max}-c},if\ c\le x<a_{\max}\\ 0,\ if\ x\ge a_{\max}\end{cases} \qquad (1)$$

This expression simulates the following expert position: "parameter A is approximately equal to c and uniquely enters to the segment $[a_{min}, a_{max}]$ ".

In practice, an analytical representation of the membership function is often used [5], for example, the Gaussian membership function has the form:

$$\mu_A(x)=e^{-\frac{(x-c)^2}{2\sigma^2}}, \qquad (2)$$

where c is a center and σ is a standard deviation (width).

Special algorithms for the fazzification of expert data have been developed [8], which are described not in numerical but in a rank (ordinal) scale. For example, it can be described as a sequence of ordered linguistic variables $x_1,..., x_n$, where one of the ranks t ($t = 1, ..., t_k$, for example, low, below average, medium, above average, and high) can be assigned to each of the observations x_k.

For the automatic formation of a rule base, a classification fuzzy decision tree based on a set of characteristics has been developed; it can be viewed as a hierarchical network with direct signal propagation [9]. Therefore, to adapt its parameters (centers and width of membership functions, as well as parameters of leaf identification), it is possible to use a hybrid algorithm for back propagation of the error by reversing each leaf node from the root node of the tree. Figure 1 presents a scheme of the parametric identification of a model. This procedure allows us to synthesize a decision tree with a small number of elements, which has a high level of generalization, with the possibility of a linguistic interpretation of the result by deriving diagnostic decision rules. Because of the optimized network scheme and the reduced number of calculations, the developed hybrid adaptive neural-fuzzy algorithms, which implement the method of classification and fine-tuning the decision tree parameters, increase the speed of the classical algorithm and preserve the hierarchical structure of the tree without any damage to interpretation.

In addition to the classification problem, the problem of multidimensional approximation with a logical conclusion that implements the mapping $F : \Re^n \rightarrow \Re$ is of interest. Further, we will show the possibility of solving this problem based on constructing a neural-fuzzy approximating model and fine-tuning the parameters of the rule consequents.

2.3 A fuzzy cognitive model (level 3)

The third level is associated with fuzzy cognitive modeling, which solves a number of forecasting problems [10]. Currently, one of the developing areas of cognitive sciences is cognitive modeling, which is an interdisciplinary scientific direction that unites the theory of cognition, cognitive psychology, neurophysiology, cognitive linguistics and a theory of artificial intelligence. The results of such synthesis, for example, in the form of the intellectual decision support system described, have a pronounced synergistic effect.

Traditionally, the cognitive model is a cognitive map [11] in the form of a sign-oriented graph (digraph) $G = <V, E>$, where:

V is a set of concepts (vertices),

$V_i \in V$, $i=1, 2,..., k$ are the elements of the study system;

E is a set of arcs, and the arcs $e_{ij} \in E$, $i,j=1, 2, ...,N$ reflect the relationship between the concepts.

The cognitive map allows analyzing both qualitative and quantitative data, and the overall goal of cognitive modeling is to generate and test hypotheses about the functional structure of the observed situation. Iterations are conducted up to a certain state of cause-effect relations, the structural scheme of which will enable us to understand and analyze behavior of the system.

3 Results and Discussion

Based on the conducted foresight research, experts evaluated the degree of interaction of innovative technologies in conditions of transition to the sixth technological order from the point of view of influence on the process of conjugation and formation of the load-bearing sectors and clusters of economy of a new technological order.

Figure 2 shows a neutrosophic cognitive map (*NCM*) of the technologies interactions generated based on the results of foresight studies (numbering corresponds to their technology list in the questionnaire).

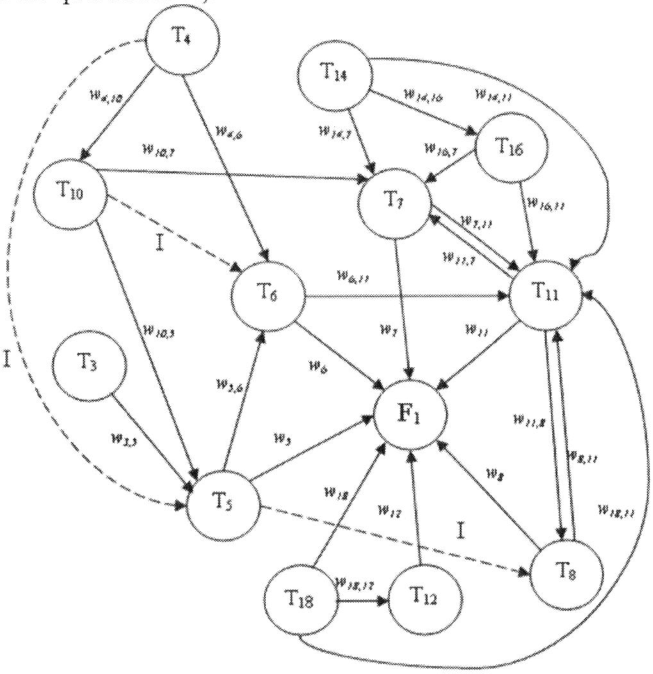

Fig. 2. Neutrosophic cognitive map of the conjugation technologies.

The adjacency matrix obtained from the Neutrosophic cognitive map has the following form:

0	0	$w_{3,5}$	0	0	0	0	0	0	0	0	0	0
0	0	I	$w_{4,6}$	0	0	$w_{4,10}$	0	0	0	0	0	0
0	0	0	$w_{5,6}$	0	I	0	0	0	0	0	0	w_5
0	0	0	0	0	0	0	$w_{6,11}$	0	0	0	0	w_6
0	0	0	0	0	0	0	$w_{7,11}$	0	0	0	0	w_7
0	0	0	0	0	0	0	$w_{8,11}$	0	0	0	0	w_8
0	0	$w_{10,5}$	I	$w_{10,7}$	0	0	0	0	0	0	0	0
0	0	0	0	$w_{11,7}$	$w_{11,8}$	0	0	0	0	0	0	w_{11}
0	0	0	0	0	0	0	0	0	0	0	0	w_{12}
0	0	0	0	$w_{14,7}$	0	0	$w_{14,11}$	0	0	$w_{14,16}$	0	0
0	0	0	0	$w_{16,7}$	0	0	$w_{16,11}$	0	0	0	0	0
0	0	0	0	0	0	0	$w_{18,11}$	$w_{18,12}$	0	0	0	w_{18}
0	0	0	0	0	0	0	0	0	0	0	0	0

The purpose of the analysis is to calculate the degree of exposure to one or more of the concepts (technologies) taking into account their condition on the target by meansof *NCM* .

Let us consider the impact the concept of the T_5, on the system, which corresponds to the respective genomic, proteomes and post-genomic technologies. Based on definition 5, we show how to find is a fixed point under the condition that the system is affected by the state vector A_1=(0 0 1 0 0 0 0 0 0 0 0 0 0).

First, we multiply the vector A_1 by the neutrosophic adjacency matrix W, and build a new state vector in accordance with the result. Then we multiply the vector A_2 obtained on the previous stage by the neutrosophic adjacency matrix W, pre-adding a unity to the third component of the vector (this is necessary in order to consider the initial single exposure). Further, in a similar way, we write a state vector to the result of multiplication obtained. Iterative procedure is conducted until the state vector at n step will be equal to the state vector at n-1step. This means that we managed to reach the fixed point as a result of the iterative procedure:

$$A_1W=(0\ 0\ 0\ w_{5,6}\ 0\ I\ 0\ 0\ 0\ 0\ 0\ 0\ w_5) \rightarrow A_2=(0\ 0\ 1\ 1\ 0\ 0\ 0\ 0\ 0\ 0\ 0\ 0\ 1)$$

$$A_2W=(0\ 0\ 0\ w_{5,6}\ 0\ I\ w_{4,10}\ w_{6,11}\ 0\ 0\ 0\ 0\ (w_5+w_6)) \rightarrow A_3=(0\ 0\ 1\ 1\ 0\ 0\ 1\ 1\ 0\ 0\ 0\ 0\ 1)$$

$$A_3W=(0\ 0\ w_{10,5}\ (w_{5,6}+I)\ (w_{10,75}+w_{11,7})\ (I+w_{11,8})\ 0\ w_{6,11}\ 0\ 0\ 0\ 0\ (w_5+w_6+w_{11})) \rightarrow$$
$$A_4=(0\ 0\ 1\ 1\ 1\ 1\ 0\ 1\ 0\ 0\ 0\ 0\ 1)$$

$$A_4W=(0\ 0\ 0\ w_{5,65}\ w_{11,7}\ (I+w_{11,8})\ 0\ (w_{6,11}+w_{7,11}+w_{8,11})\ 0\ 0\ 0\ 0\ (w_5+w_6+w_7+w_8+w_{11})) \rightarrow$$
$$A_5=(0\ 0\ 1\ 1\ 1\ 1\ 0\ 1\ 0\ 0\ 0\ 0\ 1)=A_4 \qquad (3)$$

In this case, we managed to reach the fixed point in *4* iterations. The value A_4W represents the system state vector, each element of which describes the impact applied to the corresponding concept in the result of a single impact T_5.

Let us consider a special case of the simultaneous action of several concepts T_5, T_7, T_8, that is an effect of the vector B_1=(0 0 1 0 1 1 0 0 0 0 0 0 0) :

$$B_1W=(0\ 0\ 0\ w_{5,6}\ 0\ I\ 0\ (w_{7,11}+w_{8,11})\ 0\ 0\ 0\ 0\ (w_5+w_7+w_8)) \rightarrow$$
$$B_2=(0\ 0\ 1\ 1\ 1\ 1\ 0\ 1\ 0\ 0\ 0\ 0\ 1)$$

$$B_2W==(0\ 0\ w_{10,5}\ w_{5,6}\ w_{11,7}\ (I+w_{11,8})\ 0\ (w_{6,11}+w_{7,11}+w_{8,11})\ 0\ 0\ 0\ 0\ (w_5+w_6+w_7+w_8))$$
$$B_3=(0\ 0\ 1\ 1\ 1\ 1\ 0\ 1\ 0\ 0\ 0\ 0\ 1)=B_2 \quad (4)$$

According to the results obtained, we can conclude that the change of the influence on the target indicator depends on the concepts, which were influenced.

The combination of neutrosophic cognitive modeling with models of neuro-fuzzy inference provides the opportunity to carry out a deep analysis of semi-structured problems. Another advantage is an ability to enter additional primary data based on the expert assessment, which in certain cases may lead to an increase in the reliability of the conclusions to a decision-maker.

4 Conclusion

1. Based on the synthesis of interdisciplinary research strategies the authors have presented a structural and functional scheme of an intellectual multilevel system of analysis and forecast of scientific and technological, innovative research projects, innovative facilities, and areas. Research strategies included expert foresight methods, which represent a system of peer review of strategic prospects of innovative development, identification of medium and long-term technological breakthroughs capable to have the greatest impact on the economy and society, fuzzy inference models, identification of parameters of neural

network technologies, and fuzzy cognitive models. The core of the system is formed by the designed multicriteria models of adaptive neuro-fuzzy inference and fuzzy cognitive analysis aimed to solve a number of problems and forecast evaluation scenarios of multidimensional attribute of objects and processes. The system developed has the following distinctive features:

- distributed iterative information system that operates in a real time-mode;
- possibility to make a foresight forecast of the future;
- adaptability to goals, tasks and problems because of the possibility to adjust to the neuro-fuzzy and cognitive model;
- logical transparency, which allows you to accumulate knowledge in the sphere of analytical and forecast problems of innovative development;
- universality, which means the possibility to apply various innovative studies, projects and objects (including enterprises, industries, territories (cities, regions, innovation complexes) for analyzing and forecasting the development.

2. The use of fuzzy cognitive maps as a tool for modeling semi-structured systems allows representing the system under analysis clearly and flexibly, as well as to interpret causal relationships between concepts.

3. Combining neutrosophic cognitive modeling with the results of neuro-fuzzy conclusion resulted in a new intelligent foresight tool, which provides logical transparency of the forecasting studies. It gives the opportunity for deep and quick analysis of scientific, technological, and innovation development taking into account many signs, their significance, mutual influence, and contingency with multiplier effects. It also allows carrying out a situational analysis depending on the target factor. On the other hand, the methods give the possibility to introduce additional expert data, which in general case can increase the reliability of the conclusions.

4. The results obtained have the practical value, which includes the following main factors. They allow improving the quality of forecasts of the most important areas of scientific and technological development of the state objects and territories. The results also allow identifying factors determining the prospects for economic modernization and the enhancement of the effectiveness of scientific and technical policy aimed at outstripping the development of basic production and transition to a new technological order.

The correspondence of the results obtained with the world level of research makes it possible to use them in the economic, scientific, technological, and social spheres with the purpose of scientifically substantiated support of decision-making procedures. Prospects for further research are related to the development of algorithms and methods for parametric and structural identification of the models constructed.

Acknowledgements

The article is prepared within the framework of Competitiveness Enhancement Programme of National Research Tomsk State University and grant of the Russian Foundation for Basic Research No 16-29-12858.

The authors are grateful to Tatiana B. Rumyantseva from the National Research Tomsk State University for her assistance in preparing the paper.

References

[1] S.K. Glazyev, *Theory of long-term technical and economic development* (FIZMATLIT, Moscow, 1993)

[2] T.V. Abramova, E.V.Vaganova, S.V. Gorbachev, M.V. Gribovskiy, V.I. Syryamkin, M.V. Syryamkin, T.V. Yakubovskaya, *Cognitive systems for monitoring and forecasting the state's scientific and technological development* (Publishing house of Tomsk State University, 2012)

[3] J. Forrester, *World Dynamics* (Science, Moscow,1978)

[4] S.V. Gorbachev, T.V. Abramova, Telecom. **7**, 43 (2014)

[5] R. Popper, *Methodology: Common Foresight Practices & Tools* (Edward Elgar, 2007)

[6] S.V. Gorbachev, V.I. Syryamkin, *Neuro-fuzzy methods in intelligent systems for processing and analyzing multidimensional information* (Tomsk State University, 2014)

[7] D. Rutkovskaya, M. Pilinsky, L. Rutkovsky, *Neural networks, genetic algorithms and fuzzy systems* (Moscow, 2006)

[8] S.V. Gorbachev, *Fazzification of expert linguistic data in the ordinal scale* (Kursk, 2016)

[9] S.V. Gorbachev, *Designing a Hybrid Neural-Fuzzy System Based on Soft Computing* (Tomsk State University, 2014)

[10] A.A. Kulinich, *Methodology of cognitive modeling of complex ill-defined situations* (ISP RAS, Moscow,2003)

[11] L.A. Ginis, Scient. Rev. **8**, 219 (2015)

A COMPETENCE-BASED APPROACH TO TRAINING SPECIALISTS IN THE DIGITAL SOCIETY

Evgenia Liventsova*, *Tatiana* Rumyantseva, *Ekaterina* Syryamkina

National Research Tomsk State University, 634050, Tomsk, Russia

Abstract. The article analyzes the influence of digital technologies on transformation of the labor market in social services. Modern professions require workers with advanced digital skills as well as technical and engineering specialists with advanced soft skills. The article analyzes official documents and global trends in the future labor market and presents the experience of National Research Tomsk State University (TSU) in developing transprofessional skills of students in different specialties. The authors conclude that modern society and the labor market demand new forms and methods of education aimed to develop interdisciplinary competence.

1 How digital technologies influence contemporary society

Digital technologies are a set of processes that are based on the representation of signals by discrete bands of analog levels, but not in the form of a continuous spectrum [1]. In comparison with analog technologies, digital technologies allow you to transfer arrays of data easier and faster. Today, this type of technology is widely used in computers, robotics, and in digital computing electronics.

The President of the Russian Federation Vladimir Putin in his message to the Federal Assembly on December 1, 2016 noted that a new level of economic and social development requires focusing on areas where the powerful technological potential of the future is accumulated. These areas include digital and other breakthrough technologies, which now influence all spheres of life [2]. Therefore, digital technologies can be classified as breakthrough (cutting-edge), that is, those used in all sectors and in the social sphere as well.

Currently, information technologies cause active transformation of the society. New forms of interaction between the state and society, as well as between different groups of society greatly simplify certain processes and affect improvement of quality of life of people. For example, in the sphere of governmental social services, every citizen has an opportunity to share his problem in the online mode; and the representatives of the government services react online on these problems and submit reports on their resolutions. The on-line government services are becoming more popular; they allow getting necessary

* Corresponding author: evg.liv@mail.ru

information and documents without queues and paperwork. The use of information technology also allows timely accumulation of civic activity, volunteer movements, and charitable organizations in order to solve certain social problems and provide more targeted assistance to a particular family, child, and other beneficiaries. Development of digital technologies improve quality of life of people with disabilities because of the emergence of new digital products that allow full integration into the society; these products include televisions with teletext for receiving programs with hidden subtitles, telephone devices with text output, voice-forming devices, special clothes, special devices for reading and much more.

Digital technologies significantly influence the sphere of education, which is now becoming increasingly mobile and accessible through dissemination of online learning. Scientists and engineers create special digital simulators and online learning programs, which allow people to get new knowledge and skills.

2 New professions

The long-term program of the Russian Federation, National Technology Initiative (NTI), adopted in 2015 [2], confirms the urgency of recognizing the transformation of society under the influence of new technologies. NTI is aimed to create conditions for ensuring leadership of Russian companies in the new high-tech markets that will determine the structure of the world economy in the next 15-20 years. However, developing new technologies demands new qualified specialists who can correctly apply these technologies. NTI defines nine markets of the future, and some of them are closely related to the social sphere. These include NeuroNet (a market for human-machine communications), HealthNet (a market for personalized medical services and medicines), and FoodNet (a market for production and sale of nutrients and final types of food products, as well as related IT solutions).

Key segments of NeuroNet's market include neuro-assistants, neuro-education, neuro-medical devices, neuro-entertainment and sports, neuro-communications and marketing, and neuro-pharma. A key segment for HealthNet is information technology in medicine, which means design and implementation of devices for monitoring and correction of human health. FoodNet's market is mainly focused on intellectualization and automation of technological processes throughout the production cycle and is aimed at expanding the capacity to produce high-quality products. All the above-mentioned segments directly relate to the social sphere and demand not only developing new technologies but also new specialists who can work with these technologies.

In the rapidly changing conditions of modern life, not only IT professionals but also social workers need skills in digital technologies or the so-called digital skills. Prof. Dr. Edward V. Galazhinskiy, Rector of Tomsk State University, notes that "non-digital" professions (doctors, lawyers, actors, teachers of literature, and others) need digital skills in a standard package of competence because representatives of these professions are "members of the modern network digital society" [3].

We can better understand how specialists in the social sphere will apply digital skills if we analyze *The Atlas of New Professions*, developed and published by the Agency for Strategic Initiatives and the Skolkovo School of Management [4]. The *Atlas* describes new professions that will be in demand in the future and a list of transprofessional competencies that a "specialist of the future" should possess.

According to the *Atlas,* future professions in the social sphere will include that of a specialist in crowdsourcing societal problems, a social worker for the Internet-based adaptation of people with disabilities, a specialist in adaptation of immigrants, and a moderator of a platform for personal charitable programs. Education will demand coordinators of educational online platforms, mentors for startups, coaches for mind fitness, developers of educational trajectories, and developers of tools for teaching states of consciousness. These new professions will require such competencies as programming IT solutions, managing complex automated complexes, and working with artificial intelligence (Table 1).

Table 1. Future professions in social and educational spheres according to the *Atlas of New Professions*.

<table>
<tr><th>Profession</th><th>Transprofessional competencies and skills</th></tr>
<tr><td>Specialist in crowdsourcing societal problems</td><td rowspan="8">Programming IT solutions/ managing complex automated complexes/ working with artificial intelligence
Ability to manage projects and processes
Customer-oriented focus, ability to work with customer requests
People skills (ability to interact with others)
Multilingual and multicultural development
Project management
System thinking</td></tr>
<tr><td>Social worker for the Internet-based adaptation of people with disabilities</td></tr>
<tr><td>Specialist in adaptation of immigrants</td></tr>
<tr><td>Moderator of platforms for personal charitable programs</td></tr>
<tr><td>Coordinator of educational online platforms</td></tr>
<tr><td>Mentor for startups</td></tr>
<tr><td>Coach for mind fitness</td></tr>
<tr><td>Developer of educational trajectories</td></tr>
</table>

On the contrary, if we refer to the section of the *Atlas* devoted to the IT sphere, we can find jobs directly related to the social sphere and requiring such transprofessional competencies as client-oriented work with people (Table 2). These are such future professions as an architect of information systems, interface designer, network lawyer, digital linguist, personal security consultant, and cyber-investigator. The tendency toward "society virtualization" is clearly identified: the virtual space will be organized according to the same principles as the national government. In this regard, there is a need to transfer some standard professions (such as lawyers, investigators, insurance agents, and others) to a virtual space.

Table 2. Future professions in IT-sphere according to the Atlas of New Professions.

<table>
<tr><th>Profession</th><th>Over-professional competency and skills</th></tr>
<tr><td>Architect of information systems,</td><td rowspan="6">System thinking
Programming / Robotics / Artificial Intelligence
Customer orientation
Project management
Inter-industry communication
People skills (ability to interact with others)
Multilingual and multicultural development
Ability to work in conditions of uncertainty</td></tr>
<tr><td>Interface designer</td></tr>
<tr><td>Network lawyer</td></tr>
<tr><td>Digital linguist</td></tr>
<tr><td>Personal security consultant</td></tr>
<tr><td>Cyber-investigator</td></tr>
</table>

3 Examples of educational programs and competitions to improve the social sphere by developing people's digital skills sponsored by Russian companies

Speaking about opportunities for innovative creativity of young people, we can give some examples of programs organized by educational organizations in conjunction with socially responsible business. Let us consider an example of Rostelecom Company; it sets provision of social environment as one of the priorities of social policy. The company supports educational projects, develops new interactive programs and improves existing directions. The main attention in the company is paid to training of specialists in telecommunications, information security of university students and schoolchildren, as well as work on digital inequality. When developing educational projects, the company takes into account the interests of socially unprotected sections of the population [5].

Another example is the complex program of social investments "Native cities" realized by Gazprom Neft Company [6]. The program is aimed to increase the level of life of people in the cities, to create infrastructure for children and mass sports, and to implement initiatives in the field of culture, health and education. According to the program regulations, non-profit organizations, universities, social institutions, and even initiative groups of citizens can participate in the program. The competition is aimed at finding ways to solve problems of society or certain social groups.

Since 2017, a special nomination Digital Technologies has been allocated within the framework of the "Native cities" program [6]. The nomination includes internet projects (portals, services), applications /services for mobile devices and other gadgets, games using virtual and /or augmented reality technologies, digital-installation, interactive modules, multi touch technologies; digital technology-based projects and activities; and video campaigns.

Appearance of this nomination is associated with the successful experience of the Federal Competition of Social Projects "People Need You" organized by St. Petersburg National University of Information Technologies, Mechanics and Optics [7, 8].

This competition represents a mechanism for stimulating creativity of students in the field of social design and entrepreneurship and developing students' skills in project management that can be successfully used in future social and professional activities.

4 Experience of National Research Tomsk State University (TSU)

Program of TSU strategic development as a national research university[†] for 2010-2019 highlights social and humanistic knowledge and technologies as one of the main direction in modernization of the economy and the social sphere. Moreover, among provincial universities with the status of national research university, only TSU has announced humanistic studies as one of the priority research areas. Based on the great experience of training specialists in humanities and taking into account global trends Tomsk State University has started to pay more attention to developing soft skills of all students including students of engineering and technical specialties. This includes ability for effective team communication, learning of the basic principles of behavior on the labor market, and business activity in the social sphere, which leads to the development of interpersonal skills, system thinking, formation of an active lifestyle and professional positions.

[†] In 2010, TSU became a winner in the federal competition between Russian universities and gained a status of National Research University.

Realizing that the concepts of innovation potential and entrepreneurial resource are closely connected with the individual's creativity, TSU has created a structural unit – the Park of Social and Humanistic technologies (hereinafter, the Park) aimed to implement the above-mentioned technologies of work with students of both humanities and engineering sciences [9]. Another aim of the Park is to involve students in social and entrepreneurial activity, as well as to attract professional experience of members of the local community to the university [10, 11].

The Park's goal is to organize environment for innovative development of university students in order to enhance humanistic technologies and social entrepreneurship. Therefore, the Park's activities include commercialization of social innovative projects, involvement of students and employees of TSU in creation, development and implementation of innovative social projects, involvement of TSU students and employees in entrepreneurial activities, and coordination of interaction between society, government and business organizations. During 2012-2015, the Park has implemented the strategy for developing entrepreneurial competencies of students of various specialties.

Table 3 shows annual events of the Park.

Table 3 Annual events of the Park of Social and Humanistic Technologies of TSU.

Events of the Park	**Transprofessional skills and competencies developed**
Competition of creative ideas "Student's Tomsk" (selection and assistance in implementing students' creative ideas aimed to improve the university and the city environment)	Creative thinking People skills (ability to interact with others) Ability to work in a team
Special educational program for training social and entrepreneurial skills of students	Skills for entrepreneurship in social sphere System thinking
Competition of social projects "Important Deal" (selection and assistance in the implementation of social and entrepreneurial projects of schoolchildren and students aimed to solve problems existing in social sphere)	Ability to work in a team Project management People skills (ability to interact with others)
Social and investment exhibition "Important Deal" (presentation of projects /products to the representatives of business and city authorities)	Skills of product /project presentation Skills of self-presentation Communication skills
Intensive educational program "StartUpSocial" for acceleration of students' entrepreneurial projects in the social sphere	Skills for entrepreneurial activities in the social sphere Ability to work in a team Project management People skills (ability to interact with others) Programming skills
Business Club of the Park of social and humanitarian technologies of TSU (monthly discussions between students, representatives of business, government, NGOs and educational institutions)	Development of system thinking Communication skills
Program of additional education "TSU Students are Specialists of the Future" (a new program launched in 2017)	System thinking Critical thinking Digital skills

4 Conclusion

Rapid development of technology leads to transformation of society and change of the labor market and demand renewal of educational approaches as well as integration of information and innovative methods of education.

The totality of the abovementioned forms of educational technology allows motivating students to obtain new professional skills because it is important for a student to identify the result of his education as a set of professional knowledge and skills, which can be successfully applied in practice and will allow being competitive on the labor market.

Today, university graduates need transprofessional skills; and an important fact is that these skills can be gained within the curriculum if universities organize additional programs of professional training for students of various specialties. These programs will have greater effect if they are implemented based on deep analysis of the current situation and future trends of the labor market and in close cooperation with partner-organizations.

The competence-based approach (orientation to transprofessional competences) may facilitate overcoming a lag between the theoretical training of students within the basic university curriculum and practical experience because any activity loses sense if it does not meet the needs of a contemporary society.

Acknowledgements

The research presented in this paper was supported by the Russian Foundation for Basic Research (grant No 16-29-04388/17).

References

[1] NTI Dictionary URL: http://www.nti2035.ru/nti/reference/
[2] The President's Address to the Federal Assembly URL: http://kremlin.ru/events/president/news/53379
[3] Triumvirate «skills» as a formula of demand for a university graduate URL: http://www.tsu.ru/university/rector_page/triumvirat-skillz-kak-formula-vostrebovannosti-vypusknika-vuza/
[4] The Atlas of new professions URL: http://atlas100.ru/
[5] Principles and priorities of social policy URL: http://www.rostelecom.ru/social/politics/
[6] Grant Competitions of social initiatives URL: http://www.grant.rodnyegoroda.ru
[7] Regulations on the Federal competition of social projects «People need you!» URL: http://socialprojectspb.ru/wp-content/uploads/2017/02/regulations_tnl-2017.pdf
[8] Expanding the reach and impact of the humanities for the public good URL: http://www.cdh.ucla.edu
[9] Park of Social and Humanitarian Technologies of Tomsk State University URL: http://parksgt.tsu.ru
[10] The Fourth Industrial Revolution URL: https://www.foreignaffairs.com/articles/2015-12-12/fourth-industrial-revolution
[11] Youth social entrepreneurship advancing the field URL: http://www.sundancefamilyfoundation.org/yse-toolkit/yse-white-paper

LOCAL SYSTEM OF POSITIONING USING A WIFI NETWORK

Petr Rempel*, *Alexey* Borisov

Altai State Technical University, 656038, Barnaul, Russia

Abstract. The WiFi technology is widespread. To launch the WiFi network is very simply, thus, they meet so often. Knowledge of the location of the antennas Wi-fi is very important for internal localization (position fix of an object). System of positioning of Wi-fi (WPS) or WiPS/WFPS are used where GPS and GLONASS are unavailable because of different reasons including multipath and signal blocking indoors. Such systems include as well internal systems of positioning. Use of Wi-fi not for creation of a network, and (for example, positioning of devices on the WiFi network) is urgent for other purposes because there is a rapid growth at the beginning of the 21st century of wireless access points in urban areas. This article discusses practical application of a method of localization with use of WiFi of access points which is based on measurement of intensity of the received signal (indication of the received signal or RSSI). Here the device which executes the measurements necessary for system operation is described. Some formulas for computation of layout of the device are given. Results of system operation are considered with different layouts of access points in the given space and different environments.

1 Introduction

Nowadays WiFi networks are so widespread that with their help one can easily create a network consisting of a large number of devices which are unattached to a certain place and capable to change the position in space. At present almost any corporate client has WiFi coverage, therefore the infrastructure of this wireless network can be used for performing a number of tasks, in particular, for local positioning in the organization. Accuracy of such systems will depend on density of points of access tied to specific points in the plan of buildings, constructions and territories [1].

Positioning of objects in WiFi networks can be used for achieving a number of purposes:

- automation of accounting and placement of objects;
- control of personnel and clients of the organization;
- ensuring technical safety;
- navigation in the building;
- data acquisition for information and analytical work.

* Corresponding author: petyarempel@outlook.com

2 Principle of operation

Positioning in wireless networks can be carried out in several ways:

1. The method of recognition of a template demanding a preliminary research of the network and control of the system.
2. The method of using the coordinates of the nearest access point which allows detecting the presence of the client.
3. The triangulation method involving determination of force of a WiFi signal and calculation of a possible arrangement concerning each visible point of access.
4. An angulation or positioning with definition of the angle of the incoming signal.

The method of a triangulation is rather simple and effective, and it doesn't demand high costs to apply it as well, therefore it is used in this system.

The system consists of several points of access, a WiFi-tag and a device for information collection and data management (see Figure 1).

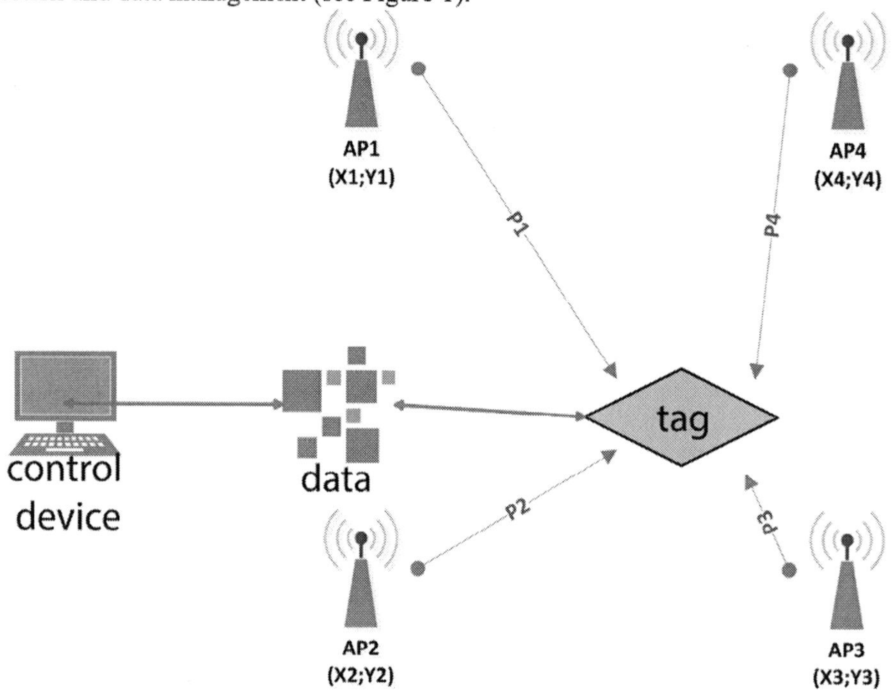

Fig. 1. The principle of operation of the local positioning in WiFi network.

The system operates as follows: the device (tag) searches for the given points of access and measures the power of the accepted signal, then sends data to the device where the location is calculated with some accuracy in the set system of coordinates [2].

Data on the control unit are transmitted "by air" through WiFi technology and can be easily intercepted. For their closing the known encryption methods in WiFi of network can be used (if it is necessary).

The coordinates were calculated by means of the algorithm Weighted centroid (the center of masses) which computes the agent's coordinate as a linear combination of coordinates of points of access taking into account the power of signals as the characteristic of weight [2]. The way this method calculates the AP position determines the situations in which the method provides good location estimations: If there are measurements taken around the real AP position or the measurements are close to the AP [3].

To calculate the location the following formulae are used:

$$\begin{cases} X_0 = \sum_{i=1}^{N} \mu_i X_i \\ Y_0 = \sum_{i=1}^{N} \mu_i Y_i \end{cases} \quad (1)$$

$$\mu_i = \left(P_i \sum_{j=1}^{N} \frac{1}{P_j^2} \right) \quad (2)$$

where X_0 and Y_0 – required coordinates, X_i and Y_i – coordinates of i-th point of access, P_i – the power of a signal of i-th point of access, P_j – the power of a signal of j-th point of access, μ_i-characteristic of weight, N – quantity of points.

The characteristic of weight of a signal μ_i from i-that access point is calculated by a formula (2). After each scanning it is necessary to calculate the characteristic of weight on a formula (2). After that tag coordinate on a formula (1) is calculated.

As a tag using the Wemos d1 mini microcontroller – a programmable board on the basis of the ESP8266EX microcontroller with WiFi support with an access to UART, SPI, I2C, GPIO interfaces (see Figure 2).

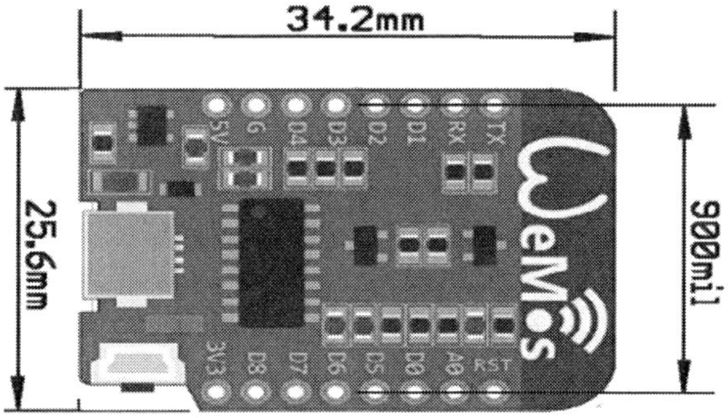

Fig. 2. Wemos D1mini.

In that this microcontroller has GPIO interface, connection to them of different sensors make the device is possible. Such sensors can serve for determination of falling, violation of integrity of an object, and also to collect object other information. Data transfer is carried out on the WiFi network.

The board is programmed so that at first to scan WiFi of a network, and then to transfer them to the server. After successful transmission the device starts over again scanning.

The accumulator of the small size and with a capacity of 1050 mAh which provides rather long operating time of the device is connected to the controller. Such operation mode in case of which the device will periodically work is possible. Then the accumulator will be enough for longer time frame

As access points WiFi-routers, spot jotas on mobile devices and notebooks were used. All of them broadcast a WiFi-signal at a frequency of 2.4 GHz.

3 Testing results of the device

This device had many tests. Testing took place in different conditions and different places. Testing was carried out both in the open spaces and indoors where distribution of a signal is interfered by overlappings, furniture and other objects.

In (see Figure 3) the results of testing the device in the room measuring 12×7 meters are presented. In the corners of the room 3 points of access are placed.

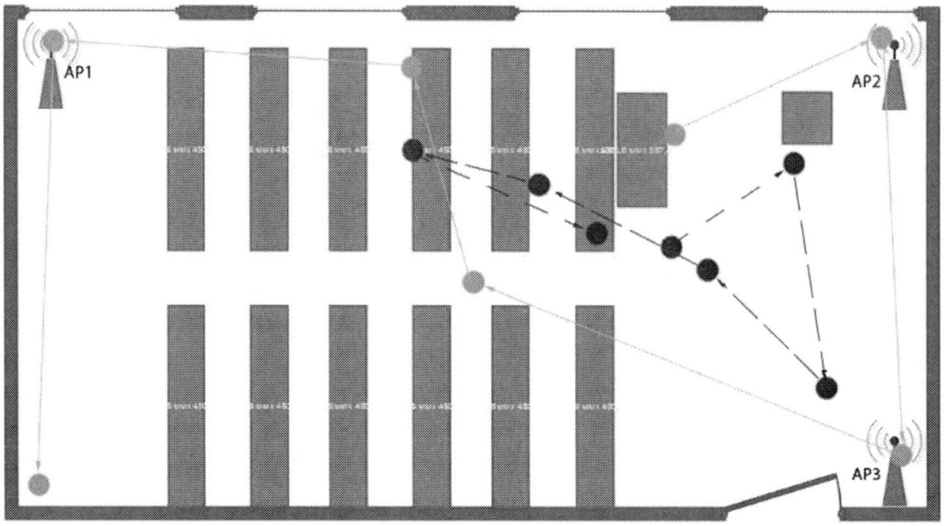

Fig. 3. Testing results with 3 points of access indoors.

Dark lines display the location calculated by the computer. Light dashed lines display actual location of a tag on the card. Gray rectangles show furniture located in the room and influencing signal propagation. The error in such conditions comes to 5 meters, therefore the number of points of access (see Figure 4) has been increased.

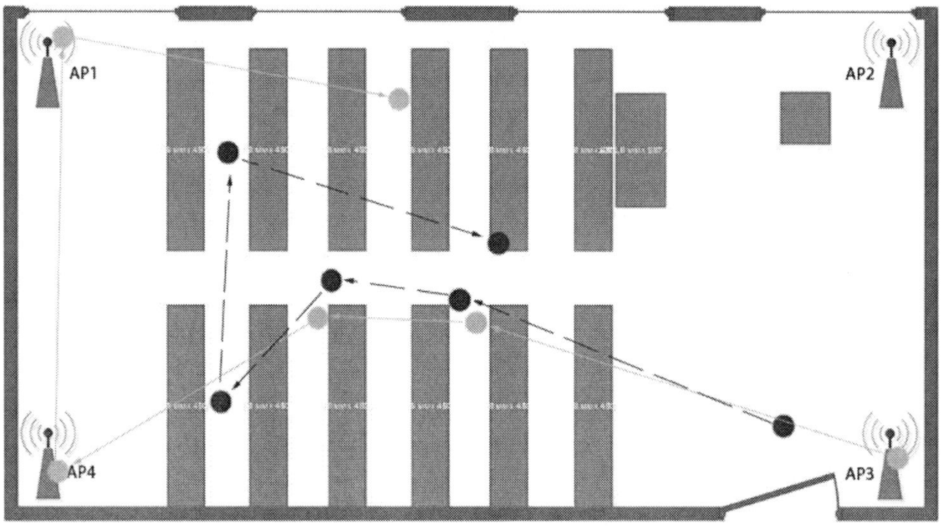

Fig. 4. Testing results with 4 points of access indoors.

Figure 4 illustrates that the accuracy of positioning has significantly improved with an increase in the quantity of points of access. The error doesn't exceed 2.5 meters.

In (see Figure 5) the results of testing the device in the open space measuring 10×10 meters with the use of four points of access are presented.

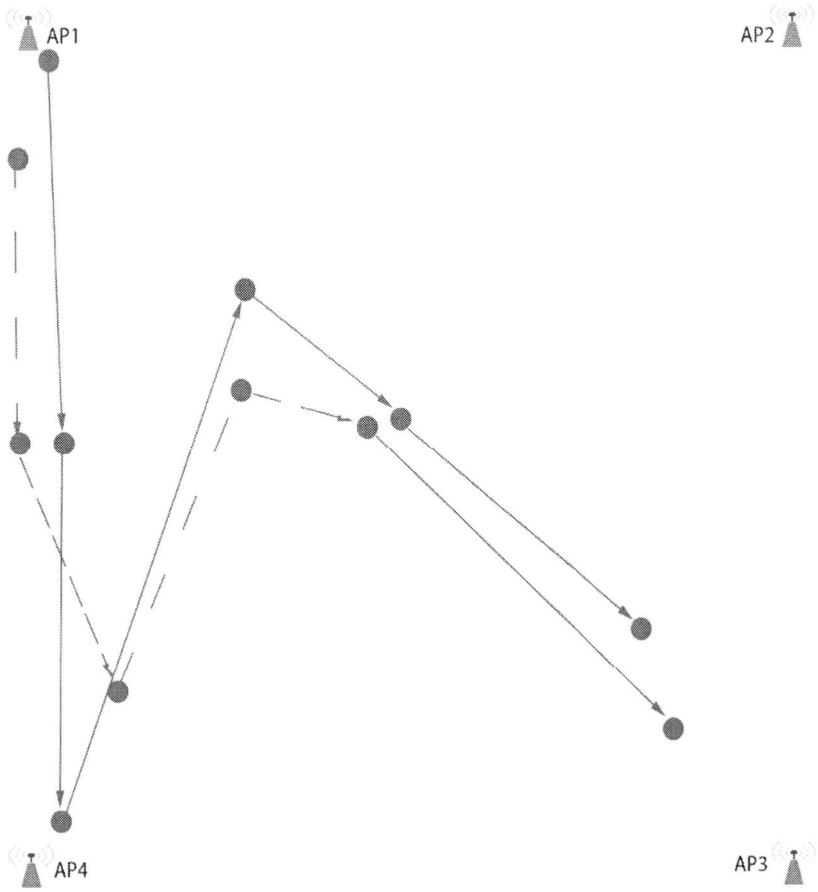

Fig. 5. Testing results with 4 points of access in the open space.

As can be seen from Figure 5 in the open space, where the propagation of the signal is not obstructed by foreign objects, the accuracy is much better.

The system also took tests in premises in which there was furniture and interroom partitions. As is well-known walls very strongly worsen WiFi signal, but even it hasn't strongly influenced operation of the device. The result was satisfactory. On (see Figure 6) results of testing of the device indoors 14×14 meters in size are presented. On corners of the room 4 points of access are placed.

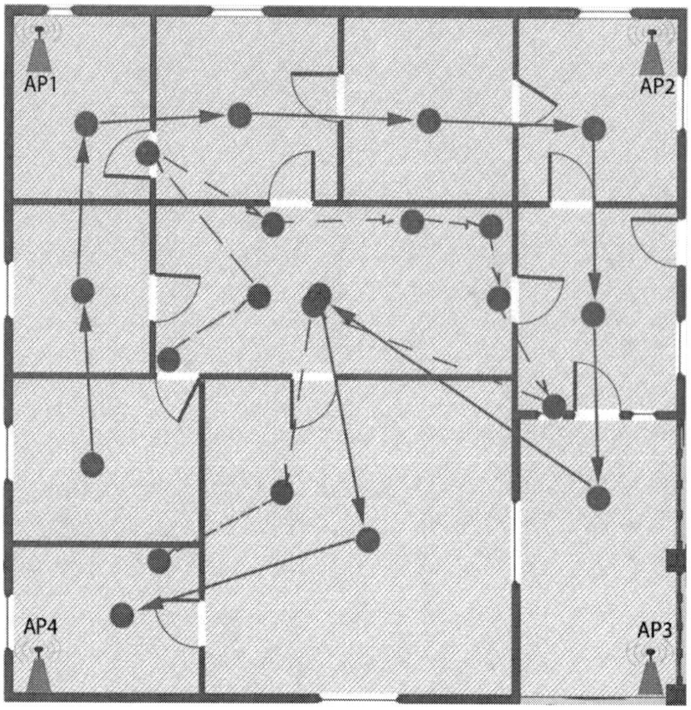

Fig. 6. Testing results with 4 points of access indoors.

As can be seen in the Figure, the walls affected the accuracy of the device slightly. In some cases the deviation was strong. It can be because on the way of a signal furniture, and also people meets.

As a result of testing the device with the use of 6 points of access the error in calculating the location has decreased to 1.5 meters. Figure 7 shows the results of testing the device in two rooms whose size totals 17×12 meters.

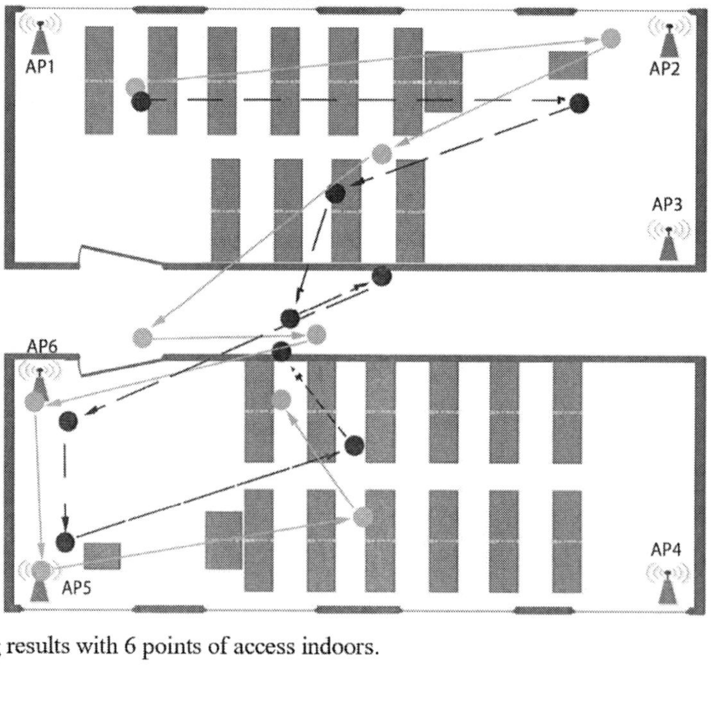

Fig. 7. Testing results with 6 points of access indoors.

Testing of the device in the presence of 6 points of access has shown good result. However, in this case points of access settled down quite close to each other. Such arrangement of access points rarely meets.

In (see Figure 8) the results of testing the device in the open space measuring 18×18 meters are presented.

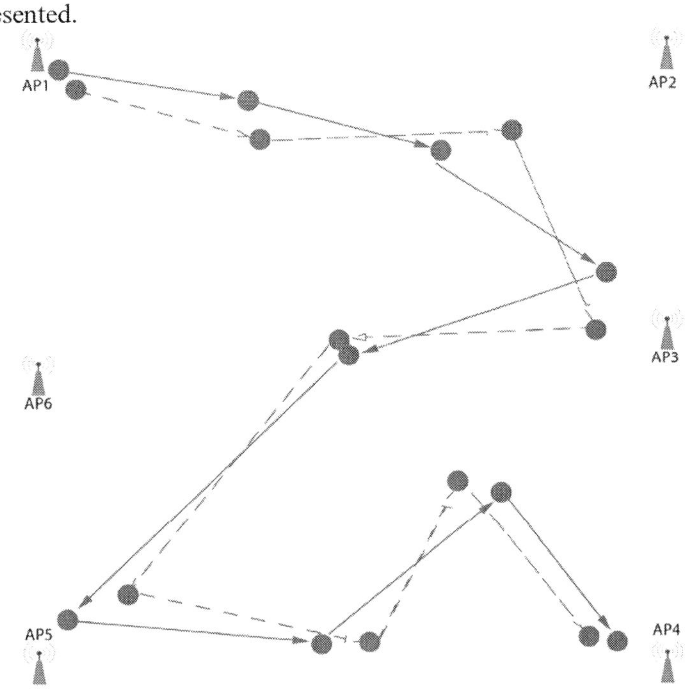

Fig. 8. Testing results with 4 points of access in the open space.

In the open space with 6 access points the device showed a good result. The error in determining the location decreased to 1 meter.

4 Conclusion

The results of testing have led us to conclude that the calculated position of a tag doesn't go beyond the bounds of the Figure formed by access currents in the set system of coordinates. Accuracy of calculation depends on the quantity of points of access simultaneously available to a tag. Our test show, that you require an anchor every 10–30m to achieve good results indoors (open space).

The device works fine even if there are obstacles in the path of the WiFi signal. However, to achieve high accuracy, the signal attenuation level must be taken into account when calculating the location.

Such a system even at the current stage of development is suitable for performing a number of tasks. Further it is planned increases in accuracy of positioning thanks to more exact algorithms of calculation of coordinates.

References

[1] P. Rempel, A. Borisov, M.C.I., XVIII I.S.T.C.**32**, (2017)
[2] I. Averin, A. Semenov., R.R.C., IV All-R. C. **475**, (2010)
[3] F. Alsehly, R. Mohd Sabri, Z. Sevak, T. I.T.M. **131**, (2010)

SOME ISSUES OF THE QUALITY CONTROL SYSTEM DEVELOPMENT FOR TELECOMMUNICATIONS SERVICES

Alexey Yurchenko[1], *Tansaule* Serikov[2], *Aliya* Alkina[1], *Yelena* Neshina[1a]

[1]National Research Tomsk Polytechnic University, 634050, Tomsk, Russia
[2]Karaganda State Technical University, 100000, Karaganda, Kazakhstan

Abstract. In connection with the rapid development of telecommunications technologies, the provision of communication quality parameters and its reliability are very relevant. In this regard, the new approaches to the operation and control of telecommunications networks are required which are inconceivable without an integrated approach to solving this problem and developing a control system not only for the network state, but also for the quality indicators of its operation. The idea of the work is to develop the control system that is able to determine the principles of strict monitoring over the telecommunications network of Kazakhtelecom JSC in Karaganda. The analysis of the key performance indicators of the telecommunications network showed that after the implementation of the Network Management System hardware and software complex, all values of quality indicators meet the regulatory values. It allows monitoring and controlling the telecommunications network at various levels of operation of basic and auxiliary equipment in a wide range of work.

1 Introduction

Modern telecommunications systems should have the high quality service level provided to their customers. The quality of these communication services is the main criterion for assessing the telecommunications network which is especially important for the national operator Kazakhtelecom JSC. The concept of "quality" for this operator is present throughout the technological cycles of the production organization of the communication service.

At present, in connection with the rapid development of telecommunications technologies, the provision of communication quality parameters and its reliability are very relevant. In this regard, the new approaches to the operation and control of telecommunications networks are required which are inconceivable without an integrated approach to solving this problem and developing a control system not only for the network state, but also for the quality indicators of its operation. In the search for various kinds of faults and failures that arise in modern information transmission systems, the greatest time is spent on identifying the causes that are related to the characteristics of the channels and communication lines, intra- and inter-system interactions, as well as unintentional or

[a] Corresponding author: 1_neg@mail.ru

deliberate actions of personnel. The identification of specific reasons for the failures of certain communication directions is extremely important for the rapid, coordinated and complete restoration of the network. At the same time, for a qualitative solution of this problem, it is necessary to use a rather large amount of information obtained from geographically dispersed objects of the network, as well as from specially conducted measurements. Information obtained in this way is characterized not only by diversity, but also by time difference which greatly complicates its generalization and processing.

Based on information from sources [1, 2] it can be concluded that the entire cycle of information transmission within the framework of a single production process requires, as a rule, the interaction of not one but several telecommunications enterprises (divisions). Moreover, each division takes part in the process of information transmission only at a certain stage.

Therefore, at the present time the monitoring systems are increasingly used to control and diagnose the various network aspects, which make it possible to solve these problems more effectively. And although such systems cover a wide range of issues being solved by them, still many of them are characterized by a certain "specialization" and do not solve the main problem such as control and diagnosing the entire network. This is precisely why the relevance of this present work is justified. The idea of the work is to develop the control system that is able to determine the principles of strict monitoring over the telecommunications network of Kazakhtelecom JSC in Karaganda.

2 Experimental study

While analyzing the management system of the backbone telecommunications networks, it is apparent that the system is built on the basis of Tivoli Netcool product family. Tivoli Netcool is an integrated software platform used to build the Operations Support System (OSS) and Quality Control System of the services provided. The structure of the telecommunications network management system (TNMS) of Kazakh telecom JSC is presented in Figure 1.

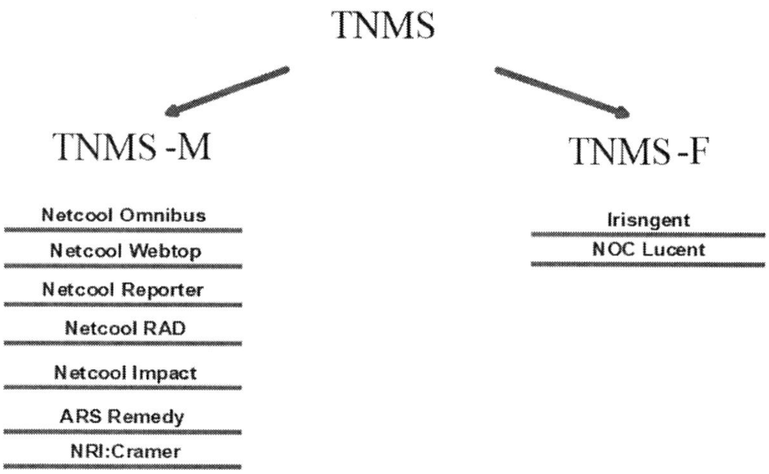

Fig. 1. Structure of the Telecommunications Network Management System (TNMS) of Kazakh telecom JSC.

The results of the network operation show that there is no control over the telecommunications network health on the network. The qualitative indicators adjusted by the company's management are not met, and the regulatory indicators are several times lower. To achieve this goal there is a need to develop a system that will not only monitor the progress of the emergency elimination, but also prevent the occurrence of an emergency on the network. In a word, it would monitor the network and its progress of operation.

Our task is to create a hardware and software complex with the name Network Management System assigned to it, which is able to function as a quality control system in efficient interaction with NETCOOL - Problem Tickets Management System (PTMS) (REMEDY). We use a systematic approach to creating a powerful multi-level system for controlling and diagnosing the telecommunications operations. The flexibility of the system will allow controlling both one and several networks simultaneously, from tens to several thousand objects. The monitoring objects connected to the management system are integrated with Netcoolupper-level telecommunications networks management system existing in Kazakh telecom JSC through the transmission of message flows using SNMP interface. All messages must contain:

- the physical and/or logical address of the equipment on which the fault occurred;
- identifier or internal report number;
- type of report (problem/resolution/information);
- priority rank for troubleshooting (critical/minor/ major, etc.);
- date and time when the fault occurred;
- a copy of the full text of the event formed by the monitoring object;
- information on correlation of opening and closing events;
- other additional information that technicians should use to eliminate and troubleshoot the equipment faults.

The interactions of components of the Telecommunications Networks Management System (TNMS) of the branch with components of the upper-level management system:

– the emergency report from the examination system (ES) is submitted for primary processing to the SEB controller, where it is reduced to a unified form, transmitted to the database of the Network Management System server and displayed on the workplace monitors of the branch personnel;

– the processed emergency report is transmitted by the Network Management System to TNMS Netcool and displayed on the monitor of the personnel of the main telecommunications network management center; Netcool system further carries out the analysis of the alarm message on the need to create a problem ticket in the PTMS for this type of emergency;

–the problem ticket is automatically created in the PTM Sand dispatched by REMEDY SW to the workplace of the responsible division for the organization of work to eliminate the damage on the network;

Figure 2 shows the interaction pattern between the Network Management System and the upper-level management systems. It is also used the notion of loss of accuracy which is estimated by the value of the error rate or the error probability rate.

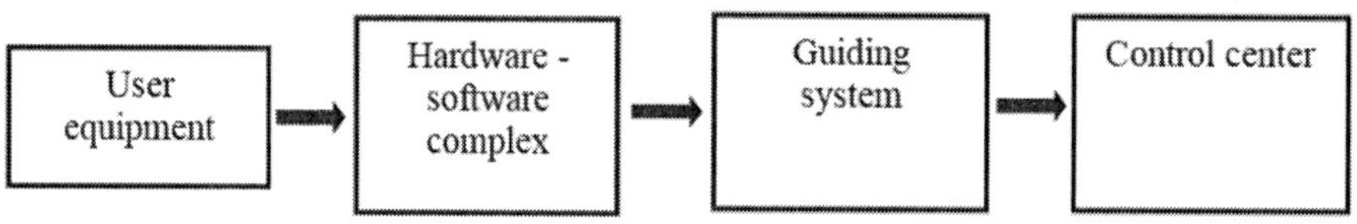

Fig. 2. Network Management System Hierarchy. Structural diagram.

The error rate Rerr. is defined as the ratio of the number of erroneously received symbols Nerr. to the total number of transmitted symbols NΣ in the measurement time interval Tmeasur.:

$$Rerr. = \frac{Nerr.}{N_{\sum Tmeasur.}}$$

When the binary signals are transmitted to the DTS (digital transmission system), the error rate is numerically the same as the error probability rate Perr.= Rerr. Since the digital flow transmitted through the digital line path (DLP) is always affected by distortion and interference which lead to digital errors. This means that some part of the binary symbols will not be accepted correctly: at the location "1" there may be "0" and vice versa. That is, the error probability is always different from zero Perr.$\neq$ 0.

The main quality assessment of the binary information transmission over the digital linear path is accuracy. The accuracy is the degree of conformity of the received information with transmitted one. The accuracy assessment is the probability of the correct reception, equal to the ratio of the number of correctly received message symbols to the total number of transmitted symbols for a certain period of time [3].

The system is built on a hierarchical basis and ensures the collection and processing of information, and then redirects it to the server that performs its final processing and analysis. The flexibility of the system allows it to be combined with any telecommunications equipment and specialized devices, and at the same time their own diagnostic capabilities can be used. The measurement monitoring elements are connected through an intermediate SEB server which perform the intermediate function of collecting and processing information and transferring it to a central server. If necessary, there is the transmission of corrective and control signals or in another word of effects to the objects from the SEB server. The whole management process happens in real time.

In addition to telecommunications equipment, Network Management System is able to manage any devices, including, for example, the contact elements of safety-alarm devices or SNMP. This means that it is possible to control any devices in the network which allows expanding the range of services provided. The information about the state of objects can be additionally sent via SMS, fax, e-mail or SNMP message [4].

Network Management System allows optimizing the amount of information output to the monitors of the dispatcher or operator, provided that the logical linking of events is logically removed at one of the objects. For example, to ignore the repeated voice messages caused by the failure of the automatic telephone exchange (ATE) and process them combining in one message about the failure of the ATE. The reports of events and faults are easily created using the interface and menus, while the issuance of reports can be programmed or executed on request to the server. If necessary, the summary of important events and emergencies can be received by e-mail or fax. The event filtering allows the operator to focus on important events and automatically configure the system's response to certain events.

Network Management System provides the visualization of processed information and data about the state of objects, presents their logic scheme or other scheme up to topographic one with the possibility of detailing to equipment or port elements which allows effectively monitoring the state of the object. It is possible to automatically prioritize the events. In the absence of an appropriate response the additional messages about important failures are made taking into account the status, importance and situation.

Types of controlled equipment using the network management system are shown in Figure 3.

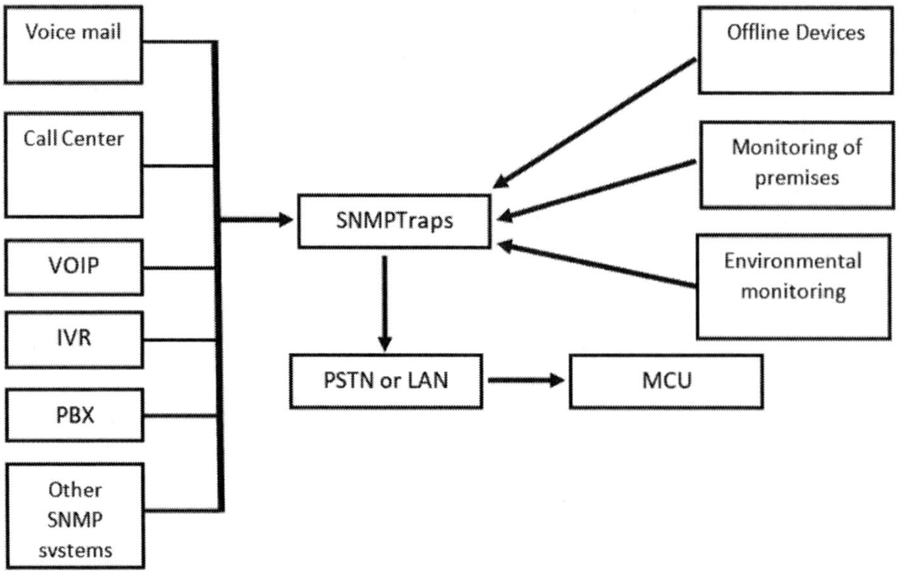

Fig. 3. Types of monitored equipment using Network Management System.

In the absence of staff reaction to critical situations and events, the priority of messages will be increased by sending them to the management units, for example, the chief engineer.

3 Developed methods

The reliability of the operation of our system is provided by different methods. For example, the standby power supply from the storage batteries ensures the uninterrupted operation and collection of information on the server after disconnecting the mains supply for several other hours accumulating information about the events. An important point is the hardware and software complex that allows processing information on the central server and making the necessary hardware settings, as well as performing the corrective operations. The collection of information about events and notification happens in real time with access to the objects serviced. The discrepancy between the information received and transmitted can be caused by the following reasons:

1) imperfection of methods of converting the transmitted message into a signal;
2) imperfection of signal transmission and reception methods;
3) imperfection of the methods of converting the received signal into a message;
4) features of signal propagation along the communication line;
5) insufficient noise-immunity of the signal.

The analysis of the result of the implemented software and hardware complex can be the reduction of key indicators. The control system with Network Management System was implemented in the network of Karaganda Regional Directorate of Telecommunications of the branch of Kazakh telecom JSC in early 2015. The results of the work after the implementation of the Network Management System hardware and software complex for the first half of 2016 are given in Table 1 [4, 5]. The interaction template: the network management system - NETCOOL - PTMS (REMEDY) is shown in Figure 4.

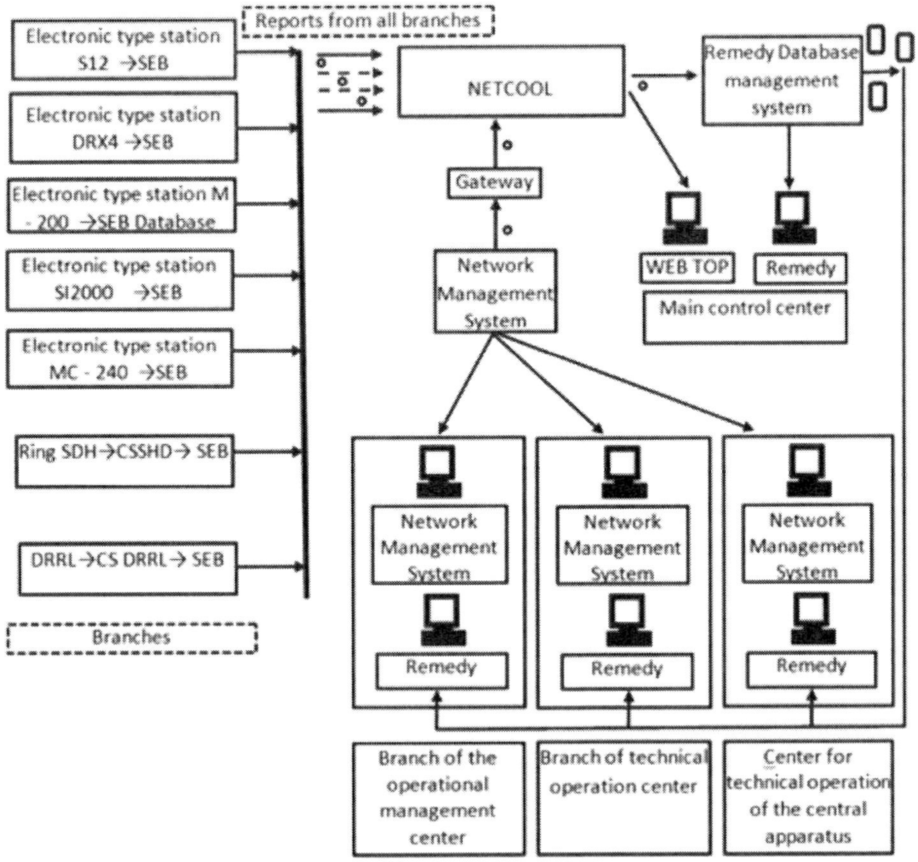

Fig. 4. Interaction Pattern: Network Management System - NETCOOL – PTMS (REMEDY).

While comparing the main indicators before and after the implementation of the Network Management System hardware and software complex concerning regulatory indicators adjusted by the company's management, it is possible to visually analyze the information (Figure 5, 6). Figure 5 shows one of the main indicators of the quality of the network; this is the network availability factor, which clearly shows the result of the Network Management System hardware and software complex.

Table 1. Main Quality Indicators.

Indicators	Time required to identify the fault of the automatic telephone exchange, h	Time required to eliminate the fault of the automatic telephone exchange, h	Time required to identify the network fault, h	Time required to eliminate the network fault, h	Network availability factor, %	Percent indicators for establishing connections,%
Regula-tory	3	3	3	4	99	50
Actual	1.3	2.1	0.5	3.2	99.97	56

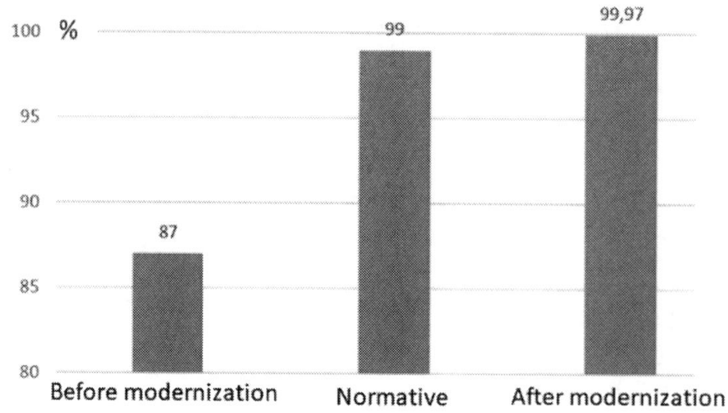

Fig. 5. Quality indicators for network availability factor.

Figure 6 shows a percent graph of the connection of network subscribers of Kazakh telecom JSC in Karaganda through communication channels. According to the results of the data, as a result of the work of the company's branch, it was not only completed, but was exceeded the adjusted standard by 22% after the implementation of the Network Management System hardware and software complex.

The analysis of the key performance indicators of the telecommunications network showed that after the implementation of the Network Management System hardware and software complex, all values of quality indicators meet the regulatory values. It allows monitoring and controlling the telecommunications network at various levels of operation of basic and auxiliary equipment in a wide range of work. The use of the Network Management System hardware and software complex when interacting with NETCOOL - PTMS (REMEDY) can be considered as a system approach to the creation of a powerful multi-level system for monitoring and diagnosing the telecommunications operations.

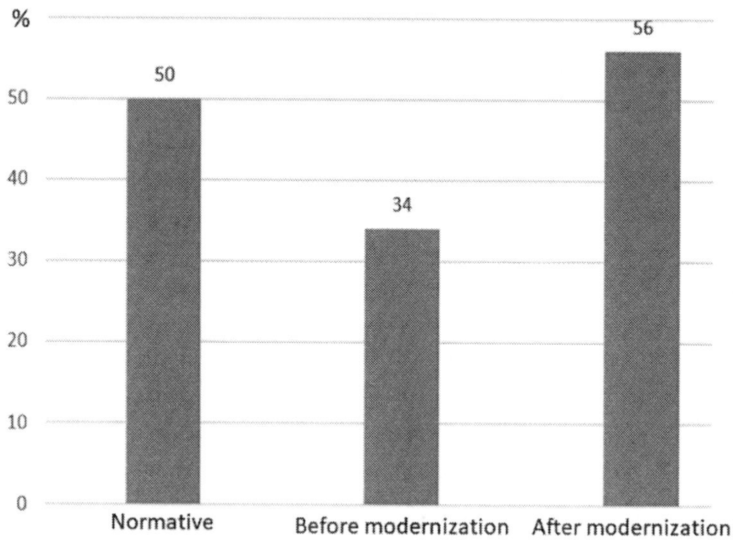

Fig. 6. Graph of the percentage ratio of connection of network subscribers of Kazakh telecom JSC in Karaganda through communication channels.

4 Conclusion

The flexibility of the system will allow controlling both one and several networks simultaneously, from tens to several thousand objects. The Network Management System hardware and software complex includes a central server and SEB control buffers. Today it is already impossible to manually and locally manage the network and at the same time to monitor all parameters, quickly correct errors and implement new services, against the background of increasing speed and the implementation of all new technologies and services. The use of the Network Management System hardware and software complex will allow ensuring in practice high level of control, efficiency of operation and development of telecommunications networks; and the implementation of a unified control system will reduce material costs. The second positive moment will be the concentration of specialists on the minimum number of objects, and not on each object which can be several thousand. This is very effective from the economic positions of costs and it solves an important problem of highly qualified personnel, since the system allows monitoring the objects remotely andthusthe need for the presence of specialists at the sites is eliminated.

References

[1] A.B. Ivanov, *Kontrol' sootvetstviia v telekommunikatsiiakh i sviazi* (Kompaniia Sairus Sisteme, Moscow, 2000)

[2] M.P. Matekin, D.S. Mironov, A.K. Skuratov, Trudy X Vserossiiskoi nauchno-metodicheskoi konferentsii "Telematika-2003", 253 (2003)

[3] T.G. Serikov, M.Z. Yakubova , A.D. Mekhtiev, V.V. Yugay, A.K. Muratova, V.P. Razinkin, A.V. Okhorzina, A.V. Yurchenko, A.D. Alkina, IFOST 2016, 510 (2017)

[4] M.Z. Yakubova, A.D. Mekhtiyev, T. G. Serikov, A.D. Alkina, A.V. Yurchenko, E.G. Neshina, X International IEEE Scientific and Technical Conference "Dynamics of Systems, Mechanisms and Machines" (Dynamics), 315 (2016)

[5] S. Shipilov, R. Satarov, I.Fedyanin, E. Balzovsky, V.Yakubov. MATEC Web of Conferences 7 01079 (2016) doi: 10.1051/matecconf/201679010

METHODS AND ALGORITHMS FOR DETECTING OBJECTS IN VIDEO FILES

Cuong Nguyen The[1,2*], *Dmitry* Shashev[1]

[1]National Research Tomsk State University, 634050, Tomsk, Russia
[2]Vietnam Maritime University, 484 Lach Tray, Le Chan, Hai Phong, Viet Nam

Abstract. Video files are files that store motion pictures and sounds like in real life. In today's world, the need for automated processing of information in video files is increasing. Automated processing of information has a wide range of application including office/home surveillance cameras, traffic control, sports applications, remote object detection, and others. In particular, detection and tracking of object movement in video file plays an important role. This article describes the methods of detecting objects in video files. Today, this problem in the field of computer vision is being studied worldwide.

1 Introduction about video files

Video files are files that store motion pictures and sounds of real-life. Video is a form of animation that gathers many images appearing continuously in sequence. Usually one second of video has 24 (or maybe more) still images and 24 or more photos, the human eyes (with retinal image retention mechanism of 1/24 sec.) will feel the motion image as continuous, smooth and beautiful.

2 The application for searching objects in video files

Detecting and tracking moving objects in video files has wide range of applications in real life. If a person just watches a video file, human eyes will not be able to give specific details about the time and the process of changing location of objects, especially those with a fast change process (such as detection of human face, smile detection when taking pictures, cars on the road, the path of a moving tennis ball and others), objects with complicated orbits (such as soccer players running on the pitch), or objects with a process of change slowly compared to the background (objects between the sea and sky, daylight changes, and tides). In addition, the object detection in the video file at the moment will help a lot in real life, for example, to determine if there was a goal or not (in football), if a tennis balls is in/out of court (in tennis), or which athlete has finished first (in speed races).

* Corresponding author: cuongntit@vimaru.edu.vn

3 Overview about searching objects

As we all know, video is a collection of discrete images that are constantly displayed to create motion effects. Thus, object detection in video files is also based on the idea of detecting objects in an image file. We need to perform the image-partitioning step to determine where the area of objects is, and where the area of the background is. Image segmentation is an important step in image processing, and this phase analyses the image into components of the same nature based on the boundaries or interconnected regions. An image is a detail, an object in panorama. An image area describes the surface properties of an image; this area is surrounded by a boundary and points with a relatively uniform grey level. Based on the physical properties of the image area, we define a number of partitioning techniques. The main partitioning methods include:

- Classification or threshold-based method.
- Structure-based method.
- Boundary-based method [1].

Object detection in video files should be based on the object's history of motion over discrete images over time. In function **I(x, y, t)**, 'I' is an image, 'x', 'y' are the coordinates of the object's location in two-dimensional space, and 't' is the time of the object at that coordinate. An approach based on the history of motion pictures is a view based on the method of pattern detection over time. This is a simple but effective way of displaying motions; it is used in a number of studies related to motion recognition, motion analysis, and other related applications.

An overview of real-time video detection, classification, and mark up of motion shows that this system can implement the following operations: (1) distinguish between transient or stationary objects and static background objects in the scene; (2) detect and distinguish objects, which move and disappear; (3) classify objects that have been discovered in different groups such as people, groups, vehicles, and others; and (4) mark the objects and generate information about the paths in the video image.

Video processing-based applications have common characteristics. Commonly used technologies for detecting motion objects are background subtraction, static methods, time difference, optical flow, and others.

3.1 Background subtraction method

Background subtraction is a part of motion-moving technology in static scenes. It tries to detect motion regions by subtracting pixels from the current image to a background image that was created by the average background image for a period of one initialization cycle. There are several approaches, which can be presented as follows.

Let $I_n(x, y)$ be a representation of the greyscale intensity value at a pixel location (x, y) with a value of [0 ÷ 255] in the $\mathbf{n}^{\text{th}}$ case of the video sequence I. Let $B_n(x)$ be a corresponding background magnitude value for a pixel at time (x, y) estimated from the video image I_0 to I_{n-1}. A pixel in position (x, y) in the current image belongs to the dominant component (highlighted) if

$$| I_t(x,y) - B_t(x,y) | > T_n(x, y) \tag{1}$$

is satisfied, where $T_n(x, y)$ is a predefined threshold.

The B_t background image is updated by using the Infinite Impulse Respone (IIR) filter as follows:

$$B_{t+1} = \alpha.I_t + (1- \alpha).B_t \tag{2}$$

This background is updated for all types of pixels. In the background subtraction methods, the source background image is updated only for pixels in the background.

3.2 Static methods

The W4 method uses a static background pattern, where each pixel is represented again with the smallest value (*M*) and its maximum value (*N*) in terms of intensity and maximum intensity difference (*D*) between any consecutive frames observed during the training initialization cycle, in which the scene does not contain moving objects. One point in the current image is classified as a prominent feature if it satisfies:

$$| M(x, y) - I_t(x, y) | > D(x, y) \quad \text{or} \quad | N(x, y) - I_t(x, y) | > D(x, y) \tag{3}$$

Every pixel is modelled by a blend of the Gaussian methods and is updated online by the input image data. Gaussian distributions will evaluate whether a point belongs to the dominant point processing or to the underlying processing.

3.3 The difference over time

Difference over time tries to detect motion regions using the difference between pixels in consecutive frames (two or three) in a sequence of videos. This method is highly adaptable to the scene; however, it often causes errors in detecting moving objects. The single-color area of the left-hand side causes the time difference algorithm to extract all pixels of the human motion.

A theory of image identification and representation was proposed by Bobick and Davis [2]. In particular, this theory analyses motion-based identification along with description of the spatial element and describes how the object moves. It represents formation of the Motion Energy Image (MEI) or the Binary Motion Region (BMR), and we can see that it represents a position of the transfer available in a sequence of images. MEI represents the shape of movement and the spatial division of motion. Then the MHI (motion history image) is made. The intensity of each pixel in the MHI represents an interval of the number of times that can be encoded by each frame, and this MHI method will hook up the time proportion of human gestures. If to put MEI and MHI together, we will see that MHI and MEI can be viewed as two component versions of the sample over time. In a vector image, each component of each pixel is a number of functions that move at that pixel location. Specific specimens that are grafted will have competition with the storage patterns of the known motion view. Unification of MEI and MHI models is constructed using the MHI method with the historical motion function $H_T(x, y, t)$ calculated from the difference in image and optical flow.

At the simplest level, each object is identified and represented by a boundary box as shown in Figure 1. As the object moves, the boundary box moves around. The difference in image is done by the entire image in the area of the bounding box. Combination with existing patterns will give the type of object to be identified.

Fig. 1. Illustration of the object in the bounding box [3].

At a more granular level, each object is identified by an image area that is distinct from the background (the boundary will have a shape of the object). Movement of the object is monitored by subtracting the difference in optical current. Illustration is made by converting to binary image as shown in Figure 2.

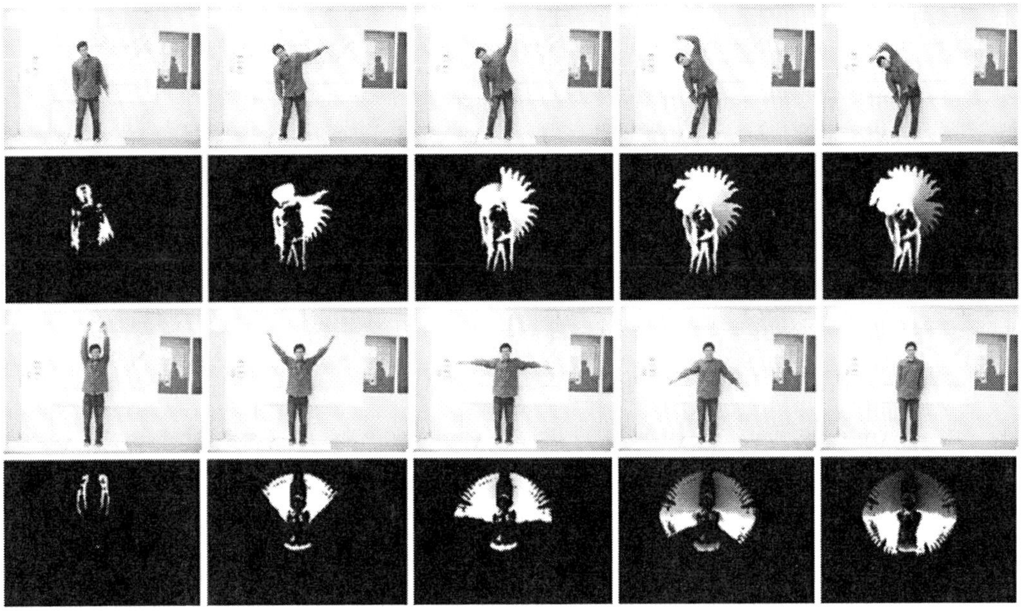

Fig. 2 Development of the MHI images for two different actions. The MHI images produced are shown under the actions sequentially [4].

4 Issues of detecting and tracking an object in video files

Detecting and tracking objects in video files encounters seven following issues: (1) Brightness change. The brightness of an image can change in space and time; some parts of an object may be brighter and brightness of objects in the current image may become darker on the next day. (2) Ability to expand or shrink an object. An object can move close to or move out of the camera's frame. (3) Rotations of an object. An object can rotate relative to the camera in all three axes of the three-dimensional space. (4) Screening of objects. An object can be overlapped by other objects. (5) Video has distorted objects or noise. Noise in the image can occur because of natural and artificially created reasons. The natural work can be attributed to the incorrect operation of the photo (video) camera and all the natural conditions such as snowfall or rain, under which the image (video) is obtained. Artificially created reasons include all human actions aimed to change the image quality such as changing colour space, compression of the image with loss of quality, blurring of the image, and others. (6) Change of the geometric shape of an object. The observed object can be deformed because it can turn to the camera with a part, which differs from the form observed before. An object can consist of small inter-connected objects that can move in space along different trajectories. For example, the desired object is a person whose hands can move in different directions. (7) Change of the colour histogram of an object. An object can have different colour histograms on different sides of its body [5].

Scalability and rotations refer to affine transformations [6, 7]. According to [8], projective transformations include affine transformations and take into account the distortions associated with the slope of the camera.

5 Methods and algorithms for searching objects

The main methods and algorithms for searching objects include the following:

- Dense optical flow. These algorithms help to estimate the motion vector of every pixel in a video frame.
- Sparse optical flow. These algorithms, like the Kanade-Lucas-Tomashi (KLT) feature tracker, track the location of a few feature points in an image.
- Kalman Filtering. This is a very popular signal-processing algorithm used to predict location of a moving object based on prior motion information. One of the early applications of this algorithm was missile guidance! Moreover, the on-board computer that guided the descent of the Apollo 11 lunar module to the moon had a Kalman filter.
- Meanshift and Camshift. These are algorithms for locating the maxima of a density function. They are also used for tracking.
- Single object trackers. In this class of trackers, the first frame is marked using a rectangle to indicate location of the object we want to track. The object is then tracked in subsequent frames using the tracking algorithm. In most real life applications, these trackers are used in conjunction with an object detector.
- Multiple object track finding algorithms. In cases, when we have a fast object detector, it makes sense to detect multiple objects in each frame and then run a track finding algorithm that identifies, which rectangle in one frame corresponds to a rectangle in the next frame [9]. Such algorithm is shown in Figure 3.

Fig 3. Tracing multiple objects [9].

6 Conclusion

Currently, anti-jamming algorithms (and methods) are not suitable for actual finding objects in the video files. The method proposed is based on finding the main characteristics of the object.

In comparison with other methods, the algorithms proposed remain unchanged for projective transformations, are resistant to noise and brightness change, and are characterized by less computational costs. The searching method requires immutability to object transformations and minimally achievable computational complexity in a real-time mode [10–12].

The further research is oriented to find an algorithm for detecting fast-moving objects in video files in order to use it to for detecting a potentially dangerous person. These algorithms can be used for detecting dangerous people who carry suspicious weapons or other things that must to be checked. Such situations include, for example, a person sitting but suddenly standing up and running at high speed, or an object, which suddenly moves to another object at high speed. The authors will study the main characteristic features and colour charts in the process of determining an object (as a person) in order to detect object deformation in video files (because rotation of objects is divided into different parts: head - body - leg).

Acknowledgements

The research presented in this paper was supported by the Russian Foundation for Basic Research (grant No 16-29-04388/17). The authors are grateful to Tatiana B. Rumyantseva from Tomsk State University for her assistance in preparing the article.

References

[1] S. M. Bryan, *Lecture 15: Segmentation (Edge Based, Hough Transform)* (Brigham Young University, 2000)
[2] A. Bobick A, J. Davis, An appearance-based representation of action, International Conference on Pattern Recognition (1996)
[3] Motion detection in OpenCV, URL: https://www.youtube.com/watch?v=z1Cvn3_4yGo (2013)
[4] Md. Atiqur Rahman Ahad, J. K. Tan, H. Kim, S. Ishikawa, Machine Vision and Applications **23**, 255 (2012) doi :10.1007/s00138-010-0298-4
[5] A. V. Pastushkov, Izvestiya Vysshikh Uchebnykh Zavedenii. Physics **56**, 334 (2013)
[6] A. A. Potapov, *Newest methods of image processing* (FIZMATLIT, Moscow, 2008)
[7] D. Shashev, S. Shidlovskiy, V. Syriamkin, A. Yurchenko. International scientific conference on radiation-thermal effects and processes in inorganic materials **81,** (2015)
[8] I.S. Gruzman, *Digital imaging in information systems* (Publishing house of the NSTU, Novosibirsk, 2002)
[9] Satya Mallick, Object Tracking using OpenCV (C++/Python), URL: https://www.learnopencv.com/object-tracking-using-opencv-cpp-python/
[10] Diego Cavalca, Object Detection with Tensorflow API, URL: https://www.youtube.com/watch?v=_zZe27JYi8Y
[11] A. V. Pastushkov, Dissertation for the degree of candidate of technical sciences (2017)
[12] S.V. Shidlovskii, J. Comp. Sys. Sci. Int **45**, 282 (2006)
[13] D.V. Shashev, S.V. Shidlovskiy, Opt. Inst. D. Proc. **51**, 227 (2015)

CONTROL OF LEVITATING PARTICLE IN ULTRASOUND FIELD

Dmitry Sukhanov[1], *Igor* Minin, *Oleg* Minin, *Ivan* Kuzmenko, *Timur* Muksunov, *Evgeniy* Sivkov, *Fedor* Emelyanov

Tomsk State University, 634050, Tomsk, Russia

Abstract. The experimental setup for ultrasonic controlled levitation in the air has been developed. Two phased arrays made of 91 ultrasonic radiators placed in front of each other are used. Arrays are focused in the region of particle levitation. The length of the focus area allows us to move the particles along it, controlling the standing waves through the phase difference of the two arrays.

1 Introduction

Ultrasonic levitation in air for small-sized particles is ensured by the displacement of particles in the extremes of the field. Acoustic levitation in the air finds more and more applications [1–3]. In particular, the traction beam technology [1] is developed, bearings with friction minimization [2, 3]. In [4], the possibility of levitation of a large flat object (disk) with dimensions much larger than the wavelength was demonstrated. Ultrasonic levitation is also used to clean surfaces from small particles [5]. With increasing power and focusing radiation, levitation of large objects of arbitrary shape is possible [6, 7].

2 Acoustic levitation setup

In this paper, we propose a method for levitation of particles by focusing the field from a set of low-intensity emitters. Due to the in-phase interference of waves in a certain region, the intensity of the field is sufficient to allow the particle to levitate. To keep a particle at a certain point in space, it is necessary to create an extremum of the field in it. In this case, the particle is pushed out to the equilibrium point corresponding to the extremum of the field intensity. We consider two arrays of ultrasound sources in front of each other. The arrays consist of 91 elements placed with a certain shift to produce the focusing effect (Figure 1). We place emitters in a hexagonal grid with a step of 11 mm as shown in Figure 2. The phase of the emitters is selected in such a way that the field distribution at a distance of 180 mm is a localized maximum with diameter 20 mm.

[1] Corresponding author: sdy@mail.tsu.ru

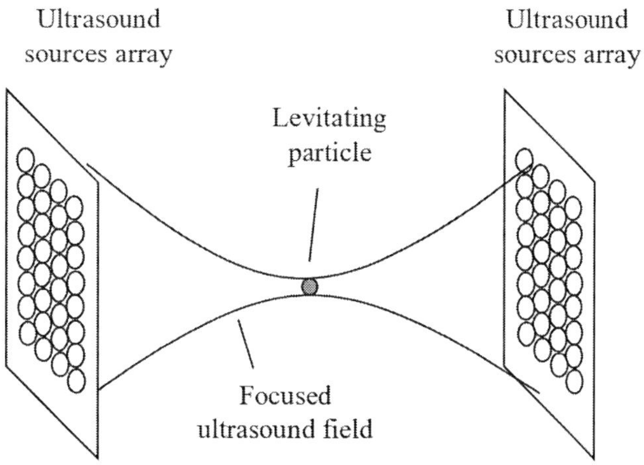

Fig.1. Ultrasonic levitation by means of phased arrays of transmitters.

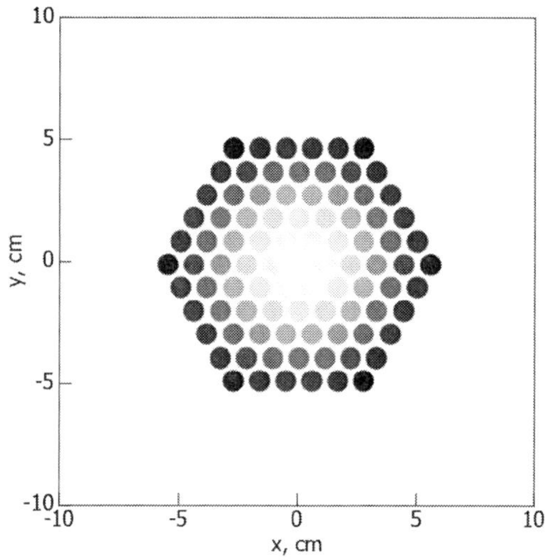

Fig. 2. Placement of source elements in array (grayscale color shows a relative depth shift of each source).

The radiation frequency is 40 kHz, which corresponds to a wavelength of 8.25 mm in air. To form an arbitrary field without secondary maxima, it is necessary to place the emitters at a distance of no more than 4 mm from each other. However, in our case, the pitch of the emitters is 11 mm, which leads to the appearance of side maxima. Nevertheless, in our problem, the presence of lateral peaks is not a difficulty, since the levitating particle is located in the center.

For experimental studies, an array of radiators from the elements MA40S4S was created (Figure 3). All the radiators were connected in parallel to the powerful amplifier basen on the TDA7297 chip. The phase of each radiator was regulated by the height of the placement in the hole. Radiators are shifted according to a spherical surface with radius of 180 cm. The required arrangement of 91 emitters was synthesized to form the distribution of ultrasonic waves in the form of an ultrasonic trap at a range of 180 mm. The laboratory model of the installation for ultrasonic levitation in the air based on the radiator array at a frequency of 40 kHz is shown in Figure 3.

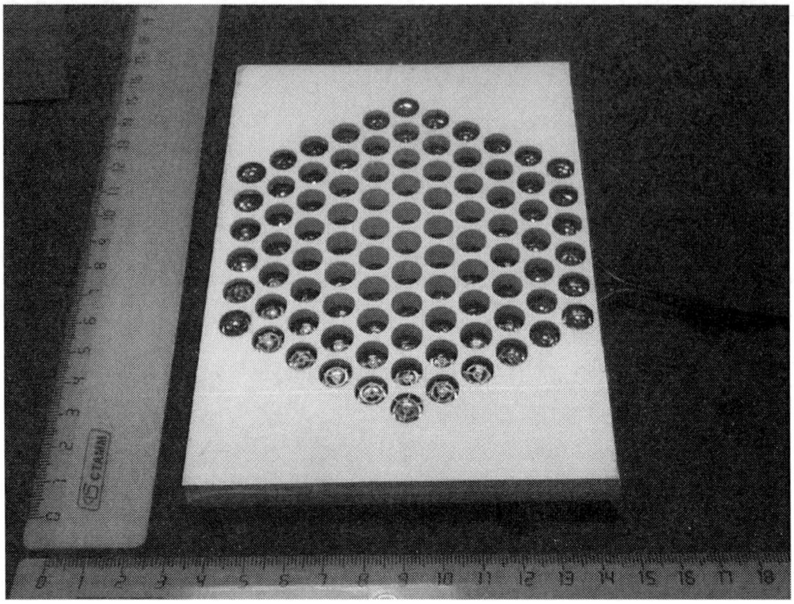

Fig. 3. Photo of sources array made of 91 elements.

To ensure the focusing properties of the source array we have made a measurements of sound field dependent on distance and horizontal shift. The resulting image of the measurement of sound field is shown on Figure 4. We can see the maximum of magnitude on distance 180 mm. The diameter of the focusing spot is about 20 mm.

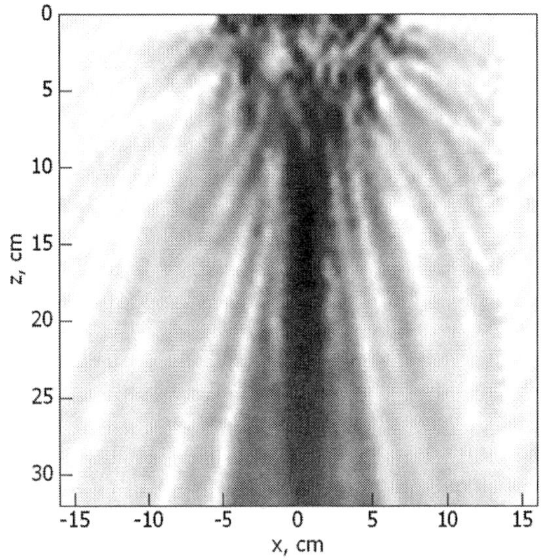

Fig.4. Measured ultrasound field magnitude.

We placed two arrays in front of each other on distance 360 mm. Arrays was controlled by separate signals, generated by digital-analog converters with amplifiers. In the photo, in the central part, you see two white levitating particles of foam. In this case, levitation is provided in the area of focusing of both arrays. By controlling of the phase difference between two signals we able to shift particles horizontally. Hence the focusing area is extended by range the interval of particles shift is comparable with the half of distance between arrays. On Figure 5 you can see photos of levitating particles between two arrays.

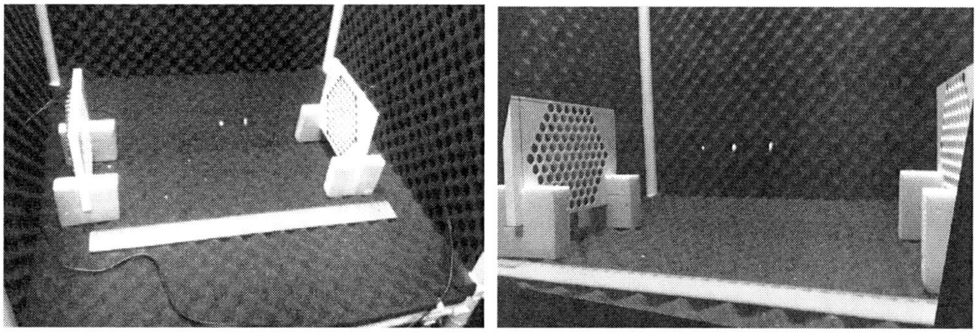

a b

Fig. 5. Photos of levitating particles: *a*- two particles; *b*- three particles.

Thus, the levitation of particles of polystyrene at a range of ~100 mm near the focusing point of the radiator arrays is shown experimentally. Levitation is provided for a long time due to the influence of pressure oscillations of ultrasonic waves, as a result of which particles tend in the region of the maximum of oscillations. To shift the position of the particle, a radiator phase difference is changed, which resoling in shift of standing wave maximum.

3 Conclusion

Application of two radiators towards each other allows us to ensure the levitation of particles in the interference field in the focus region. The change in the height and lateral shift of the emitters led to a change in the height of the particle. Thus, the possibility of controlling the position of the particles in the range by means of adjusting the phase difference between the direct and the counter emitters is demonstrated.

Acknowledgments

The scientific research was carried out with the support of the grant of the Russian Science Foundation No. 17-79-20051.

References

[1] C.E.M. Démoré, P.M. Dahl, Z. Yang, P. Glynne-Jones, A. Melzer, S. Cochran, M. Macdonald, G.C. Spalding, Physical Review Letters **112** (17),174302 (2014)
[2] J.-S.Wang, C. Chen, G.-C. Chen, X.-J. Yan, Proc. of the 2013 Symposium on Piezoelectricity, Acoustic Waves and Device Applications, SPAWDA 2013, 6841135 (2013)
[3] C. Chen, J. Wang, B. Jia, F. Li, Journal of Intelligent Material Systems and Structures **25**, 755 (2014)
[4] S. Zhao, J. Wallaschek, Archive of Applied Mechanics **81**,123 (2011)
[5] A. Fuhrmann, J.S. Marshall, J. Wu, Applied Acoustics **74**, 535 (2013)
[6] M.A.B. Andrade, F.T.A. Okina, A.L. Bernassau, J.C. Adamowski, Journal of the Acoustical Society of America **141**, 4148 (2017)
[7] M.A.B. Andrade, A.L. Bernassau, J.C. Adamowski. Applied Physics Letters **109**,044101 (2016)

IDENTIFICATION OF A PERSONALITY

Denis Kuznetsov*

Tomsk State University, 634050, Tomsk, Russia

Abstract. Biometric technologies are based on biometrics, measurement of the unique characteristics of a certain person. These are the unique signs received by a person from birth and acquired characteristics that can change under the influence of time or external environment. This article discusses the principles of operation, types of biometric identification systems, describes examples and areas of their application.

1 Biometric technologies

Are based on biometrics, measurement of the unique characteristics of a certain person. They may include unique signs received by a person from birth, such as DNA, fingerprints, iris; as well as the acquired characteristics that can change under the influence of time or external environment, such as handwriting, voice or gait [1, 2].

Principle of operation. All biometric systems work almost identically. First, the system remembers a sample of the biometric characteristic (this is called as a recording process). During the recording, some biometric systems may ask to make several samples in order to make the most accurate image of the biometric characteristic. Then the received information is processed and transformed into a mathematical code. In addition, the system may ask to make some additional actions in order to "ascribe" the biometric sample to a certain person. For example, a personal identification number (PIN) is attached to a specific sample, or a smart card containing a sample is inserted into the readout device. In this case, a sample of the biometric characteristic is made again and is compared with the presented sample:

by fingerprints (dactylography);
by iris;
hand geometry;
by speech characteristics;
by face;
by handwriting.

Biometrics is the technology of identification of a personality that uses physiological parameters of the subject (fingerprints, iris, etc.). Biometric parameters include static fingerprints, hand geometry, iris, etc. Dynamic parameters include the dynamics of reproduction of a signature or a handwritten keyword, voice, etc.

* Corresponding author : novorostomsk@gmail.com

2 Fingerprints

This method is the oldest and the most common among identification methods using the physiological characteristics of the organism. It is used in all countries by the law enforcement agencies for identification of criminals [3].

Nowadays, dactyloscopy (identification of fingerprints) is the most developed biometric method for identification of a personality. The method is very well developed because of its wide use in criminology in the 20th century.

Each person has a unique papillary fingerprint pattern that makes identification possible. Usually algorithms use characteristic points on fingerprints such as the end of the pattern line, the branching of the line, single points. In addition, information about the morphological structure of the fingerprint is used: the relative position of the closed lines of the papillary pattern, of the "arched" and spiral lines. Specific features of the papillary pattern are transformed into a unique code, which preserves the information value of the print image. "Fingerprint codes" are stored in the database used for search and comparison. Transferring the fingerprint image into the code and its identification depends on the database size and usually does not exceed 1 second. The time of taking fingerprints is not taken into account.

3 Speaker recognition systems

This is one of the oldest biometric technologies. Currently, its development has become more active; specialists think that it will have a great future and wide use in building "intelligent constructions". There are many ways to construct the identification code by voice; as a rule, they include different combinations of voice frequency and statistical characteristics (Figure 1).

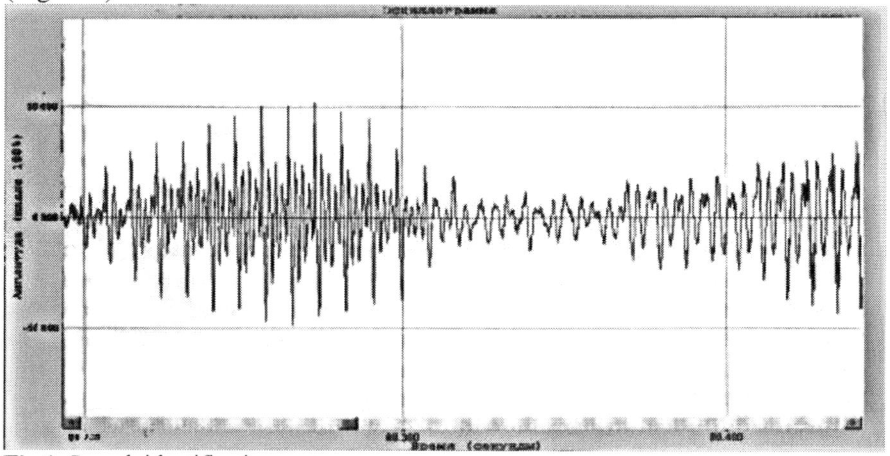

Fig 1. Speech identification systems.

4 Iris recognition systems

This is the most simple and convenient method, but less reliable. Iris is a unique characteristic of a person. The iris pattern is formed on the eighth month of intrauterine development; it finally stabilizes at the person's age of two and practically does not change during life, except because of severe injuries or sudden pathologies. The method is one of the most accurate among other biometric methods.

The iris recognition system is logically divided into two parts: the device for image capture, its preprocessing and transmission to the calculator, and the calculator that compares the current image with the images from the database, and transmits the admission command to the executive device [4].

The time of image preprocessing in modern systems is approximately 300…500 milliseconds, the speed of comparison of the obtained image with the images from the data base is about 50000–150000 comparisons per second (on an ordinary PC). Such speed of comparison does not impose restrictions for application of the method in systems for access in large organizations.

5 Facial recognition systems

When using this method of recognition, a three-dimensional image of a person's face is constructed. The contours of eyebrows, eyes, nose, lips, etc. are outlined on the face, the distance between them is calculated, and not a simple image is constructed, but many possible variants for cases of face rotation, flexion, and change of expression. The number of images varies depending on the purpose of application of this method (for authentication, verification, remote search on large territories, etc.).

6 Hand geometry recognition systems

In respect of technological structure and level of reliability, the method of identification by hand geometry is quite comparable with the fingerprint identification method.

7 Signature recognition systems

Typically, a person's signature or writing a code word (sometimes) is used for a person's identification. Depending on the required degree of protection and availability of equipment (graphic tablet computer, screen of a handheld Palm computer, etc.), two types of the digital identification code can be formed:

• By the signature itself (just the degree of coincidence of two images is used for identification);

• By the signature and dynamic characteristics of writing (identification is conducted based on the convolution, which includes information about signature, time characteristics of making signature and the statistical characteristics of the pressure dynamics).

8 Conclusion

Biometric technologies are actively used in many areas related to protection of access to information and material objects, as well as in the tasks of unique identification of a personality in order to provide:

- access control;
- information protection;
- identification of clients.

Comparison of the cost of identification systems:

- Fingerprint identification system (dactyloscopy)–827 rubles.
- Iris recognition system–17400 rubles.
- Hand geometry recognition system–47200 rubles.
- Speaker recognition system–803 rubles.
- Facial recognition systems–27900 rubles.

Currently, our team is developing a method for secretive collecting and analyzing biological information. We will present it in detail in the future publications.

Acknowledgments

The article is prepared within the framework of Competitiveness Enhancement Programme of National Research Tomsk State University and grant of the Russian Foundation for Basic Research No 16-29-04388.

References

[1] M. Yazov Yu.K., V.I. Volchihin, A.I. Ivanov, V.A. Funtikov, I.G. Nazarov, *Neural network protection of personal biometric data* (Radio technics, Moscow, 2012)

[2] Akhmetov B.S., Ivanov A.I., Funtikov V.A., Bezyaev A.V., Malygina E.A. *The technology of using large neural networks to convert fuzzy biometric data into the code of access key, (*LEM Publishing house, Kazakhstan, Almaty, 2014)

[3] Yu.S. Bekhtin, S.A. Klestov, M.S. Kutsov, V.I. Syryamkin, D.V. Titov, *Theoretical foundations of digital image processing in build-in vision systems,*(STT Publishing house, Tomsk, 2016)

[4] V.I. Syryamkin, *Intellectual robotic and mechatronic systems: textbook,* (STT Publishing house,Tomsk, 2017)

SOLUTION OF ACOUSTIC TOMOGRAPHIC PROBLEMS ON THE BASIS OF NUMERICAL SIMULATION OF PARTICLE DYNAMICS

Angela Kuzovova[1], *Ivan* Kuzmenko

Tomsk State University, 634050, Tomsk, Russia

Abstract. A method is proposed for imaging of scattering heterogeneities in continuous media on the basis of numerical modeling of forward and backward wave propagation. It is shown that the combination of a solution for backward propagation of waves and direct wave propagation allows us to visualize scattering heterogeneities. The results of numerical simulation are presented.

1 Introduction

Acoustic tomography is widely used in flaw detection. To obtain tomographic images, it is necessary to solve the inverse problem – determine its shape and characteristics of object by means of the measured signal scattered by heterogeneity. Inverse problems do not always have a unambiguous solution, and are extremely sensitive to measurement noise and inconsistencies in the mathematical model and sounding conditions of the medium. Most of the existing methods of ultrasonic tomography use the locational principle of sounding [1, 2], in which objects are visualized by back scattered waves. There are also transmission schemes for ultrasonic tomography to determine the flat shadow images of the objects under study. Often, the approximation of geometrical optics is used to the solve inverse tomography problems [3, 4]. The most universal method of processing ultrasound sounding data is the method of spatially-matched filtering. By measuring the field at a variety of different points in space and it is possible to synthesize the aperture and obtain images with a resolution comparable to the wavelength. This method works in the frequency domain and for ultra-wideband signals it is demanding of large computing resources. However, there are particular cases of sounding schemes where rapid processing of signals and obtaining of three-dimensional images is possible, for example, the Stolt method. To process signals in short-pulse sounding, a migration method in the time domain is used, or the method of diffraction hyperbolas summation. This method is quite universal, and can take into account any combination of positions of the emitters and receivers. Such methods are fast-acting, but do not allow to take into account for the influence of background heterogeneities and nonlinear distortions.

In this paper, we propose a method for solving the inverse problem of acoustic tomography based on numerical simulation of the propagation of acoustic waves in solids.

[1] Corresponding author: Ang_kuz93@mail

2 Method of acoustic tomography

Solid is represented as an array of particles in a tetrahedral equidistant grid. The dynamics of particles is described by Newton's equations of motion. To reconstruct the image of scattering heterogeneities, the time reversal method is used in the numerical model under consideration. In this case, it is assumed that background media heterogeneities are known, but it is necessary to detect unknown defects, for example, in the form of a gap in the crystal lattice.

Let us consider a numerical model of the propagation of acoustic waves in a solid in which the scattering object (heterogeneity) has the form shown in Figure 1, and the background medium is the equivalent of brass (sound velocity 4430 m/c, density 8485 kg/m^3). Heterogeneity it is a gap of the crystal lattice, through which acoustic waves do not pass. We consider the source of the acoustic signal in the form of a plane wave radiated in the forward direction of the y axis. The shape of the original signal is a monopolar pulse in the form of a Gauss curve with a duration of 2.4 μs (at half-amplitude level). A plane wave, propagating downward in a medium, collides with a scattering heterogeneity, diffracts and reflects. Part of the scattered field is registered by receiving sensors, in which are measured the particle velocity and their deviation in the crystal lattice depending on time. Receiving sensors are located on the plane $y = 0$ with a step equal to the step of the particles placement.

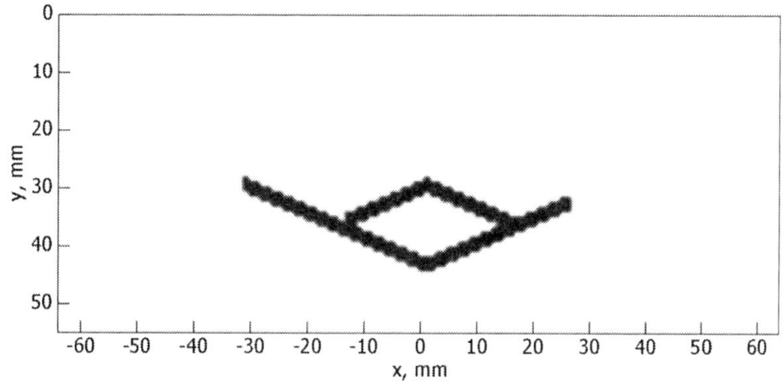

Fig.1. Image of test heterogeneity.

In Figure 2 the image of the field of the scattered wave on the heterogeneity under consideration is presented. One can observe the reflection of the acoustic pulse of a plane wave from the heterogeneity. The incident wave does not pass through the upper part of the heterogeneity, but is completely reflected, hence, the lower part of the heterogeneity will not scatter the field and will not be detected. The all scattered field toward to the side the plane $y = 0$ is recorded, namely the velocity vectors and the coordinates of the upper particles.

The solution of the inverse problem is based on the method of backward wave propagation. In work [5] this approach was used to reconstruct images of acoustic wave sources. The method of backward propagation of waves is based on the same process of numerical modeling as the direct problem, but here the time step is taken to be negative. At the same time, the receiving points become emitters, and they radiate precisely that one the signal that was accepted at the appropriate moment of time.

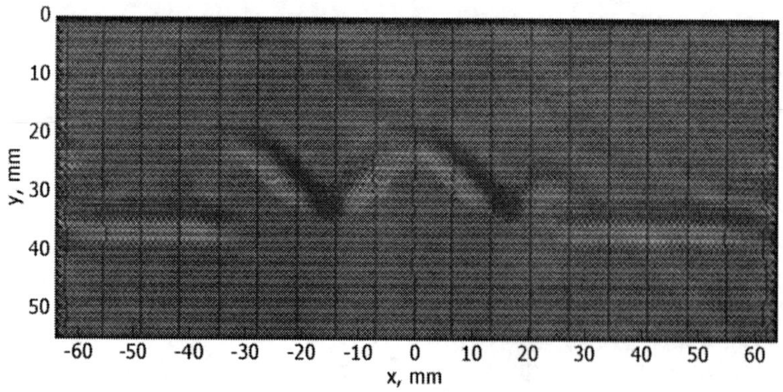

Fig. 2. The solution of a direct problem at a certain moment in time.

The denser the receivers will fill the whole space, the more accurate it will be possible to restore the complete initial state of the system at past moment times. So it will be possible to restore the cause of the appearance of the scattered field. In our case, the receivers are located at the top, so the backward propagation wave will pass from the top to bottom. At the point where the point scatterer is located, the backward propagation wave converges in a point (with a localization region of the order of the pulse length). If continue the reverse course of time already focused field in the scattering point will begin to diverge. It is not known in advance at what point in time it is necessary to stop reverse propagation, so we can not exactly guarantee the restoration of the scatterer. To stop the reverse propagation makes sense exactly at the moment when to the considered point in space reached by a forward sounding wave.

It is proposed to reconstruct the image of scattering heterogeneities by combining the numerical simulation solutions with a positive time course and with a negative time course (Figure 3). Initially we assume that the medium is homogeneous and does not contain scatterers, and we model in it the propagation of a plane wave from a source. Next, we simulate the backward propagation of waves from receivers, reversing the time. We scalar multiply the velocities of particles from two models at coinciding instants of time. In the region where the scatterers are located, the forward and backward waves overlap, when they are superimposed (by scalar multiplication of the particle velocities), it is possible to visualize the scattering object. Since the wave does not penetrate into the heterogeneity, which is a gap of the crystal lattice, we cannot visualize hidden parts.

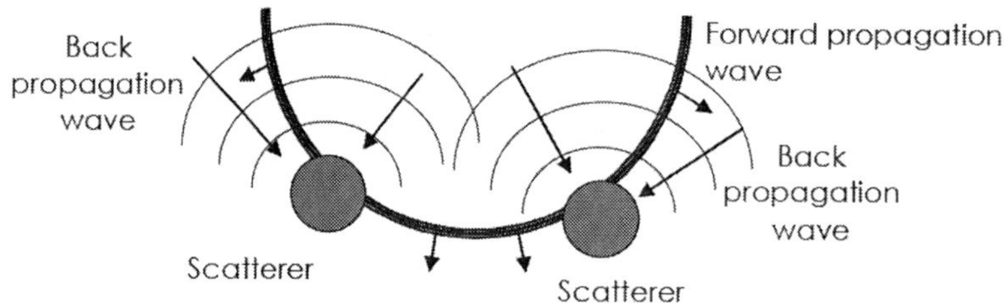

Fig.3. Explanation of the method for solving the inverse problem.

On (Figure 4) is presented the result of the plane wave simulation at back propagation of the field. Red indicates the velocity of the particle along the X axis, green along the Y axis, blue along the Z axis. The initial perturbation was directed along the Z axis. It is seen that in the region where the test heterogeneity is located, the field scattered from the object is focused.

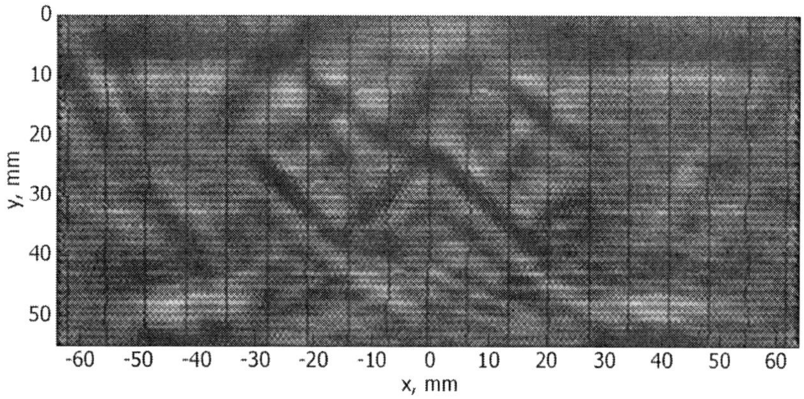

Fig.4. Backward propagation of the field.

Figure 5 shows the reconstructed image of the test heterogeneity. According to the restored result we can determine the shape of the object. The proposed method of image restoration allows to take into account the effect of diffraction. It should be noted that the concave regions and internal voids of objects will not be visualized, since the wave field does not penetrate into these regions.

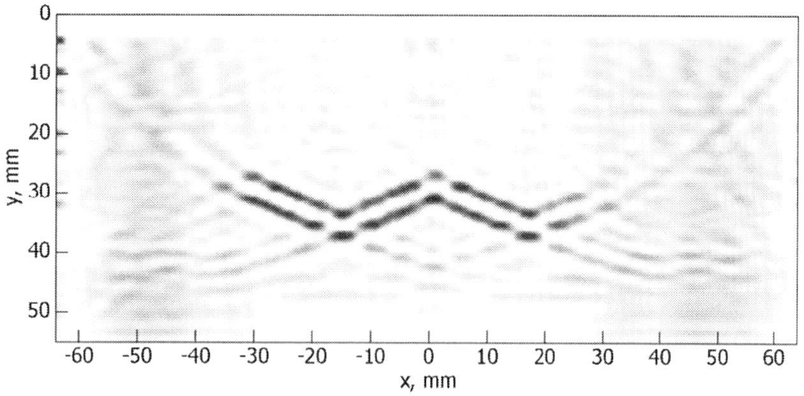

Fig.5. Refurbished image of heterogeneities.

To test the possibility of visualizing objects of more complex shapes, the scatterer was modeled in the form of a circle (Figure 6).

As a result of solving the inverse problem by the proposed method, we have restored the image of the scatterer, shown in Figure 7. It can be seen that only the upper part of the circle is visualized. This is due to the fact that most part of the waves scattered on the object do not return to the side of the receivers. Thus, objects of not any shape can be visualized, it is necessary the presence of waves scattered to the side of the receivers.

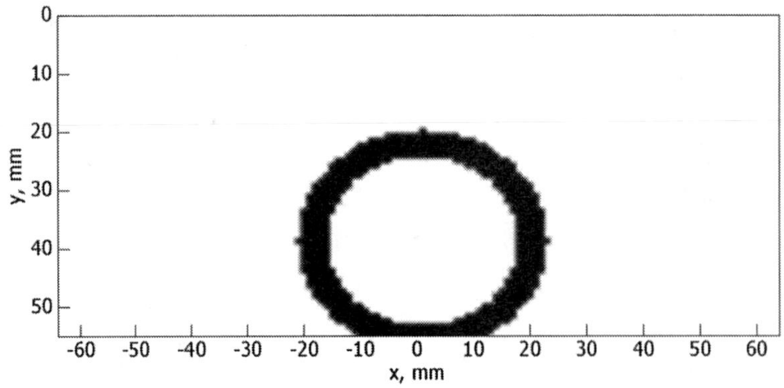

Fig.6. Image of modeled heterogeneity.

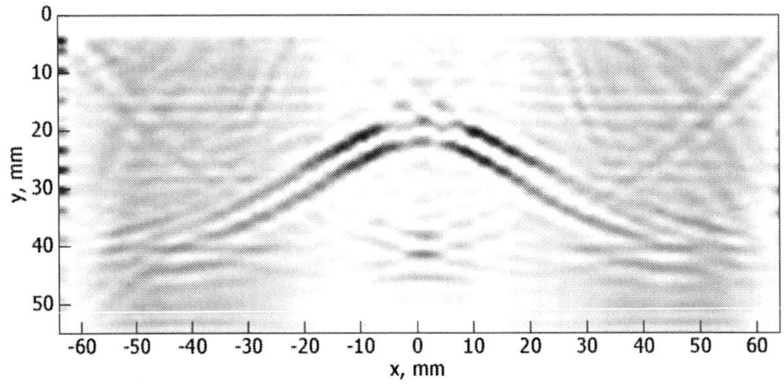

Fig. 7. Refurbished image of heterogeneity.

As a result of the work, the new acoustic topography method is proposed. The method is based on forward and backward time - domain numerical simulation of particles dynamics. It is possible to visualize scattering heterogeneities in a solid media.

Acknowledgments

The work was supported by the Ministry of Education and Science of Russia within the framework of the project "Creation of a high-tech complex of ultrasound surgery" (Unique identifier of the project RFMEF157517X0163).

References

[1] A.V. Osetrov , Acoust. Journal **37**, 528 (1991)
[2] A.V Osetrov, S.N. Samolenkov, Acoust. Journal **42**, 679 (1996)
[3] K.J. Opielinski, T. Gudra, Ultrasonics, 295 (2006)
[4] I. Peterlik, R. Jirik, N. Ruiter, J. Jan, Radioengineering, 125 (2008)
[5] D.Ya. Sukhanov, M.A. Kalashnikova, Acoustical Physics **60**, 304 (2014)

APPLICATION OF THE E – LAYER MODEL FOR SOLVING THE PROBLEMS OF PARAMETRIC ESTIMATION IN MEASURING DEVICES

Alexey Yakunin* and *Larisa* Suchkova

Altai State Technical University, 656038, Barnaul, Russia

Abstract. The article proposes a technique for estimating the maximum possible errors in measuring the parameters of the signals observed against a background of interference that do not exhibit statistical properties on the measurement interval and of which only the region of their possible change is known. This region is a layer within which the interference -free signal can vary due to the variation in the values of the parameters of the model function describing it. The technique is also proposed that makes it possible to estimate the measurement error of the monitored parameter for a specific implementation of the signal described by the proposed model with a confidence probability close to unity. An example of the use of the offered techniques for estimating the coordinate of a light stain created by a point source is shown. The offered approach is actual, when each measurement represents special worth and there is no possibility to improve the measurement result by averaging over the set of signal realizations.

1 General definition of the problem

At calculation, designing and use of control and measuring devices the estimation of their potentially achievable characteristics of accuracy and measurement errors is of great importance, as defines the basic metrological characteristics of devices and the demands shown to their hardware representation [1]. In particular, level of internal noise and maximum deviations from rating values of electronic components, capacity of analogue-to-digital converters, capacity of a memory for a data storage, type of variables used for a data handling.

For an estimation of errors of measurement mainly statistical approach now is used [2–7]. Its application is justified, when value of parameter inspected by the measuring device is selected from a quasi-determined signal containing additive stationary ergodic noise. However not for all factors influencing a controllable signal their laws of distribution can be set. Moreover, sometimes the informative signal is presented by unique realization of nonstationary acyclic process with the limited number of references that excludes a possibility of use of classical statistical methods of its processing. Therefore in cases when in the course of measurement statistical properties of controllable process are not exhibited, the possibility of carrying out of repeated measurement is absent, each separate measurement has special value and it is necessary to define a range of possible values of outcome of

* Corresponding author: yakunin@agtu.secna.ru

measurement, more expediently to find an interval estimation of controllable parameter and potential accuracy of such estimation with application of methods of the interval analysis [8–10]. Such methods are well developed with reference to problems of an estimation of errors of outcomes of joint and cumulative measurements in various applied areas [11, 12]. Generally the vector of required controlled parameters λ with dimension m in such cases defined by results of measurements (x_i, y_i), connected by expressions

$$y_i = F(x_i, \lambda);\ x_i, y_i \in X;\ \lambda \in \Lambda, \tag{1}$$

where $i=1,\dots,n$, and y_i and x_i are measured with accuracy $\varepsilon \in X$, X and Λ – spaces of accordingly measured and controllable values.

However in modern intellectual informational-measuring systems often it is necessary to deal not with a final population of outcomes of measurements, but with continual and in some cases also with many-dimensional multiparameter signals. It can be, for example, two-dimensional optical signals varying in time, or wavelet-transformations of acoustic signals. A part from parameters of such signals are form parameters, a part - support parameters, and only small part - the informational parameters defining estimation of required value.

For determination of an interval estimation of errors of measurement of parameters in such cases in works [13–15] it is offered to use a method based on application of model of ε – layer for exposition of signals with high degree of a priori uncertainty about conditions of their observation. We will consider at first the general approach to application of this model for calculation of potential accuracy and errors of measurement of measuring devices, and then - its application for a solution of a concrete practical problem.

2 E – layer model and its using for solving the problem of estimation of potential accuracy of measuring equipment

According to model of ε – layer the signal containing controllable parameter is represented in the form of superposition of quasi-determined component (modelling function) of the useful signal and some component including random casual noise and interference, and also other deviations of a real signal from its description by the chosen modelling function. This component in all range of definition of a signal is described by an arbitrary class continuous or even explosive and not submitting to conditions of Lipschitz functions. On this function unique restriction is superimposed. It consists in that all values in a signal range of definition laid on an interval $[-\varepsilon^-\ \varepsilon^+]$, where ε^- and ε^+ – some positive values, which estimations, as a rule, much less than maximum value of the module quasi-determined component.

Let's suppose for simplicity of reasoning, that all describing a quasi-determined component of model of a signal the set of parameters consists only of one controllable parameter $\lambda \in \Lambda$ concerning which the interval estimation is searched, and all remaining parameters $\xi \in \Xi$ among which can be both form and support parameters, and other informative parameters which also can be subject to measurement. Then for any arbitrary value $\lambda_0 \in \Lambda$ according to ε – layer model it is possible to note, that the following relation is fulfilled:

$$S(r, \lambda_o, \xi) - \varepsilon < S(r, \lambda_e, \xi) < S(r, \lambda_o, \xi) + \varepsilon, \tag{2}$$

where λ_e – estimation of controllable parameter, $S(r, \lambda_x, \xi)$ – quasi-determined component of a signal, $r \in R$ – vector in space of coordinates R on which function $S(r, \lambda_e, \xi)$ is defined. This vector can include in itself of three coordinates in three-dimensional space and the fourth coordinate - time.

Generally instead of the fixed interval [$-\varepsilon$, ε] (a layer of the fixed thickness) it is possible to use interval function [$-\varepsilon^-(r)$, $\varepsilon^+(r)$] if in the field of signal observation there are sections with various degree of accuracy of its exposition. Obviously, the lower the level of noise and interference and the less inaccuracy of the correspondence of the model function to the real signal, the thinner will be the thickness of the layer.

Let's consider now as by means of ε – layer model it is possible to find potential accuracy of the measuring device on which input signal described by the equation (2) is fed. Potential accuracy at calculation and designing of control and measuring devices is of great importance [1]. Potential accuracy shows, what minimum possible errors of measurement can be received for the given aspect of a signal by means of the most perfect algorithm of its handling, but at the same time at the most unfavorable cases of its realization. And in this sense potential accuracy is its minimax estimation. Besides, being an interval estimation, it sets a resolving power of the measuring device, and finally defines its dynamic range and volume of the information received at measurement [16, 17]. Indirectly potential accuracy allows to estimate also possibilities of the further perfection of performances of accuracy of developed equipment. By definition, the interval estimation [λ_-, λ_+] is a domain of definition with a given confidence probability of the point estimation λ_e of the controlled parameter λ, taking in concrete realization of the observed signal having parameter value λ_o. According to a model of ε – layer, such estimation should satisfy to an inequality (2) and its confidence probability aspires to unity. Then $\Delta\lambda = \lambda_+ - \lambda_-$ will define an absolute error of measurement of parameter λ for its true value λ_o. Generally this error will be function of all parameters of a signal entering into the determined component. For true value of parameter λ, equal λ_o, it is possible to define the greatest possible value of this error as

$$\Delta_\lambda(\lambda_0) = \max_{\xi \subset \Xi}(\lambda_+(\lambda_o, \xi) - \lambda_-(\lambda_o, \xi)), \tag{3}$$

and the maximum error of an estimation of parameter λ in all range of its modification can be defined as

$$\Delta_{\lambda.\max} = \max_{\lambda o \subset \Lambda}(\Delta_\lambda(\lambda_0)). \tag{4}$$

Expression (4) represents an upper bound of an estimation of an error of measurement of controllable parameter which can be received as a result of handling of signal $S\ (r, \lambda, \xi)$. Value of an error set by this expression is defined only by properties of a signal, instead of algorithm of its handling. Therefore it cannot be reduced at use even the most perfect algorithm and will define magnitude of potential accuracy which can be reached for the given signal.

Comparing expression (3) with expression (2), it is easy to be convinced, that it is possible to present it in an aspect

$$\Delta_\lambda(\lambda_0) = \max_{\xi \subset \Xi} \lambda_+(\lambda_o, \xi) - \min_{\xi \subset \Xi} \lambda_-(\lambda_o, \xi), \tag{5}$$

where value λ_+ and λ_- are radicals of a solution of the equations

$$S(r, \lambda_e, \xi) - S(r, \lambda_o, \xi) = -\varepsilon^-(r) \text{ for } S(r, \lambda_e, \xi) - S(r, \lambda_o, \xi) < \varepsilon^+(r) \tag{6}$$

and

$$S(r, \lambda_e, \xi) - S(r, \lambda_o, \xi) = \varepsilon^+(r) \text{ for } S(r, \lambda_e, \xi) - S(r, \lambda_o, \xi) > -\varepsilon^-(r). \tag{7}$$

Thus from the discovered values greatest be assigned for λ_+, and smaller for λ_-.

If the thickness of the layer is sufficiently small, then the deviations $\Delta\lambda$ and $\Delta\xi$ of the signal parameters λ and ξ from their true values λ_o and ξ_o must also be small so that the signal realization does not exceed the limits specified by the ε-layer. Then it is possible to take advantage of linear approximation of function $S(r, \lambda, \xi)$ in a point neighbourhood (λ_o, ξ_o), having presented the left parts of the equations (6) and (7) in the form:

$$S(r,\lambda_e,\xi)-S(r,\lambda_o,\xi)=\frac{d}{d\lambda}S(r,\lambda,\xi)\Big|_{\lambda=\lambda o,\xi=\xi o}\cdot\Delta\lambda+\sum_{i=1}^{n}\frac{d}{d\xi_i}S(r,\lambda,\xi)\Big|_{\lambda=\lambda o,\xi=\xi o}\cdot\Delta\xi_i \quad (8)$$

Such representation allows to linearize the equations (6) and (7), that essentially simplifies their solution.

Thus, it is necessary for determination of potential accuracy at first with application of expression (8) or it is direct from (6) and (7) to discover boundaries of an interval estimation $[\lambda_-, \lambda_+]$, and already then from (3) and (4) to search for extreme value of a difference of these boundaries.

3 Using the E – Layer Model to Estimate Interval Errors of Measurement for a of a Signal Realization

From the statistical theory it is known, that in cases when conditions of a stationarity and an ergodicity of observable process are not observed, there can be so-called superefficient estimations when the confidential interval of controllable parameter for concrete realization can appear an arbitrarily small and even to become equal to zero [3,4]. It at all does not contradict expression (4) which sets only greatest possible value of an interval estimation for the most unfavorable case, instead of its value for concrete realization of a signal which will be always less or equally magnitudes finding from expression (4).

Let we have such concrete realization $S_x(r,\lambda,\xi)$. As this realization should be in limits ε – layer, the quasi-determined component of a signal for any value r cannot appear more than $S(r,\lambda,\xi) + \varepsilon^-(r)$ and less than $S(r,\lambda,\xi) - \varepsilon^+(r)$, that is

$$S(r, \lambda_o, \xi) - \varepsilon^+(r) < S_x(r, \lambda_e, \xi) < S(r, \lambda_o, \xi) + \varepsilon^-(r), \quad (9)$$

It is clear from the comparison of (9) and (2) that the process of finding the measurement error for a particular implementation will not differ much from the case of calculating the potential accuracy, and accordingly, we can use the same algorithm to find the measurement error. The only, but rather important, difference is the substitution in (2), as a rule, of a smooth model function $S(r, \lambda_o, \xi)$ by a concrete realization of the signal $S'(r,\lambda_e,\xi)$, representing the sum describing a quasi-determined component of modelling with the inherent noise component and the model inaccuracy lying within the ε-layer defined by the boundaries ε + (r) and - ε - (r). Such a function, unlike the model one, can contain a lot of irregularities, overfalls, protuberances and hollows. This difference also allows to receive for concrete realizations of an error, essentially smaller the errors set by potential accuracy.

For acknowledgement of the given fact we will consider the elementary one-dimensional case when the quasi-determined signal component represents a constant, and ε – layer is constant and symmetric, that is $\varepsilon^+(r) = \varepsilon^-(r) = \varepsilon$, and $S(r, \lambda_o, \xi) = \lambda$. For such a signal, depending on its physical nature, r can be time, or one of the spatial coordinates. It is obvious, that for such signal the parameter λ can accept values only from an interval $[\lambda_-, \lambda_+]$, and the maximum error of its measurement $\Delta\lambda$ can reach values 2ε.

Such error will be observed when the real signal in all area of the definition has the fixed value that is improbable, but, nevertheless, probably. Besides the considered model assumes, that each realization of a signal is unique and to improve outcome of observation by its average on a series of measurements it is impossible, as the interference accompanying a signal throughout all series of measurements can not exhibit statistical properties and remain invariable. At the same time such improbable situations, when $\Delta\lambda = 0$ are possible also. The example of such situation is shown in Figure 1*a* when on an interval of observation a signal undergoes a jump from the lower boundary to the upper bound ε – layer, varying on magnitude 2ε.

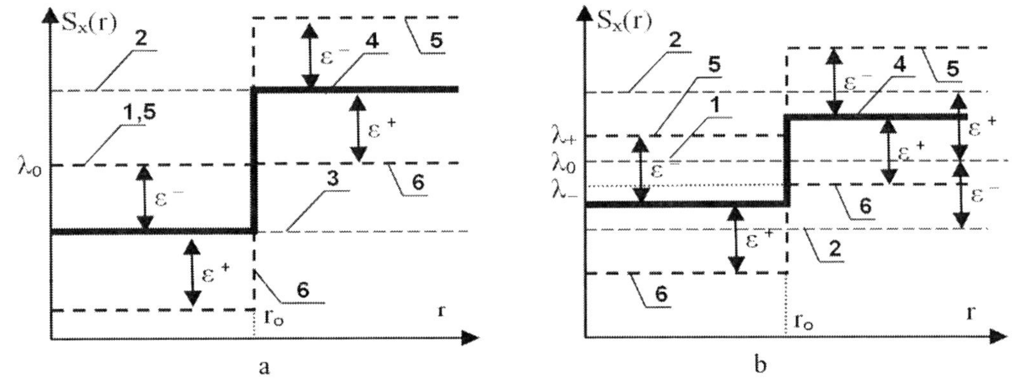

Fig. 1. Determination of an interval estimation for concrete realizations of signals. 1 – the modelling function corresponding to true value of λ_0 observable signals; 2 – an upper bound of ε – layer for modelling function with parameter λ_0; 3 – the lower boundary of ε – layer for modelling function with parameter λ_0; 4 - observable realization of signal $S_x(r)$; 5 – an upper bound of values of modelling function for an observable signal; 6 – the lower boundary of values of modelling function for an observable signal. a) rise of a signal is equal to a full width ε – layer . b) rise of a signal is less than a full width ε – layer

For this case $\lambda_+ = \lambda_- = \lambda_0$, that is at such realization it is possible to restore absolutely precisely true value of controllable parameter, when $\lambda_e = \lambda_0$. However, if the swing width is less than the total thickness of the ε-layer, the confidence interval will have a finite length equal to the difference between the total thickness of the ε-layer and the value of the drop (Figure 1*b*). We will consider now an example of application of the offered methods of determination of interval estimations not for abstract, and for the real optical signal projected on an input of the one-line converter of the image of the device for the monitoring linear movements.

4 Computational experiment for modeling the estimation procedure of the interval error as applied to the passive optical positioning system

Optoelectronic devices are widely applied in measuring technique to the solution of the most various problems, including for positioning of remote objects by non-contact methods [18–21]. For this purpose, a light source is fixed at the monitoring object or a light spot is formed on it. The light source or a spot is projected on a surface of a photodetector sensitive to movements or on the image converter in an electric signal (for example, CCD [22, 23]) and a position of a projection of a light source define an object position. Obviously, accuracy of positioning will be defined first of all by quality of formation of an optical signal an objective and presence in a controllable scene of the image or on a section between object and an optical system of additional installations, presence of background illumination and other factors, to consider which difficult or even it is impossible.

To demonstrate the application of the ε - layer model to estimate the error caused by these dominant factors, consider the case where the image of the luminous source placed on the object is projected onto the surface of a linear CCD photodetector. If to assume, that a source is dot, that in this case it is quite admissible, the signal shape on certain wave length and under using an ideal (no aberration) system will be described by a Bessel's function of first order J_1 (kr) with a scale factor *k*, proportional to diameter of an entrance pupil of objective D_o and inversely proportional to a focal length of objective F_o as for enough remote installations the image plane will coincide with a focal plane [24]. As the optical signal will be perceived by a photodetector only over the range lengths of waves of its spectral response the useful two-dimensional signal will be equal to integral from J_1 (kr) on the given range. Since in semiconductor photodetectors the upper limit of the spectral range is at least twice as high as the lower, the result is a signal that is close in shape to a gaussian:

$$E(x) \equiv E(x, x_0, R) = E_0 \cdot e^{-\frac{(x-x_0)^2}{R^2}} \tag{10}$$

where E_0 – the maximum level of light exposure created by a light source on a surface of a photodetector, x - cartesian coordinate, x_0 – coordinate of the center of a source image on a photodetector surface, and R – some parameter, proportional to effective radius of Airy scattering circle and in addition considering not ideality of an optical system [24].

As it is supposed, that concerning other components of an optical signal any information is absent, we will present them in the form of surrounding a signal ε – layer. Thus positive values will be created by the additional reflecting light installations which are in a zone of the control, and negative - by shading installations. As an instance of such installations the clouds recoating a laser beam at sounding of an aerosphere can serve, for example. Then the general model of an optical signal can be written down in an aspect

$$E'(x,R,x_0)=E(x,R,x_0)+\varepsilon\ (x), \tag{11}$$

where $-\varepsilon^-(x) \leq \varepsilon\ (x) \leq \varepsilon^+(x)$.

From here, according to (8), it is possible to note, that the estimations of any called by errors of parametres of a signal of its deviation from an initial aspect will be equal:

$$\Delta\text{E(x)}=\text{E'(x)}-\text{E(x)}=\frac{d}{dx_0}E(x,R,x_0)\cdot\Delta x_0+\frac{d}{dR}E(x,R,x_0)\cdot\Delta R+\frac{d}{dE_0}E(x,R,x_0)\cdot\Delta E_0, \tag{12}$$

whete

$$\frac{dE}{dE_0}=e^{-\frac{(x-x_0)^2}{R^2}}; \tag{13}$$

$$\frac{dE}{dx_0}=2E_0\cdot\frac{x-x_0}{R^2}e^{-\frac{(x-x_0)^2}{R^2}}; \tag{14}$$

$$\frac{dE}{dx_0}=2E_0\cdot\frac{(x-x_0)^2}{R^3}e^{-\frac{(x-x_0)^2}{R^2}}. \tag{15}$$

Then, solving the equations

$$\Delta\ E(x) - \varepsilon^+(x) = 0 \text{ with } \Delta\ E(x) \geq -\varepsilon^-(x) \tag{16}$$

$$\Delta\ E(x) - \varepsilon^-(x) = 0 \text{ with } \Delta\ E(x) \leq \varepsilon^+(x) \tag{17}$$

concerning deviations of any of parameters of function E (x, R, x_0) at the fixed values of remaining parameters in the field of their possible modification, and then, discovering for them the greatest probable values, it is possible for that to discover the greatest probable errors of measurement of all parameters entering into an optical signal.

Examples of calculation of the greatest possible and actual values of errors for the considered model of a signal are shown in Figures 2–7. For calculations the normalized values of parameters E_0=1, R=1 and X_0=0 were set, and ε – layer relied symmetric and invariable on a thickness which has been chosen equal 0.1. For such a thickness of the layer, the linear approximation (12) turned out to be unjustified and a solution for the initial model of the quasi-deficient component given by expression (9) was used. If variations of amplitude of signal E_0 and effective radius are absent, the greatest possible error of definition of coordinate of light stain X_0 makes 0.1168 (Figure 2) and if thus the effective radius increases by 10 % (for example, at the expense of an objective defocusing) this error will decrease to 0.042 (Figure 3).

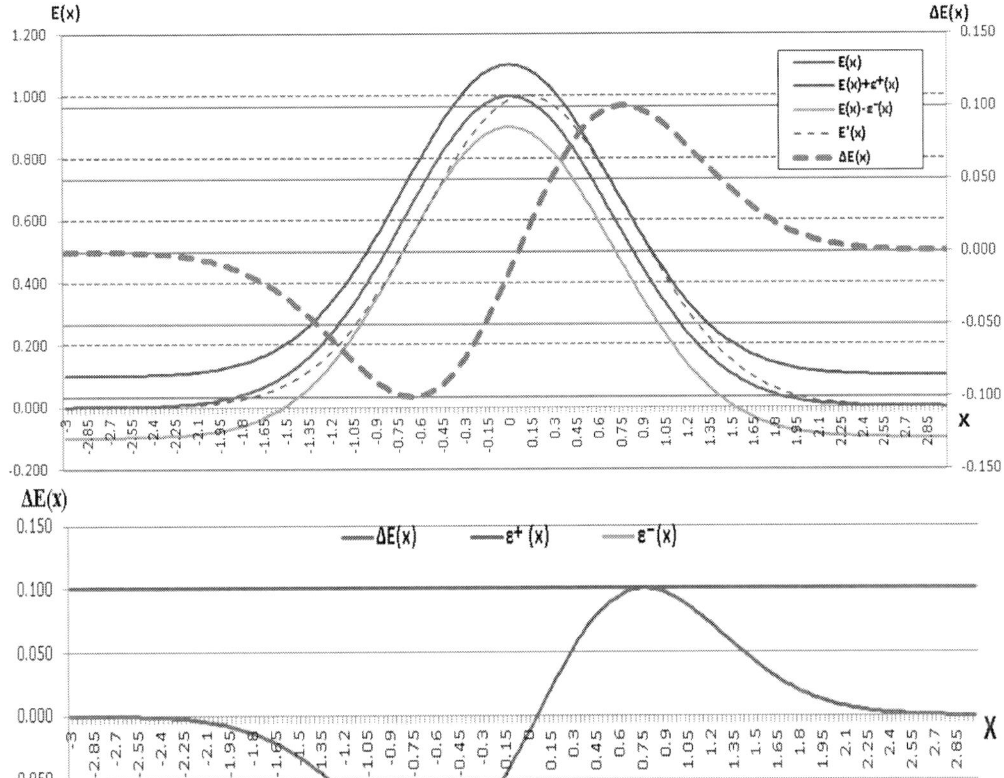

Fig. 2. Initial signal E (x), upper and lower boundaries of ε – layer $E(x) + \varepsilon^+$ and $E(x) - \varepsilon^-$, the greatest possible variation of signal $E'(x)$ on parameter x_0 and corresponding to this variation of the signal deviation $\Delta E(x)$ from its original form.

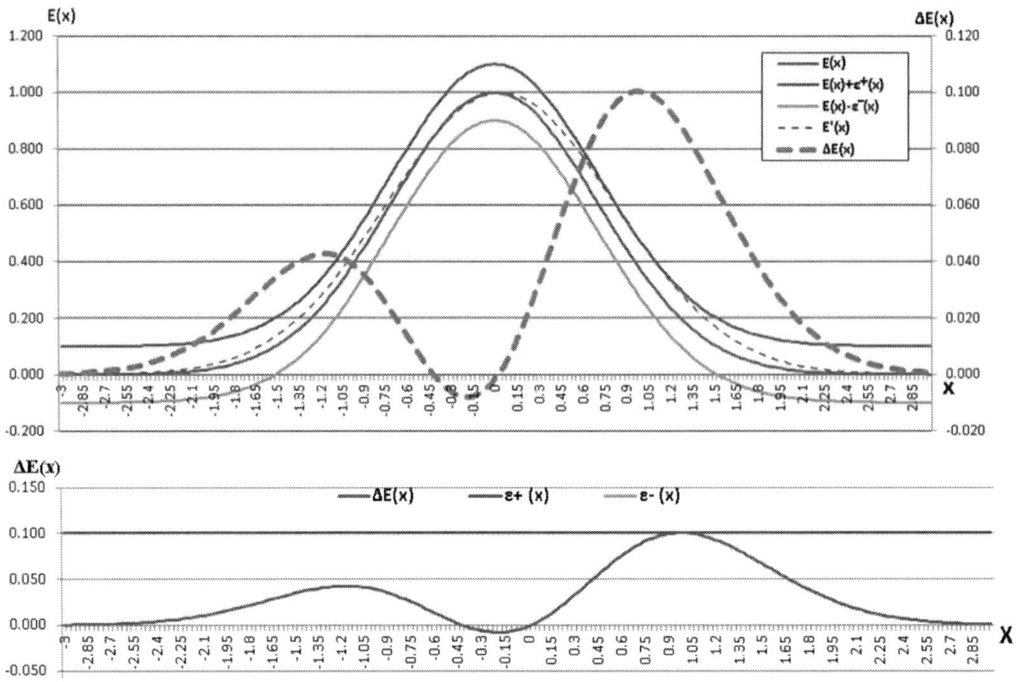

Fig. 3. Outcomes of evaluations for a case of the greatest possible variation of parameter Δx, at increasing of effective radius R on 0.1 and invariable value of parameter E_0.

On the other hand, if coordinate of a light stain to leave invariable, but thus to increase amplitude E_0 of modelling function $E(x)$ to its greatest possible value within the set thickness ε – layer, that is to 1.1 the maximum error of an estimation of parameter R will appear equal -0.16. Thus, the interval estimation for the effective radius will be sharply asymmetric with respect to R. The full correlation between boundary values of parameters ΔE_0 and ΔR is presented in Table 1.

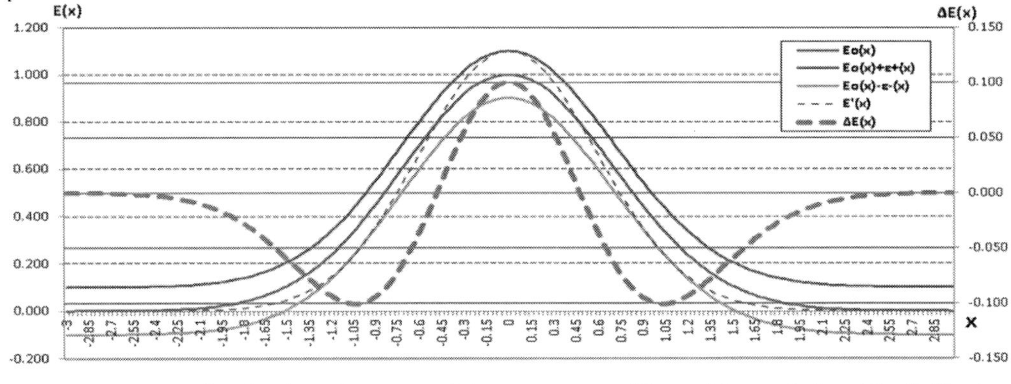

Fig. 4. Outcomes of evaluations for a case of the greatest possible positive variation ΔR of effective radius R at invariable values of remaining parameters of a signal.

Table 1. The greatest possible variations of parameters of signal E_0 and R under condition of absence of variations X_0.

ΔE_0=	0.00	0.10	-0.1
ΔR=	[-0.1270 , 0.1461]	[-0.1583, 0.0495]	[-0.0518, 0.2101]

Unlike effective radius, the measurement error Δx_0 always will be symmetric, that is obviously visible from Figures 2, 4. As a whole the correlation between possible deviations of all three parameters of an optical signal is set by three-dimensional area. Since for these studies it is of interest to analyze only the maximum possible error in measuring the position of the light spot, the dependence of this error on the parameters E_0 and R have been received. Outcomes of such calculations are presented in Figure 5 from which it is visible, that the maximum error will be observed at a choice $\Delta R = 0$ and $\Delta E_0 = 0$ and is equal 0.1169.

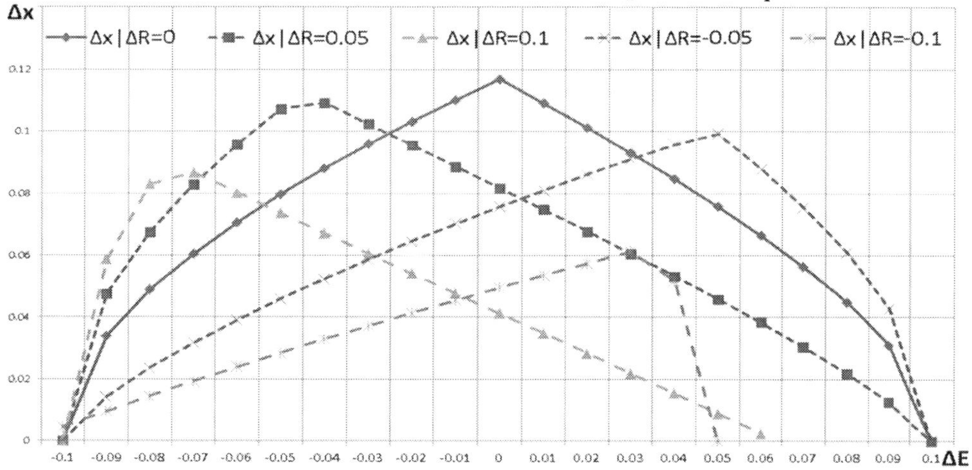

Fig. 5. Cross sections of the three-dimensional definition area of the variations of values of the signal parameters *E '(x)*, provided it is in the region of the ε-layer of the original signal *E (x)*. The parts of the cross sections in the positive half-plane, mirror-symmetric with respect to the axis ΔE is shown.

At the same time the value of given estimation is maximally achievable and for concrete realizations it can appear essentially smaller, that is obviously visible from Figures 6, 7.

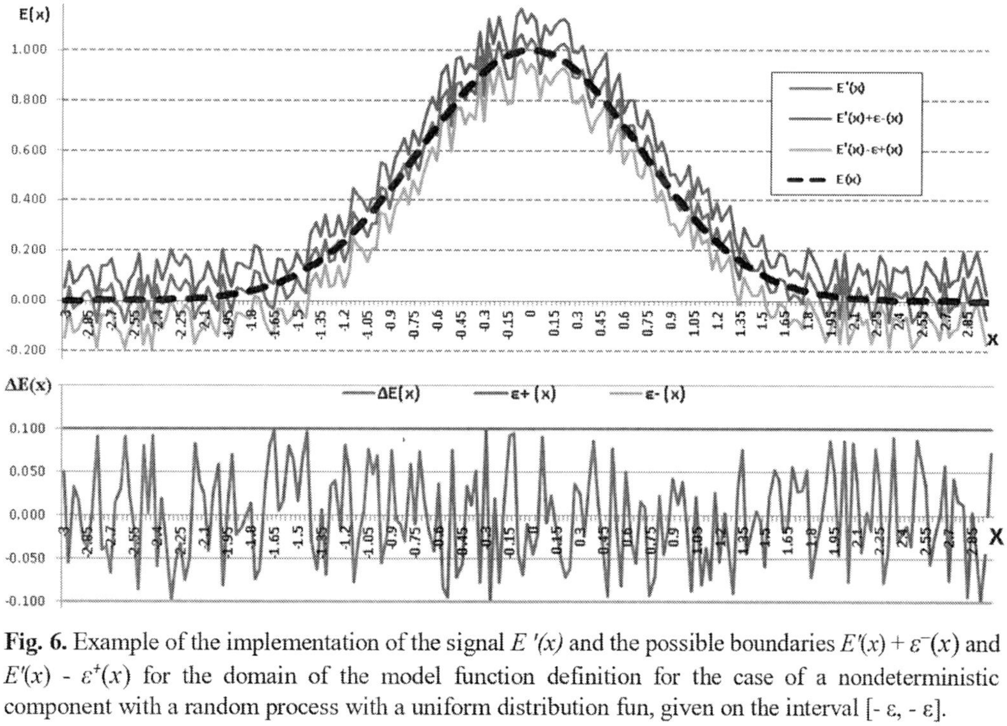

Fig. 6. Example of the implementation of the signal *E '(x)* and the possible boundaries $E'(x) + \varepsilon^{-}(x)$ and $E'(x)$ - $\varepsilon^{+}(x)$ for the domain of the model function definition for the case of a nondeterministic component with a random process with a uniform distribution fun, given on the interval [- ε, - ε].

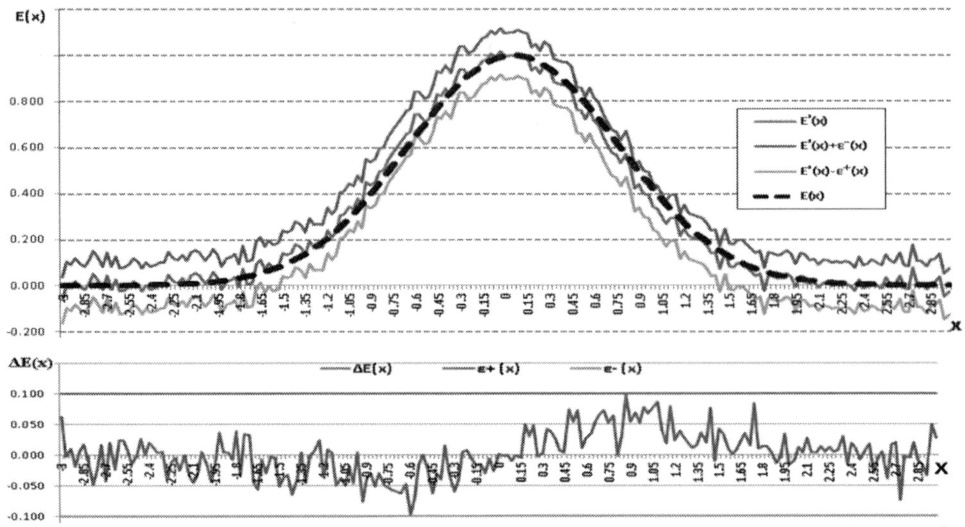

Fig. 7. Example of realization of signal $E'(x)$ and possible boundaries $E'(x) + \varepsilon^{-}(x)$ and $E'(x) - \varepsilon^{+}(x)$ of a range of definition of modelling function for a case of representation of nondeterministic component as normal casual process with value of the variance equal 1/3 ε and an additional restriction of the density function of distribution by the interval $[-\varepsilon, \varepsilon]$.

For the signal realizations presented in Figures 6, 7 these values lay within 0.05 for the normal law of distribution and 0.008 for a rectangular distribution of signal variations. Thus as can be seen from these figures, an increase in the observation interval will lead to a decrease in the interval error estimation due to an increase in the probability of appearence the signal values close to the ε-layer boundaries. The maximum measurement error is reached at displacements x_0 by a value close to the value of the effective radius, when the model function touches the boundaries of the ε-layer. And the increase in the interval of observation does not allow to reduce the magnitude of this error, which is evident from Figures 6.7 However it is not necessary to forget, that the given approach was developed just for those cases when noise and other interference cannot be described casual process. If, for example, to suppose that there is a linearly varying background illumination in the observation zone, then the measurement error x_0 will increase at once, that obviously illustrates Figure 8.

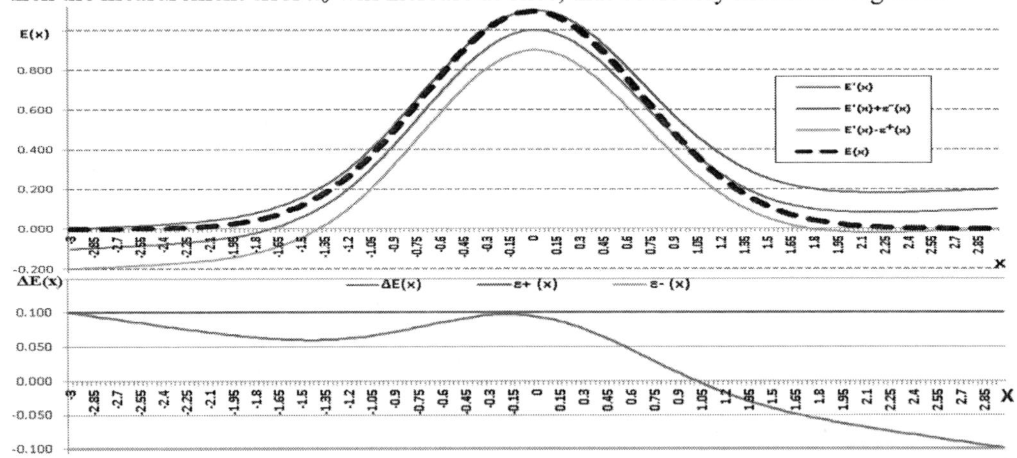

Fig. 8. Example of realization of signal $E'(x)$ and possible boundaries $E'(x) + \varepsilon^{-}(x)$ and $E'(x) - \varepsilon^{+}(x)$ of a range of definition of modelling function for a case of linearly varying on an interval of observation of a background illumination from -0.1 to 0.1. $\Delta x \in [-0.942, 0.942]$ with $\Delta E_0=0$ and $\Delta R=0$.

5 Conclusion

The executed experiments have shown, that the offered methods for calculation of greatest possible and actually observable interval errors are very simple and useful to applications when noise and interferences during conducting of procedure of measurements do not manifest statistical properties. The considered concrete example of the application of the method for solving the particular problem of positioning objects by optical methods has shown that the use of the ε - layer model makes it possible, without complex calculations, to analyze the influence of various factors on the accuracy of measurements and to obtain both qualitative and quantitative results.

Acknowledgement

The author expresses his deep gratitude to the leadership of the Altai State Technical University of I.I. Polzunov (AltSTU) for financial support of the publication of the article and OOO Promavtomatika for technical assistance in the preparation of the article.

References

[1] V.A. Balalaev, V.A. Slaev, A.I. Sinjakov, *Potential accuracy of measurements. Scientific issue* (Scientific professional unity «Professional», St.Peterburg, 2005) (in Russian)
[2] B.R. Levin, *Theoretical bases of a statistical radio engineering* (Soviet Radio, Moscow, 1975) (in Russian)
[3] S.E. Phalkovich, A.N. Homjakov, *The statistical theory of measuring radio systems* (Radio&Communication, Moskow, 1981) (in Russian)
[4] J.R.Taylor, *An Introduction to Error Analysis: The Study of Uncertainties in Physical Measurements* (University Science Books, 1999)
[5] R.B. Ash, A.Cathterine, Doléans-Dade, *Probability & Measure Theory, Second Edition* (Academic Press, 2000)
[6] K.B. Athreya, S.N. Lahiri, *Measure Theory and Probability Theory* (Springer Science + Business Media, 2006)
[7] P. Billingsley, *Probability and Measure, Third Edition* (John Wiley & Sons, 1995)
[8] R.E. Moore, *Interval analysis* (Englewood Cliffs. Prentice Hall, 1966)
[9] R.E. Moore, R.B.Kearfott, M.J.Cloud, *Introduction to interval analysis* (SIAM, Philadelphia, 2009)
[10] R.E. Moore, *Methods and applications of interval analysis* (SIAM, Philadelphia, 1979)
[11] O.Ye Rodionova, A.L.Pomerantsev, In: «Progress in Chemometrics Research», 43 (2005)
[12] S. Markov, K. Okumura, In: «Developments in Reliable Computing», 167 (1998)
[13] A.G. Yakunin, Scientific Pract.Conf. «Optical scanners and devices on their basis», 169 (1986) (in Russian)
[14] L.I. Suchkova, A.G. Yakunin, The bulletin of Dagestan State Technical University. Technical Sciences, 4 11(2011) (in Russian)
[15] L.I. Suchkova, A.G. Yakunin, Proc.of TSU of Control Systems and Radioelectronics 2(28), 147(2013) (in Russian)
[16] P.V. Novitskij *Bases of the information theory of measuring devices* (Energy, Leningrad,1968) (in Russian)
[17] P.V. Novitskij, I.A. Zograph, *Estimation of errors of results of measurements* (Energoatomizdat, Leningrad, 1991) (in Russian)

[18] S. Donati, *Electro-Optical Instrumentation: Sensing and Measuring with Lasers 1st Edition* (Pearson Education, New Jersey, Inc.,2004)
[19] P.C.D. Hobbs, *Building Electro-Optical Systems: Making It all Work* (Wiley series in pure and applied optics) (John Wiley & Sons Inc., USA, 2009)
[20] P. Fiala, T. Jirku, R. Kubasek, P. Drexler, P. Konas, PIERS ONLINE, 2, 685 (2006)
[21] S.P. Mahulikar, H.R. Sonawane, G.A. Rao, Progress in Aerospace Sciences, 43(7-8), 218 (2007)
[22] E.R. Fossum, IEEE Transactions on Electron Devices, 44, 10, 689 (1997)
[23] H.S. Wong, IEEE IEDM Technical Digest, 8.5.1(1997)
[24] M. Bom, W. Emil. *Principles of Optics, 6lh edition* (Cambridge University Press, reissued, 1998)

NEAR-FIELD INTERFERENCE MICROWAVE DIAGNOSTICS OF CULTURAL PLANTS AND WOOD MATERIALS

Viktor Belichenko*, *Andrey* Zapasnoy, *Aleksandr* Mironchev

National Research Tomsk State University, Radio Physics Department, 634050 Tomsk, Russia

Abstract. A schematic solution of the near-field interference microwave microscopy technology is discussed. This solution is implemented in the form of a maximally simplified microscope structure. Testing was conducted to determine the capabilities of this microscope. It is shown that technology can be used to solve a number of hygroscopy and defectoscopy problems.

1 Introduction

In modern devices of near-field microwave microscopy one resonant or nonresonant probe (usually as an open end of coaxial lines of different cross sections) is used for probing [1–3]. The independent studies of several scientific groups have shown that to increase the resolution and sensitivity of near-field microwave microscopy, the so-called "wave" spatial scale should be used, along with "quasistatic" spatial scale (see, for example [4]). However, this conclusions concern to the technology of "one-sided" near-field sounding. At the same time, our study [5] of probing by overlapping near (evanescent) fields of several active probes revealed the presence of a new, "interferential" spatial scale.

2 The Prototype and the Technique of Experimental research

One of this work purposes is to confirm given conclusion on the numerical simulation basis using CST Microwave Studio software package. The probing system model includes two identical conical coaxial probes which have 360 mm long and located coaxially. The apertures of these probes are turned by towards each other. The outer diameter of the probe central conductor in the aperture plane is 30 mm and the inner diameter of the external conductor is 70 mm; the interval between apertures is equal $L = 92$ mm. Excitation of the probes is produced by means waveguide ports with identical characteristics. But for the signals on the ports phase shift is introduced.

The real and imaginary parts of the Poynting's vector longitudinal component P were calculated in the interval ($0 \le z \le L$) between apertures of probes, along the axis of their symmetry. Radiated frequency was accepted equal 2 GHz (wavelength – 15 cm). The re-

* Corresponding author: bvp@mail.tsu.ru

sults of calculations processed with use of the mathematical Mathcad package are presented in the Figure 1. They show the nature of change Re(P) (Figure 1*a*) and Im(P) (Figure 1*b*) depending on initial phase shift ($\Delta\varphi$) of the fields in the probes: 1 – $\Delta\varphi = 0^\circ$, 2 – $\Delta\varphi = 15^\circ$, 3 – $\Delta\varphi = 45^\circ$, 4 – $\Delta\varphi = 90^\circ$.

Directly near the probes apertures there are domains with the maximum values P (module P). It is interesting to note that, for example, at $\Delta\varphi = 0^\circ$ oscillating character P is accurately demonstrated. Moreover, there are points in the neighborhood of which P has small values. This circumstance directly indicates the presence of an interference energy flux in the overlapping evanescent fields of the probes. With increase $\Delta\varphi$ values of Re(P) increase too and the displacement of its maximum in space between apertures of probes occur. For Im(P) great values near apertures of probes are generally characteristic. However change $\Delta\varphi$ leads to very significant change of Im(P) localization domain structure. In other words, it is observed considerable redistribution of the stored energy localized near apertures of probes. And the most noted circumstances are observed at $\Delta\varphi = 90^\circ$.

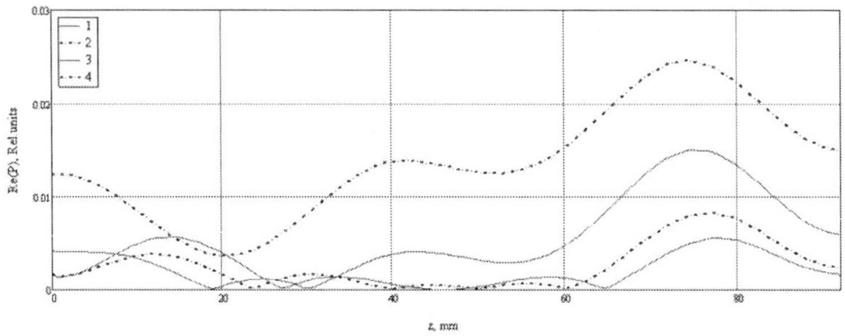

a)

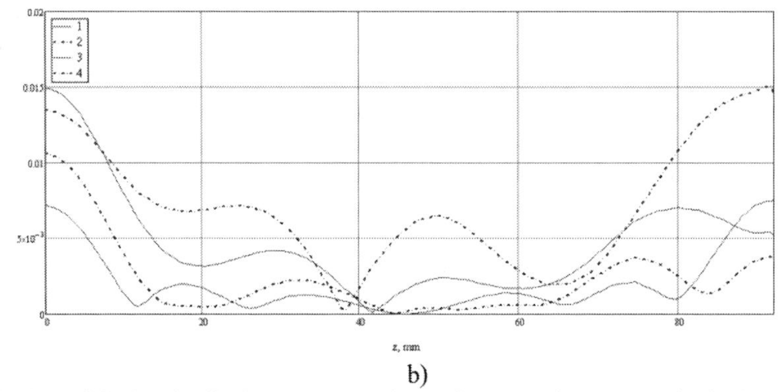

b)

Fig. 1. Distribution of the longitudinal component of complex Poynting's vector in the interval between apertures of two active probes.

It should be emphasized that a change of $\Delta\varphi$ undergoes significant changes and in Re(P) distribution. Presented here simulation results are quite consistent with the results of [5]. In this paper we prove the existence of additional interference energy flux generated in counter propagating evanescent fields of the two active probes. This energy flux provide a deeper field penetration in the medium under study, as well as improve sensitivity compared to conventional technology "one-sided" sensing. In addition, the detailed theoretical study of the structure of the near-field of a few simple systems radiators has shown that effective control the parameters of said interference energy flows is possible.

This feasibility has also been confirmed in the present work based on rigorous electrodynamic modeling of the real probing system. It is important to emphasize that the parameters of the interference energy flux formed in the gap between the probe apertures can be controlled; moreover, they can be controlled electrically, namely, by changing the phase difference between the probe fields. This control is realized via "accentuating" one of the interference components of the Pointing vector. This circumstance means that a sufficiently small selected volume of the examined object or medium can be detected via flexible change of the regime of its irradiation by the evanescent fields created by a radiator system (moreover, the motionless system).

3 Experimental results: analysis and interpretation

By results of numerical modeling the circuit design of technology of near-field interferential microwave sensing which is most simplified on structure was developed. Based on this circuit design, a working breadboard model of the device was developed and tested experimentally to estimate its resolution and sensitivity.

The breadboard model of the near-field interference microwave microscope contains the following basic elements: a N5230C PNA-L Network Analyzer (Agilent Technologies) with the working frequency band 10 MHz – 40 GHz, two coaxial conical horns (probes), power divider, and additional sections of transmission lines (Figure 2). The network analyzer allowed measurements to be performed with high speed and high accuracy, and the correct calibration provided correction for twelve error components.

An operating procedure of the prototype is as follows. A signal generated by the network analyzer splits at the power divider which is working on the frequencies of 10 MHz – 10 GHz with the maximum power of 0.5 W. Then, signals are sent at the inputs of the coaxial conical probes. The probe apertures are faced to each other, the test object is placed in the gap between the apertures and measurements of the reflectivity from the power divider entrance are performed. Creation of required phase shift between the signals arriving on probes is reached due to inclusion in the path of transfer additional sections of coaxial cable, so, that provides the control of interference energy flows which appears in overlapping evanescent fields. The diameter of the probe inner conductor in the aperture plane is 30 mm, the internal diameter of the outer conductor is 70 mm. This provides an opportunity to control the size of overlaping zone of evanescent fields over a wide range using the 10 MHz – 10 GHz microwave radiation, sounding a medium or an object with a counter propagating evanescent fields. The breadboard model of the interference microwave microscope is presented in the Figure 2.

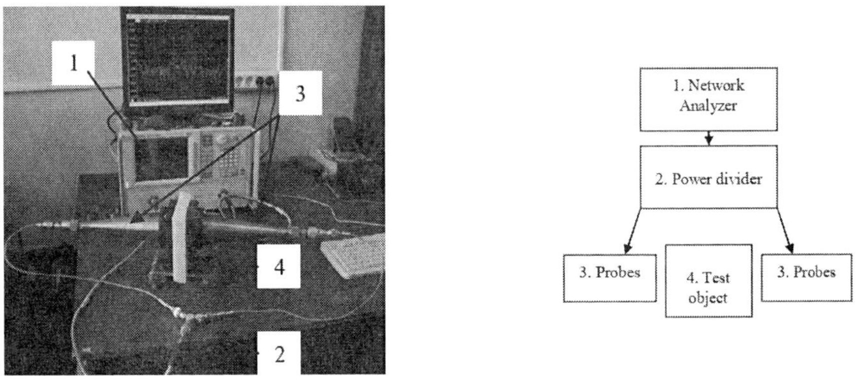

Fig. 2. Breadboard model of the microwave interference microscope.

As data registered by the network analyzer, the frequency dependence of the complex reflection coefficient (S11) was chosen. The frequency band during measurements was 10 MHz – 10 GHz. The step of sampling was 250 kHz since measurements in the chosen frequency band were performed in 38000 points. For convenience of perception during each measurement the registered signals were normalized on the reflectivity for lack of the studied samples. Further we will designate this parameter as the relative reflectivity coefficient (RRC, NS11).

For more detailed analysis of sensitivity of the breadboard model, the series of the experiments concerning researches of quality of seeds of different cultures was conducted. As the studied cultures seeds of soy, flowers of the impatiens and the amaranth were chosen. The purpose of the experiment was to determine the quality of the seeds provided (will they ascend when planting them in the ground or not). Seeds which will sprout have to possess the large supply of water, and, therefore, have big dielectric permittivity and absorb more the radio wave radiation falling on them. For the experiments seeds divided into "fertile" and "infertile" were provided, and the division was carried out by specialists in the botanical garden of the National Research Tomsk State University with the application of a standard methodology.

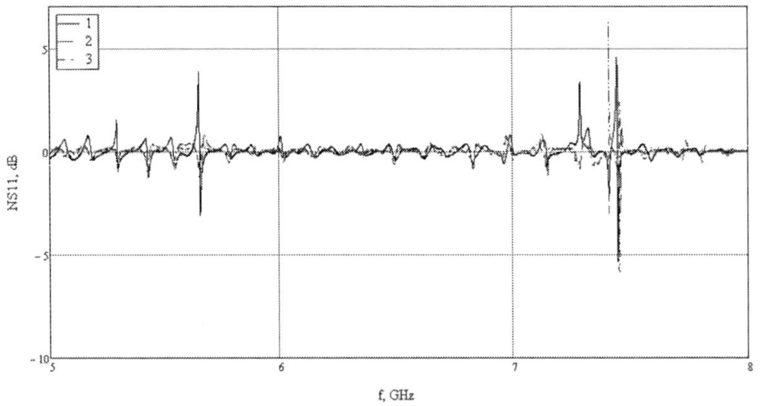

Fig. 3. Frequency dependence of RRC for seeds of three different cultures: 1 – the amaranth, 2 – the impatiens and 3 – soy.

The experiment was carried out as follows. It was made special the cuvette for seeds which was placed between microscope probes, thickness of the cuvette made 45 mm. After setup and calibration of the breadboard model the measurement data of the empty cuvette were obtained and it date set using for normalization. Then cuvette were filled up in turn by "fertile" and "infertile" seeds, the measurements were normalized and then two signals subtracted in the entire frequency range from 0.5 MHz to 8 GHz (Figure 3) was make. Thus, seeds with maximum RRC were chosen. In Figure 4 is show the frequency band from 2.5 GHz to 3.5 GHz, it can be seen that for all three types of seeds, it is possible to allocated seeds with higher dielectric permittivity and, therefore, with a higher likelihood of germination.

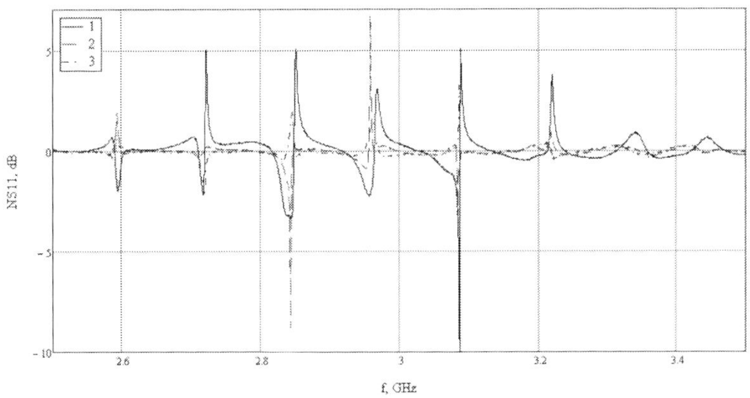

Fig. 4. Frequency dependence of RRC for seeds of three different cultures in the band from 2.5 GHz to 3.5 GHz: 1 – the amaranth, 2 – the impatiens and 3 – soy.

The presented dependences demonstrate that the breadboard model of the microwave interference microscope offered in this work, allows to register even minor changes of moisture content in seeds. As a result of data processing and comparisons of the received results with results these provided specialists of the botanical garden full coincidence is received.

Other series of experiments belonged to clarification of opportunities of the breadboard model at diagnostics of quality of wood materials. The products from the wood having these or those defects including hidden were chosen as objects of researches. As the example in the Figure 5 results of diagnostics of the board of the laminate having defects in the form of through openings with a diameter of 4 and 10 mm are presented.

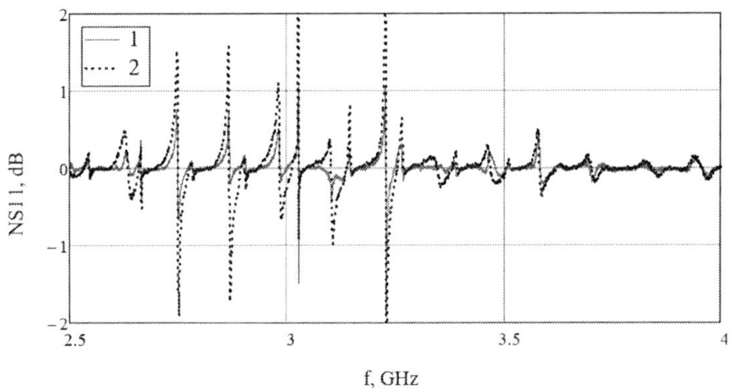

Fig. 5. Frequency dependence of RRC for the laminate 5 mm thick: 1 – with the opening with a diameter of 4 mm, 2 – with the opening with a diameter of 10 mm.

As we showned, the breadboard model allowed to find similar defects. Also it is showned as with increase in diameter of the opening up to 10 mm amplitude of the accepted signal also increases. It is essential to emphasize that defect with a diameter of 3-4 mm is fixed even at the frequency $f \approx 2.6$ GHz of used the electro-magnetic radiation (at the same time wavelength makes 11.5 cm). This demonstrates very essential overcoming the known fundamental limit on permission "in the farfields".

For demonstration differences of traditional technology of sounding from above described we made experiment on diagnostics of defect of the laminate with use evanestsent fields only of one probe (Figure 6). The fact that defect existence (even with a diameter of 10 mm) is extremely poorly displayed on this schedule attracts attention.

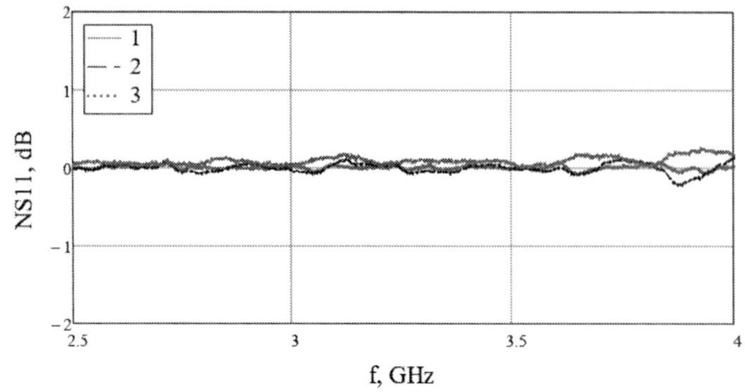

Fig. 6. Frequency dependence of RRC for the laminate 5 mm thick: 1 – the continuous laminate, 2 – with the opening with a diameter of 3 mm, 3 – with the opening with a diameter of 10 mm.

4 Conclusion

The presented results lead to the conclusion that the sensitivity and resolution of described interference microwave microscope are higher in comparison with a microscope constructed according to a conventional scheme. Indeed, our experimental studies, used the most simplified interference microwave microscope model, confirmed this conclusion. In a number of experiments, a microscope with a single probe did not reliably detect the material defect, in contrast to a microscope with two probes. This circumstance is explained by the absence of an additional interference energy flux which generated in the counter propagating evanescent fields of the two probes. This energy flux ensure a deeper penetration of the field into the investigated object, as well as an improvement in sensitivity in comparison with conventional near-field technology. In particular, this effect was manifested in experiments on the hygroscopic properties of seeds of cultural plants, as well as in the detection of defects in wood products.

Acknowledgments

The research was carried out with the financial support of the Russian Foundation for Basic Research and the Administration of the Tomsk Region within the framework of the scientific project No. 16-42-700942 of the regional competition (r_sibir_a).

References

[1] S.M. Anlage, V.V. Talanov, A.R. Schwartz, *Scanning Probe Microscopy: Electrical and Electromechanical Phenomena at the Nanoscale* (Springer-Verlag, New York, 2007)
[2] K.P. Gaikovich, P.K. Gaikovich, Inverse Problems **26**, 125013 (2010)
[3] A.N. Reznik, N.V. Yurasova, Technical Physics **51**, 86 (2006)
[4] A.N. Reznik, Journal Applied Physics **115**, 084501 (2014)
[5] V.P. Belichenko, A.S. Zapasnoy, P.V. Shestakov, Optics and Spectroscopy **118**, 633 (2015)

METHOD OF RESTORING MINE HOISTING MACHINE BRAKE UNIT LEVER TRANSMISSION BUSHING

Marat Ibatov[2], *Ali* Mekhtiyev [1,2], *Felix* Bulatbayev[2], *Oxana* Yakupova[3a]

[1]National Research Tomsk Polytechnic University, 634050, Tomsk, Russia
[2]Karaganda State Technical University, 100000, Karaganda, Kazakhstan
[3]National Research Tomsk State University, 634050, Tomsk, Russia

Abstract. In the article results of researches of brake devices of lifting machines UD of joint-stock company "ArcelorMittal Temirtau" are resulted. Defects were revealed as a result of the examination. A method and device for the unconditional control of defects and technical condition of the hinged joints of brake devices is developed in terms of the gap size and the angle of rotation between the links of the kinematic pairs "Widget-finger". For the monitoring of gaps, capacitive sensors are installed, mounted on hinged joints. It is substantiated that with the use of capacitive sensors it is possible to control the gaps in the hinged joints in real time. A device has been developed that makes it possible to estimate the magnitude of the gap at several points of the kinematic chain of the mechanism simultaneously.

1 Introduction

One of the most important elements in the technological chain of mineral resources mining at mines id hoisting-transporting equipment. Intensification of stoping and preparatory operations requires continuous increasing of mine hoisting plants capacity and productivity. With productivity increasing there increases the significance of providing normative reliability, as in these conditions freight hoisting plants failures lead to the significant loss of mineral resources mining, man hoisting failures lead to the potential danger for peoples' life. Solving the problem of increasing the technical level, wear resistance and reliability of constructional parts of hoisting-transporting equipment is one of the main priorities of mining machine building development.

A brake unit is one of the most complicated and important units of a hoisting plant which is the final element in the protection means chain. The raised standards are claimed to it, as on its perfectness there depends reliability and safety of hoisting plant operation [1].

In the process of using hoisting plants there occur significant braking forces in the brake unit elements both in normal modes and in emergency situations with the hoisting vessel sudden stops in the shaft [2]. Hinges jamming or their excessive wear can become a cause of a serious accident. Brake system hinge joints are to be timely repaired.

---

[a] Corresponding author: 1_neg@mail.ru

Mine hoisting plant brake unit hinge joints wear can be a result of constructional characteristic features, contact surface material mechanical properties, manufacturing technology and quality, friction type, existing abrasive particles on the friction surface, use and mode of operation. To prevent untimely failures in brake units operation it is necessary still at the designing stage to take into consideration all the factors effecting hinge wear.

Meanwhile, there are not used the possibilities of ensuring the hinges rated reliability due to increasing the contact surface areas. Starting from the above-said, establishing hinge joints elements rational constructional parameters is an urgent problem. To solve this problem at the first stage it is necessary to determine stresses on the hinge coupling surface. Using imitation modeling, there was carried out the analysis of occurring contact stresses depending on bushings geometrical form and contact surface area.

2 Experimental study

There were studied typical hinges with cylindrical inner surface. Besides, there were suggested bushings with partially conical boring (with different ratios of the cylindrical and conical parts) and bushings with complete conical boring on the full length [3].

The experiment was carried out with five imitation models of hinges with different constructional parameters (Figure 1).

The imitation models have the following differences:

the first variant is a typical structure with the bushing cylindrical inner surface on AK line with diameter d_2;

the second variant has a partial conical boring of the bushing inner surface for 0.1 of its length on ABC line with diameters d_2 and d_3;

the third variant has a partial conical boring of the bushing inner surface for 0.35 of its length on ABC lines with diameters d_2 and d_3;

the fourth variant has a partial conical boring of the bushing inner surface for 0.7 of its length on ABC line with diameters d_2 and d_3;

the fifth variant has a complete conical boring of the bushing inner surface for the full its length on AC line with diameters d_2 and d_3.

Due to the use of the conical surface there increases the bushing and the pin surface contact area and, as a result there reduce contact stresses in the operating condition and the bushing contact surface wear. Studying the imitation model of the "bushing-and-pin" pair using applied programs ANSYS package showed that external loads applying would change stress distribution in the zone subjected to wear.

There was experimentally established the dependence of stresses on the bushing contact surface on the boring depth relative to its general length. Due to the use of the conical surface there increases the contact area of the bushing and pin surfaces and as a result there reduce contact stresses in the operating condition. The geometrical sense of this phenomenon is demonstrated in Figure 1. The bushing wear reducing takes place due to the bushing contact area with the pin, as the contact area on the body surface with ABC generatrix is larger than that on the surface with AK generatrix. The sum of AB and BC sides is larger than AC side, consequently, the genetratrix with ABC bend is more effective than AC straight L_1 if the bushing length; L_2 is the bushing inner conical surface boring depth; d_1 id the bushing outside diameter; d_2 the bushing inner diameter; d_3 is the bushing larger inner diameter after boring; γ is minimal possible thickness of the bushing wall from the condition of crush line. In designing C point is determined from the condition of the bushing crush condition. The inclination angle $\angle$ CBK is to be larger than Morse cone angle (70). With the boring angles less than 70 there can occur the coupled surfaces catching and the hinge jamming [4].

The analysis of the obtained dependences of stresses on the bushing contact surfaces of various structures permitted to determine the rational form of the hinge bushing. The conical part boring depth makes 0.42 … 0.64 of the bushing general length.

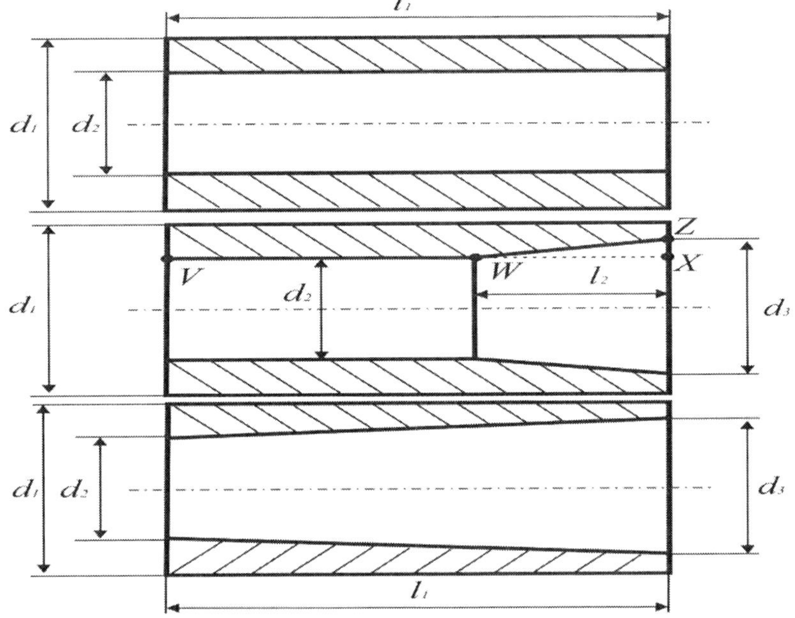

Fig. 1. Bushing longitudinal section: l_1 – length of the sleeve; l_2 – depth of boring of the inner conical surface of the bushing; d_1– outer diameter of the bushing; d_2 – inner diameter of the bushing; d_3 –larger internal diameter of the bushing after boring.

The dependence on the voltage on the contact surface of the bushing of the drilling depth relative to its total length is presented in Figure 2.

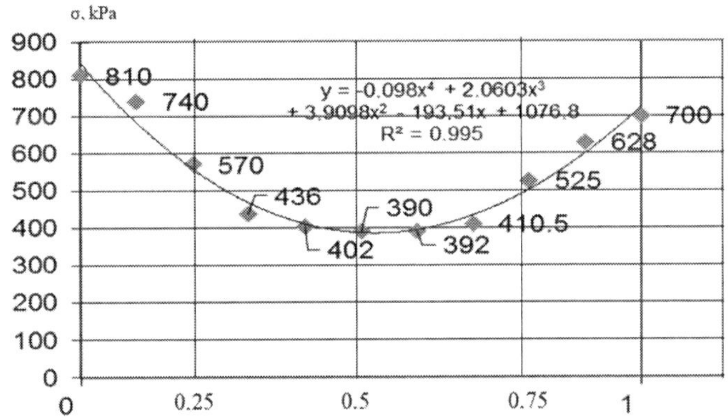

Fig. 2. Stress dependence on the bushing contact surface of the boring depth relative to its general length.

3 Developed methods

The aim of the developed method of hinge restoring is widening technological possibilities of repairing and increasing the operation resource of mine hoisting machine brake unit mechanism hinge joints [5, 6].

The technical decision essence consists in the possibility to restore a hinge joint without complete dissembling mine hoisting machine brake unit mechanism.

There are known various methods of restoring mine hoisting machine brake unit mechanism hinge joints. These methods have the drawback in the necessity to disassemble the brake system mechanism in order to press in the bushes, the bush setting complexity, the necessity of the double boring before and after the bush setting. This is connected with certain technological difficulties, as there must be disassembled the brake mechanism elements and stopped the mine hoisting machine operation for a sufficiently long period.

Typical bushing-and-pin elements of the brake mechanism hinge joints has a relatively short service life because, as a result of wear there form inadmissible clearances in hinge joints elements [2, 7]. It in turn leads to the impossibility to set normative clearances between the braking shoe and the braking rim of the rope winding element. Such state of the brake unit can lead to the mechanism jamming and damaging separate elements. To replace worn hinge joints with new ones there is performed the complete disassembling of the brake unit mechanism, the worn bushings removal and pressing in new bushings at workshops. As a result we have losses due to the hoisting-transport equipment downtime and expenses for its following adjustment.

The suggested method of repairing can be used for a typical structure of industrially manufactured bushing-and-pin elements of unit hinge joints for mine hoisting plants of NKMZ 2C-4x2.3; 2C–5x2.3; 2C-5x2.4 and 2C-6x2.4 types.

The studies showed that the main causes of the bushing-and-pin pair failure are the following factors: increasing clearances due to the bushing wear, limit contact stresses in the process of use, insufficiently effective distribution of lubricants over the contact surfaces. For the bushing manufacture there is used cast iron (SCh), for the pin – high-carbon steel of ST 45X grade. This pair possesses low wear resistance but here it is possible to use a worn bushing repeatedly within several inter-repair periods.

The developed method permits to reduce costs and time for the repair-and-restoration operations of mine hoisting machine brake unit mechanism. The hinge bushing restoring is performed without the complete disassembling of mine hoisting machine brake unit. The technology stipulates boring a cylindrical conical opening at the place of the bush setting using a mobile cutting tool. The bush opening boring along its geometrical axis is performed with a cylindrical and conical cutter. The bushing conical part makes 0.35 of the bushing full length. In the bored opening there is set a pin of corresponding configuration. It is economically expedient to use a bush in two repairing periods without its and the whole mechanism disassembling. In the repeated restoring the conical part is bored for 0.7 of the bushing total length. The conical form permits to increase the contact surfaces area, to exclude the radial skewing when setting a pin and to restrict the pin end travel in the bushing in the process of operation. This increases the brake unit mechanism hinge joint service life.

The method of restoring mine hoisting plant brake unit hinge joints is presented in Figure 3.

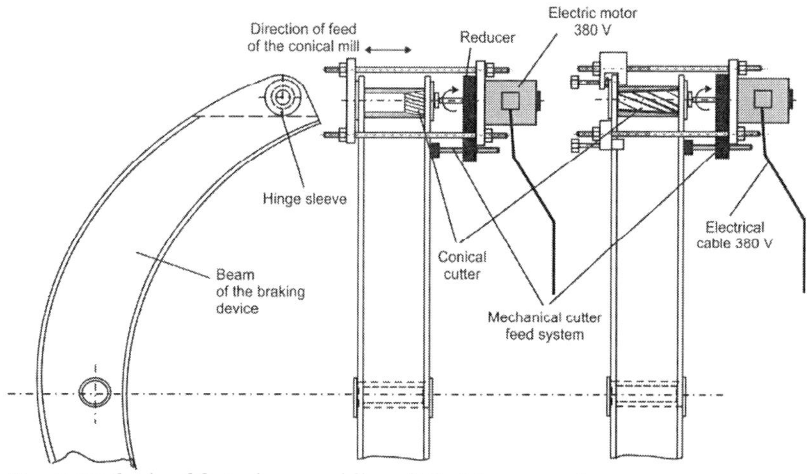

Fig. 3. Methods of fastening a mobile unit for hinge restoring.

When performing repair, the clearance value and the degree of the pair "bushing-and-pin" wear are preset in advance. Experimental studies showed that uniform wear of the total lever transmission bushing surface is characteristic, and this leads to increasing a clearance between the pin and the bushing. This excess breaks the hinge normal work and requires replacing the bushing.

A hinge with the excess of the clearance admissible value is disassembled. Using a mobile cutting plant there is performed the boring of an opening with cutter l_1 of the cylindrical and then partly conical form. Guiding units 7 and pneumatic cylinder 4 provide the necessary travel of the boring unit working element whose rotation is carried out through reduction gear 10 from electric drive 9. To restore mine hoisting machine brake unit hinge bushings working surfaces and liquidation of the admissible clearance excess, there is performed boring to the first repair size. Let's make the primary measuring of diameters d_1;d_2 and length L_1 (Figure 4). The boring depth of the bushing L_3 inner conical surface (Figure 5) is selected as a part of the total length L_1 (conical part makes 0.35 of the bushing total length). The remaining part of L_2 length is bored to the cylindrical form with the new opening diameter d_3 equal to $d_2+2\Delta_{wear}$ (Δ_{wear} is the value of the bushing inner diameter radial wear). The boring angle is: $tg\varphi = \delta/L_3$. The bushing larger inner diameter d_4 is determined from the condition of the bush crushing. The maximum angle of the bushing conical boring is restricted by its outer diameter. The minimum angle of the conical boring is restricted by the value 70 (Morse's cone). The link pin has the configuration similar to that of the bushing inner surface by the first repair size accounting allowances and fits [3].

After a certain period of operation between the working surfaces of the bushing and pin coupling there again forms a clearance affecting the mechanism normal operation. In this case it is possible to repeat the bushing inner surface restoration to the second repair size (the conical part makes 0.7 of the bushing total length), where diameter d_3 of the second repair size is equal to the sum of the first repair size cylindrical part and 2Δ.

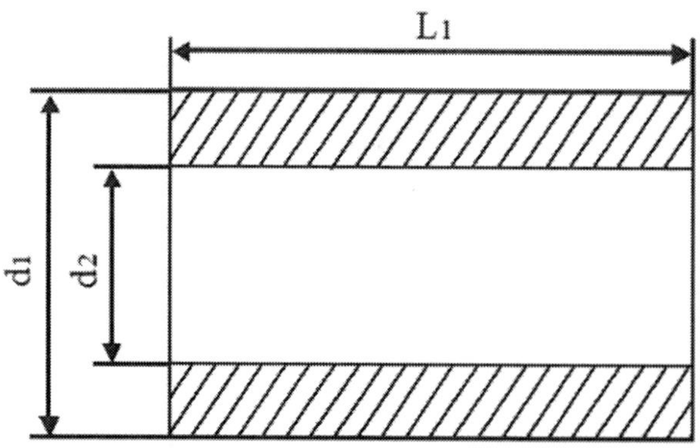

Fig. 4. Typical bushing structure.

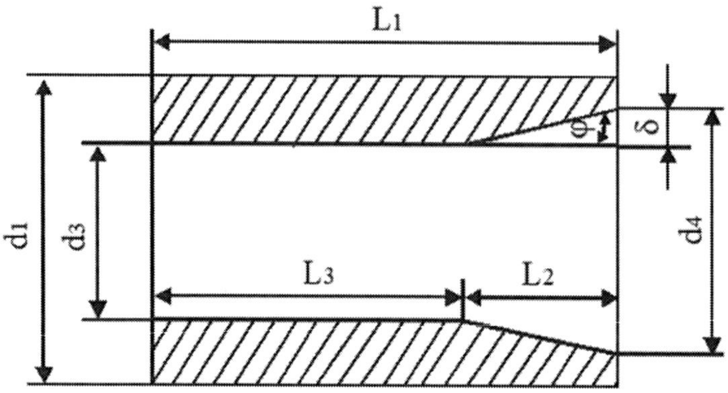

Fig. 5. The bushing structure with partial conical boring.

The conical form permits to exclude a radial skewing when assembling the pin and to restrict the ping travel in the bushing in the process of use. Taking into consideration the abovementioned, we can state the increase of the operation resource of mine hoisting machine brake unit hinge joint.

The results of theoretical and experimental studies were used in restoring the brake unit hinges on the hoisting machines of a drain well at LLP "Bogatyr Komir". The work was performed immediately on the hoisting plants within the periods of the equipment short-term stops without dissembling the brake system mechanism. In total there were restored 30 bushings of hinge joints.

4 Conclusion

The carried out experimental studies of various hinge structures stressed-and-strained state (imitation modeling) permitted to determine parameters of the bushing inner surface boring with the least stress in the zone of the bushing and the pin contact. The bushing boring interval makes 0.42…0.64 of its full length. Increasing the bushing-and pin pair contact surfaces area due to the conical boring reduces the hinge joint wear.

References

[1] V.R. Bezhok, R.Ia. Gruzutin, V.G. Kalinin, B.N. Chaika, *Neispravnosti shakhtnykh pod"emnykh ustanovok* (Nedra, Moscow, 1991)

[2] A.D. Mekhtiyev, V.V. Yugai, Scientific-analytical and production journal "Mining

equipment and electrical mechanics" **6**, 26 (2011)
[3] F.N. Bulatbayev, A.D. Mekhtiyev, V.V. Likhachoyv, University proceedings **4**, 50 (2011)
[4] A.V. Yurchenko, A.D. Mekhtiyev, F.N. Bulatbaev, Y.G. Neshina, A.D. Alkina and M.M. Kokkoz, Journal of Physics: Conference Series **881** (2017) doi: 10.1088/1742-6596/881/1/012034
[5] V.R. Bezhok, R.Ya. Gruzutin, V.G. Kalinin, B.N. Chaika, *Mine hoisting plants failures* (Nedra, Moscow, 1991)
[6] A.V. Yurchenko, A.D. Alkina, A.D. Mekhtiyev, F.N. Bulatbayev, E.G. Neshina, MATEC Web of Conferences **79,** 01085 (2016)
[7] T. R. Muksunov, N. K. Maksimova, E. Yu. Sevast'yanov, S. É. Shipilov, V. P. Yakubov, Russian Physics Journal **57**, 1287 (2015)

AUTOMATION OF THE DESIGN OF THE ANCHORAGE SYSTEM TAKING INTO ACCOUNT THE GEOMECHANICAL STATE OF THE MASSIF AND MINING DEVELOPMENT SCHEMES

Vladimir Demin, *Alexander* Tomilov[1], *Bakhyt* Sultanova

Karaganda State Technical University, 100027, Karaganda, Kazakhstan

Abstract. The article presents the system for the automation of the design of the anchorage, which regulates the calculation of the required parameters of the fasteners for the fastening of the fastening system. The main factors affecting the operation of the anchor support are grouped in the following way: mining and geological conditions, technical characteristics of the anchor support, geomechanical conditions for conducting and operating the mine workings. Mining and geological conditions for carrying out excavations include: physical and mechanical properties of rocks, the category of roof stability, fracturing, etc. Technical characteristics of the anchor support: material of the rod, filler, filling completeness, etc. Conditions (geomechanical) of carrying out and exploitation of the mine workings: the depth of the conduct, the location relative to the zone of influence of the cleaning works, the location relative to the waste zone, etc. As a result of calculations the program gives out the basic parameters of the anchor support, which coincide with the parameters adopted by the passport.

1 Introduction

In increasing the productivity of mining production, it is very important to solve the problem of improving the methods and means of fastening and maintaining the mine workings sustainably. The cost of excavation workings is quite large and amounts to 15–22% of the prime cost of coal mining.

High costs also require qualitative maintenance of preparatory workings, both for their repair before and after commissioning of cleaning faces that reach 15–22% of the cost of carrying out mine workings [1].

At existing depths of development of 600–851 m of coal basins with modern supports it is very difficult to achieve maintenance-free workings.

One of the effective ways to improve the state of excavations and save material resources is the use of a combined anchor-frame support.

[1] Corresponding author: tom44487@mail.ru

To date, the world practice has accumulated sufficient experience in applying the technology of erecting anchorage in mine workings. However, studies show that the condition of mine workings, especially excavations, is not always satisfactory.

In the solution of these issues, the weakest link is insufficient automation of the process of calculating optimal technological schemes for conducting mine workings with anchor and combined supports under different mining and geological and mining conditions of their development.

The content of the research is as follows:

1) In optimizing the mathematical apparatus by calculation:

- Combined support (anchors with frames).
- The length and density of anchoring rope anchors in the anchoring and combined fastening of mine workings.
- Anchoring of the mating of mine workings, taking into account the various conditions of the mine workings: work-up, work-in-hand, reference rock pressure, increased rock pressure, lava influence, the impact of close workings, preservation for reuse behind the bottom line.

2) In the development and creation of a computer system with:

- The possibility of taking into account the emerging strained-deformed state of the rock massif, permissible limiting deformation of the roof rocks.
- Possibility of using two-and three-level fastening, mixed mounting of mine workings.
- A modular organization of the system, the logical model of which is represented by the subsystem of the user interface, the subsystem for the formation of the information model, the subsystem of mathematical calculations, the drawing and formation of the results of calculations of the anchorage system and the output subsystem.

Improvement of the method for calculating the anchor support parameters, in terms of its automation, allows to significantly reduce the time for designing anchorage systems and increase the number of options considered, as well as to exclude the "human" factor in the calculation of technological fastening schemes for mine workings.

2 Methods of solution

The aim of the study is to increase the efficiency of the work carried out in coal mines, by reducing the time for calculating the optimal technological schemes for anchoring the anchor and combined supports for various mining and geological and mining-technological conditions for mining the mine workings of coal mines.

To achieve this goal, the following tasks were accomplished:

1) The current state of problems in the use of anchorage of mine workings in the world and in the Karaganda coal basin, such as:

- Analysis of factors affecting the effectiveness of the use of anchor support in mine workings.
- Study of stability of contours of preparatory excavations taking into account their strained-deformed state depending on mining and geological factors.
- Study of the stress-strain state of excavation workings depending on the influence of technological factors.

2) The mathematical apparatus for calculating anchor and combined technological schemes for managing the state of the array and improving the quality of anchored rocks adaptable to changing mining technological conditions is optimized:

- Anchor support.
- Combined support (anchor + frame).

- The length and density of anchoring rope anchors in the anchoring and combined fastening of mine workings.

Anchoring of mating connections of mine workings, taking into account:

- Different conditions for carrying out mine workings: work-up, work-in-progress, bearing rock pressure, increased rock pressure, lava influence, the impact of close workings, conservation for reuse behind the bottom line.
- Permissible lowering of the contours of the development, the arising strained-deformed state of the rock massif, permissible limiting deformation of the roof rocks, the possibility of two- and three-level fastening.

3) An automated system for designing technological schemes for the development of rocks for various mining and geological and mining conditions has been developed, which regulates the calculation of the required parameters of fastening passports, the construction of a fastening system, and the validation of design:

- Digital models for calculation of anchor and combined supports taking into account various mining-geological and mining-technical conditions for conducting mine workings.
- Model and information support of the automated system for designing technological schemes for mining workings, with the following capabilities:
- Taking into account the emerging strained-deformed state of the rock massif, permissible limiting deformation of the roof rocks.
- Use of two- and three-level fastening, mixed mounting of mine workings.

The architecture of the computer system for the automation of the design of the anchorage system, represented by a multi-user database and a modular structure of program modules, its logical model consists of a subsystem of the user interface; subsystems of mathematical calculations; subsystems of tracing and the subsystem of formation of results of calculating anchor support parameters. And also the module for the formation of technological passports of anchorage of excavations, taking into account mining and geological and mining conditions of operation.

3 The choice of influencing factors and the structure of the initial data

In reality, an array of rocks is a very complex environment, formed under the influence of all sorts of geological factors. Mechanical properties of rocks in the array vary among themselves in a wide range. The property of the processes occurring in the mountain mass around the mine workings, including the coal mine, apart from the mechanical properties of the rocks, is significantly affected by their stress state caused by man-made processes in mining operations.

Since it is very difficult to take into account the diversity of the properties of the rock massif, various mechanical models are used to study it. They reflect only the main essential properties of rocks and discard all other factors. The main properties include strength and deformation characteristics (elasticity, plasticity, brittleness), continuity, isotropy and homogeneity of the body under study.

In theoretical studies, the following basic types of models are used to study the interaction of rock massif with the support of rock mining: elastic, rigid-plastic, elastoplastic, elastoplastic non-uniform, elastic-viscous, viscoplastic.

The effectiveness of the use of anchor support is greatly affected by the correctness of accounting for all influencing factors. But to take into account all the factors for determining the parameters of the anchor support is almost impossible. Therefore, the main factors that have the greatest influence are used in the calculations.

The peculiarity of the anchor support is that, unlike the metal frame support, the anchor support from the moment of its installation works with the array as a single system. Bearing capacity of the array of installed anchors increase due to the cross-linking of rock layers between each other. That is why, the main load from the rocks located above the mine workings will be an array, fixed with an anchor support.

Group, the main factors that determine the work of the anchor support can be as follows [2]:

- Mining and geological conditions for conducting mine workings.
- Technical parameters of the anchor support.
- Geomechanical conditions of mining operation and its operation.

Mining-geological conditions for conducting mine workings can be grouped as follows:

- Physical and mechanical properties of rocks.
- Categories of stability of the roof of mine workings.
- Fracture of rocks.

Technical properties of the anchor support include:

- Anchor rod material.
- Filler.
- Completeness of filling, etc.

Geomechanical conditions for carrying out and operating the mine workings can be represented by the following set of properties:

- The depth of the mine workings.
- The location of the mine in relation to the zone of influence of the cleaning works.
- The location of the development with respect to the worked-out zone.

It is very difficult to take into account the last group of factors, because these factors influence the stability of mine workings. Therefore, depending on the geological and geo mechanical conditions, we will highlight the following most important types of calculations for the mine workings:

- Mine workings and mattings, which are carried out and maintained outside the zone of influence of clearing works.
- For main mine workings and mattings supported in the reference pressure zone with a width of> 0.05 m and <15 m.
- For excavation workings extinguished behind lava.
- For excavation workings, protected entirely with a width> 15 m and extinguished after adjacent lava.
- For excavation workings, protected entirely by a width of 2.5–3.0 m of the thickness of the formation, and then extinguished after adjacent lava.
- Mine workings, conducted in a cut-off to the worked out space.
- Mounting chambers.

The method of calculating the anchor is as follows.

Depending on the initial data for specific mining and geological and mining conditions for calculating the anchor support as a criterion for the intensity of the rock pressure, we take the calculated displacements of the roof rocks with the anchor support.

To determine the parameters of the anchorage and the choice of reinforcement, depending on the calculated displacements, we distinguish the following intervals of intensity of the manifestation of rock pressure:

- At estimated offsets of roof rocks <50 mm: small rock pressure.
- At estimated offsets of roof rocks> 50 and <200 mm: mountain pressure of medium intensity.
- At estimated offsets of roof rocks> 200 mm: intensive rock pressure.

In the case of a small rock pressure at the calculated displacements of the roof rocks <50 mm, the resistance of the anchor support and the length of the anchors installed in the bottom hole are taken from Table 1 [3], depending on the class of roof rocks.

Table 1. Stability classes of rocks.

Grade I Rocks of the roof are unstable	Grade II Roofs of medium stability	Grade III Roof resilient rocks
The roof rock collapses after exposure at a distance from the face to 1 m	The stability of the roof is preserved when the roof is exposed at a distance from the face of 1 to 3 m	The stability of the roof is preserved when exposed at a distance of more than 3 m from the face
Thin-layered and fissured mudstones predominate	Layered low-cracked siltstones and sandstones prevail	Non-inter bedded sandstones, marls predominate
Average design compressive strength of rocks (Rc) ≤ 30 MPa	30 <Rc ≤ 80 MPa	Rc> 80 MPa

For the II class of the roof rocks, the resistance of the anchor support is taken as for Class I, and the length of the anchors can be reduced by 21%.

For roofing types I and III classes, the reduction in the length of anchors is possible by 10% or 21%, but it is necessary to increase the density of the anchorage installation or to apply anchors of increased load capacity, and the design resistance of the anchor support must be increased by 30 and 50%.

Methods of strengthening the anchor support and calculating its characteristics of the anchor support of the mine workings are carried out in accordance with the value of the rock pressure. In turn, the rock pressure is determined taking into account the size and depth of the mine workings and the interface from the surface, the method and the parameters of their protection from the impact of cleaning operations, strength, fracturing and rock stability in the roof.

In conditions where the calculated displacements exceed 300 mm, we calculate the parameters of the anchor support of the mine workings from the conditions for ensuring maximum roof displacements not> 300 mm, while maintaining the workings on the boundary with the worked out space, conducting them in the crocheting and protecting the whole.

Depending on the behavior of the rock pressure, the installation density of the anchors in the roof is checked and, if necessary, increased from the condition of the required minimum density of anchorage installation depending on the stability of the immediate roof:

- For an unstable roof not less than 1 anchor per 1 m^2.
- For a medium-stable roof not less than 0.7 anchor per 1 m^2.
- For a stable roof no less than 0.5 anchor per 1 m^2.

At the same time, in order to avoid overloading, in all workings and mattings at design roof offsets greater than 150 mm, the anchors and reinforcing supporting supports must be compliant or with damped compliant elements crumpled under loads reaching 0.8 from the bearing capacity of the anchor.

For anchors with partial fastening along the length of the well, the compliance value of which should be:

- From 40 mm at displacements of 151 mm.
- Up to 70 mm at offsets of 301 mm.
- Fixed over the entire length – from 30 to 51 mm.

In the workings and mattings, the anchors must have a compliance not <50 mm and with displacements not> 50 mm, in the roof of which lie a coal seam and rocks that are dangerous to mountain impacts and emissions.

The main indicators affecting the parameters of the anchor support are: the calculated displacements and the expected loads on the anchor support.

Figure 1–4 shows the user interface of the program. The interface of the computer program is created in the Russian and Kazakh languages. At Figures 1–4 the version of the interface is presented in Russian (the decimal fraction is presented through a comma).

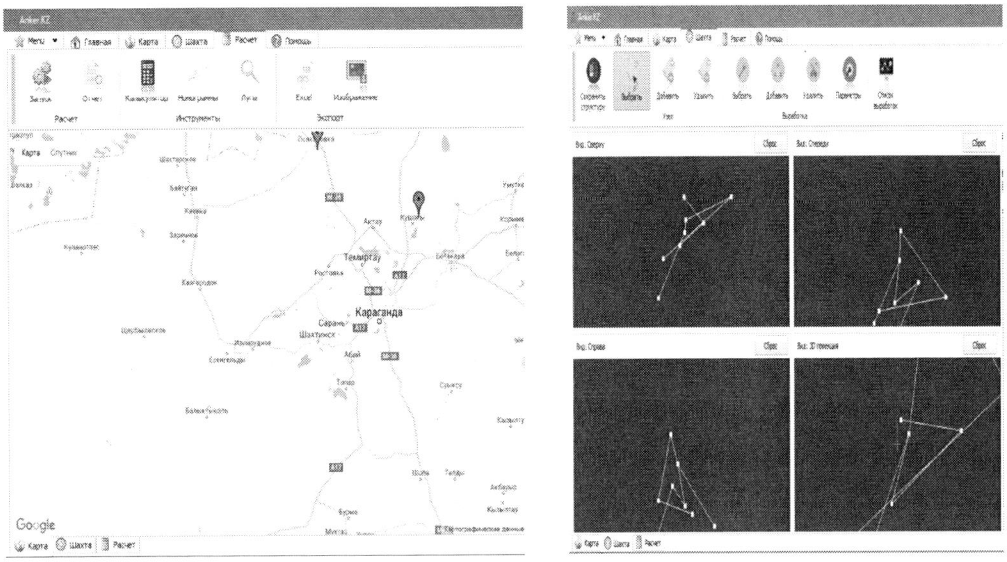

Fig. 1. Choice of mine and mine workings.

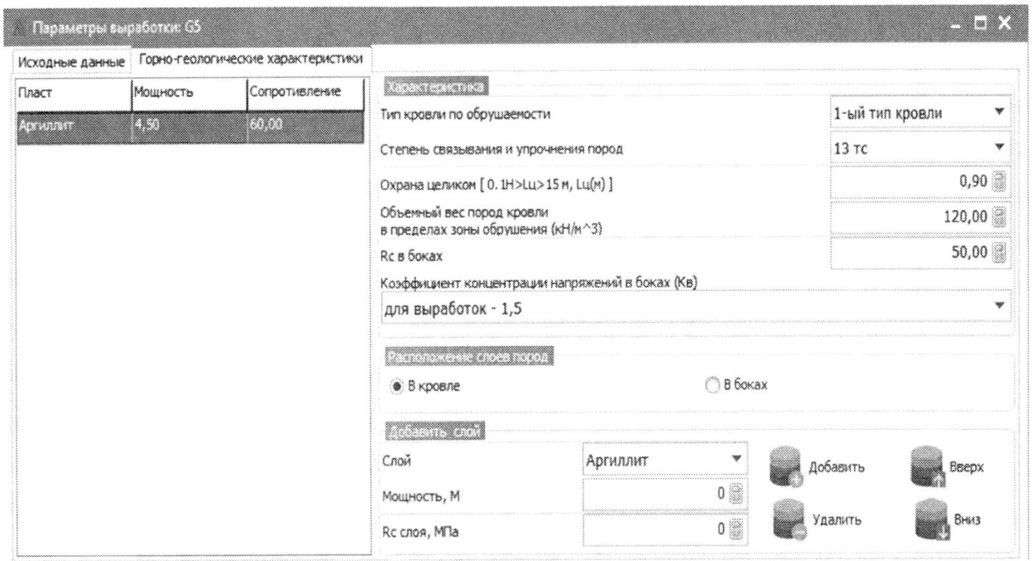

Fig. 2. Characteristics of the mine workings.

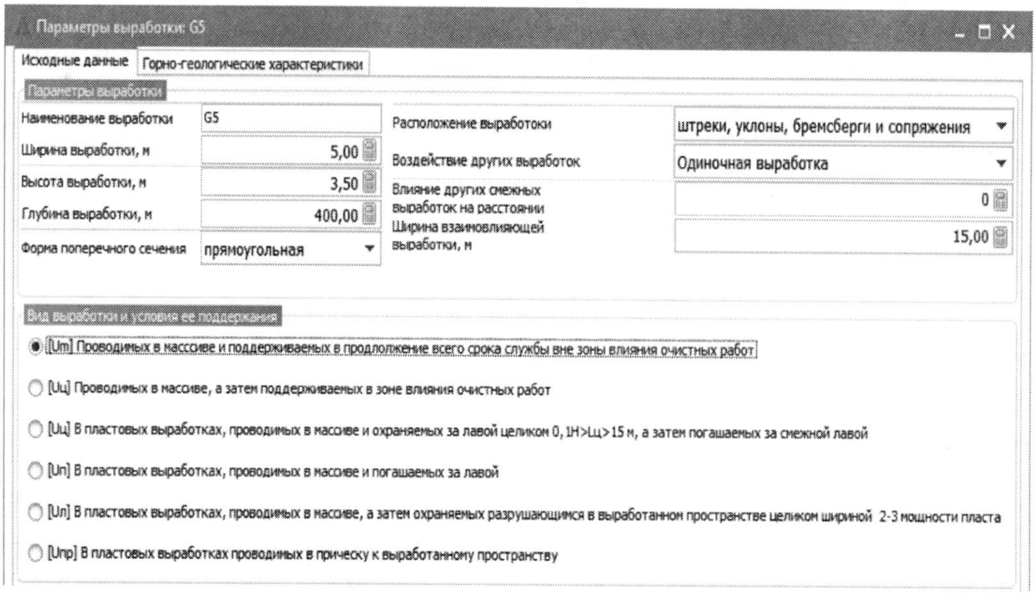

Fig. 3. Mining and geological characteristics.

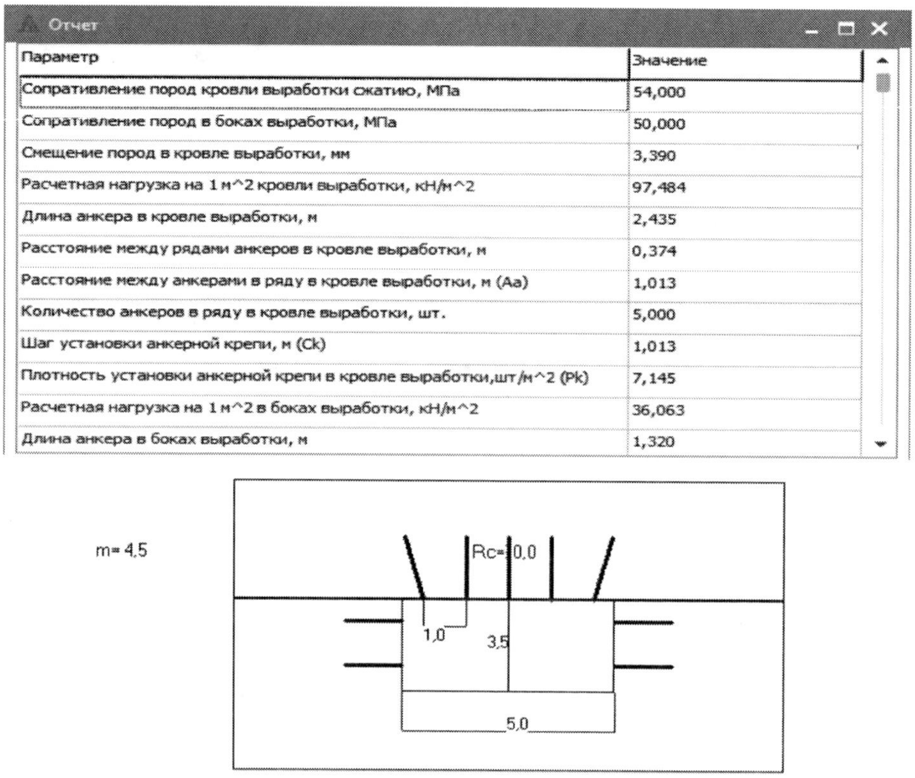

Отчет

Параметр	Значение
Сопротивление пород кровли выработки сжатию, МПа	54,000
Сопротивление пород в боках выработки, МПа	50,000
Смещение пород в кровле выработки, мм	3,390
Расчетная нагрузка на 1 м^2 кровли выработки, кН/м^2	97,484
Длина анкера в кровле выработки, м	2,435
Расстояние между рядами анкеров в кровле выработки, м	0,374
Расстояние между анкерами в ряду в кровле выработки, м (Aa)	1,013
Количество анкеров в ряду в кровле выработки, шт.	5,000
Шаг установки анкерной крепи, м (Ck)	1,013
Плотность установки анкерной крепи в кровле выработки,шт/м^2 (Pk)	7,145
Расчетная нагрузка на 1 м^2 в боках выработки, кН/м^2	36,063
Длина анкера в боках выработки, м	1,320

Fig. 4. Calculation result.

The following indicators are used as input data:

- Resistance of rocks to compression of the workings in the sides and roof of the mine to a height, Mpa.
- Volume weight of rocks, kN/m^3.

- Coefficient of structural weakening.
- Thickness of the layers, m.
- Working height and width, m.
- Shape of the section and depth of the deposit, m.
- Type of generation.
- The influence of other excavations.
- Width of the lid, m.
- Method of calculation.
- Bearing capacity of the anchor support, kN.
- The length of the protruding part of the anchor, m.
- Depth of penetration, m.
- Bearing capacity of rope anchors, kN.
- The degree of fracture of rocks.
- Type of roofing.
- The degree of collapse of the roof rocks.

4 Conclusion

In order to establish the correctness of calculation of the anchor support parameters, using the developed program "Anker.KZ", the fastenings for the following developments of the UD mines of JSC "ArcelorMittal Temirtau" (Kazakhstan) were calculated:

1) Mounting chamber 231d6 – in the mine "Shakhtinskaya".
2) Ventilation Bremsberg 4.03d6-1-Z mine them. Lenin.
3) Ventilation drift 41k12-z mine them. Kuzembaev.
4) Drainage drift 32k11-s mine "Abayskaya".
5) The second western ventilation slope K7 mine them. Kostenko.

The thickness of the reservoir is d6 – 5.4 m (fortress $f = 1$). In the immediate top of the stratum there are argillites with a thickness of 0.44 m (strength $f = 2.5$), with a capacity of 1.0 m (strength f = 3.2). Above there are siltstones 4.11 m thick (fortress $f = 4.51$).

The shape of the cross section of the mine is rectangular. The working height is 3400 mm, the working width is 5651 mm. The depth of work is 471 m.

According to the passport calculated by the technical department of the Shakhtinskaya mine, the length of the anchors is assumed to be 2.3 m, the number of anchors in the row in the roof 7 in the sides 2. The distance between the anchors in the row is 0.9 m, between the rows of anchors is 0.74 m.

As a result of calculations using the program "Anker.KZ" received the following values:

- Number of anchors in the row – 7 anchors.
- The distance between the anchors in the row is 0.85 m.
- The length of the anchor in the roof of the mine is 2.54 m.
- The distance between the rows of anchors in the roof of the mine is 0.84 m.
- The number of anchors in the row in the roof of the mine – 7 anchors.
- The length of the anchor in the sides of the mine is 1.32 m.

Comparison of the calculation results showed that the distances between the anchors calculated by the two methods are in satisfactory agreement. The length of the anchor calculated with the help of the program "Anker.KZ" is longer than the length indicated in the passport by 0.15 m.

Analysis of the geological section along the production route showed that, at a length of 2.4 m, the anchor is located in the argillite layer with a strength of $f = 2.5$.

At a length of 2.56 m, the upper part of the anchor is located in the layer of argillite with a strength of f = 3.3, which increases the efficiency of the anchor support. The number of anchors in the series, calculated by both ways is the same.

Ventilation Bremsberg 4.04d6-1-z is carried out on the upper layer of layer d6. The immediate roof of the reservoir in the area of production is represented by easily collapsible mudstones with a thickness of 0.3 to 7 m, and a strength f = 2.5. In the main roof of the layer, the siltstone is strong, difficult to fall, f = 4.

The shape of the section is arched. Type of support – combined. The working height is 3735 mm, the working width is 5570 mm. The depth of the work is 650 m.

According to the passport of fastening and carrying out for fixing the working, a combined support is adopted with the following parameters:

- Length of anchors – 2.4 m.
- The distance between the anchors in the row is 0.81 m.
- Distance between the rows of anchors – 0.74 m.
- The number of anchors in the roof – 7.7 anchors.

As a result of calculations using the "Anker.KZ" program, the following was received:

- The length of the anchor in the roof of the mine is 2.38 m.
- The distance between the rows of anchors in the roof of the mine is 0.81 m.
- The distance between the anchors in the row in the roof of the output is 0.75 m.
- The number of anchors in the row in the roof of the mine is 7.
- The length of the anchor in the sides of the mine is 1.31 m.

Analysis of calculation results showed that the distances calculated by the program are slightly higher than the corresponding values of the passport. The number of anchors in the row is the same as the number indicated in the passport.

Based on the results of testing the computer program, the following conclusions can be drawn:

1) The program has a convenient intuitive interface that allows you to easily navigate when entering the source data.

2) The program takes into account the main conditions for carrying out mine workings.

3) The computer program allows you to reduce the time for calculating the parameters of the anchor support. On average, the calculation takes about 2–3 minutes.

5) As a result of the calculations, the program issues all the basic parameters of the anchor support.

6) The results of calculations satisfactorily coincide with the parameters adopted by the passport.

7) During the testing of the program failures, no critical errors were observed.

8) The ability to enter only one type of rock in the sides of the mine, which reduces the reliability of the results obtained when calculating the lining workings carried out by the mixed face.

References

[1] V. Demin, T. Isabek, V. Melnik, *Controlling the stability of mine workings in difficult mining conditions* (KSTU, Karaganda, 2016)

[2] V. Zhurov, Perfection of a technique of calculation of parameters of fastening of developments with allowance for mining technological factors, Thesis of the candidate of technical sciences (2010)

[3] Instruction on the calculation and use of anchor support in the mines of the Karaganda basin, (2008)

PLANE SOURCE OF CONCENTRATED MAGNETIC FIELD BASED ON SQUARE COILS

Sergey Roslyakov[1], *Vladimir* Yakubov, *Ksenia* Zavyalova

Radiophysics Faculty, Tomsk State University, Tomsk, Russia

Abstract. You can see the method of the synthesis of planar sources of concentrated magnetic field. Method defines the current distribution in the plane, which is required to produce a concentrated magnetic field at a distance from the source. Carried out a numerical calculation of the current distribution to create a point field localization. It is shown that the field concentration is achieved by minimizing the level of low spatial frequencies due to antiphase currents in concentric circular coils. The resulting field distribution is extremely sensitive to the slightest changes in the current distribution, which creates difficulties for the practical implementation of the system.

1 Introduction

Variable magnetic fields have a high potential for application in the area of non-destructive quality control and inspection systems [1–8]. The disadvantages of this type of introscopy may include: the detection only of electrically conductive objects, the rapid decrease of the field because of the distance, a low resolution. These methods are used to diagnose irregularities in ferromagnetic and electrically conductive objects, as well as for detecting electrically conductive objects behind the dielectric barriers. Weak alternating magnetic fields penetrate through metal barriers and can detect objects behind them [1–8]. For the non-destructive testing of ferromagnetic materials is possible to use permanent magnetic fields. To increase the dynamic range of magnetic induction methods are applied differential magnetic field sensors [2, 4]. There are various combinations of sensors based on differential inductors. In the differential magnetic induction sensor coil sources are located with the receiving coil in a way that does not create induced currents in the receiving coil. For example, two source coils located near each other can be used, but only if source coils energized antiphase currents. The receiving coil which is placed between them cannot induce current, while a metal object appears in the medium and causing a violation of the balance between the fields of the source coils. For magnetic induction defectoscopy more preferable to use induction coils instead of the Hall sensors [3], because the coil allows obtaining higher level of signal-to-noise ratio. The transmission methods of magnetic induction tomography are being developing [8]. These methods are based on the analysis of amplitude and phase distortion of magnetic force lines which passing through the test electrically conductive medium. One of the major problems of the magnetic

[1] Corresponding author: serega_1994_0@mail.ru

tomography is a nonlocalizability of magnetic field at a distance from the source. That leads to the deterioration of resolution of reconstituted tomographic images.

This work aims to create a concentrated source of the magnetic field, which will solve the problem of low resolution magnetic induction introscopic systems.

2 System overview

It is proposed to consider the scheme of measurements shown in Figure1. Let's consider the plane current distribution which creates a magnetic field in space. It is necessary to select such current distribution that will provide a concentration of the magnetic field in the horizontal plane at distance h and around a single point. Let us assume that all the currents that exist on the plane XOY are vortex. Consequently, they can be represented as a sum of a plurality of elementary eddy currents of different amplitude. Accordingly, the field generated by the induction currents can be represented as the sum of elementary eddy current fields. Mostly we are interested in the z-component which is component of the magnetic induction vector field B_z .

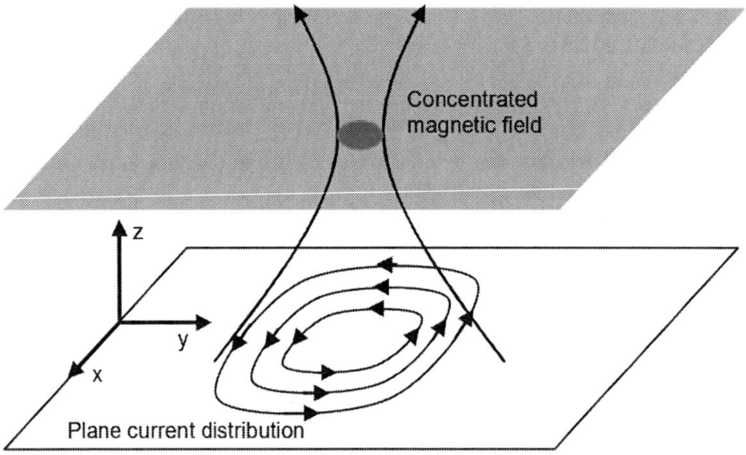

Fig.1. Measurement scheme.

Let us write the z-components of the magnetic induction vector field in the plane $z = h$ as an integral convolution:

$$B_z(x, y, h) = \iint_{XOY} p(x', y')S(x - x', y - y', h)dx'dy', \qquad (1)$$

where $S(x, y, h)$ - is the z-component of the magnetic induction field on h range and is created by elementary eddy current; $p(x', y')$ – distribution of eddy currents amplitude on the plane XOY. The field generated by the eddy current in the basic range of 10 cm is represented on Figure 2. It can be seen that the field is localized in the area of a diameter about 10 cm. It can be seen that the field is localized to the area of a diameter of about 10 cm. In fact, the resolution of magnetic induction sensing systems will deteriorate proportional to the distance.

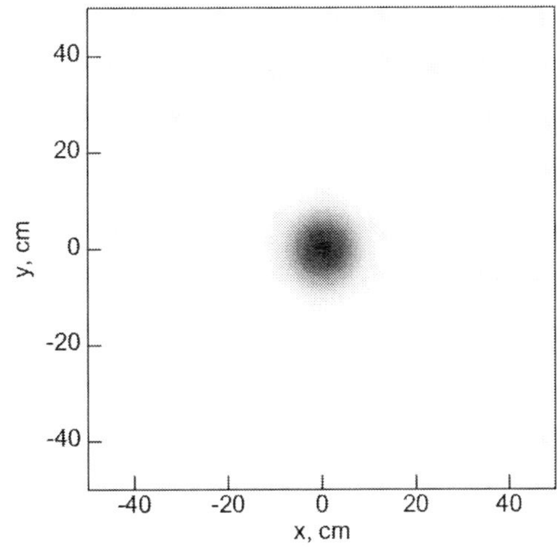

Fig.2. B_z component of magnetic induction field of elementary eddy current.

3 Solving the inverse problem

It is necessary to choose such current distribution that provides the best concentration of the magnetic field at a distance h from the plane XOY. To solve this problem suggested submitting the required field distribution in the form of a delta function, and the current distribution can be found by using the operation of inverse convolution with regularization. The spatial spectrum of the current distribution in the plane is calculated from the spatial spectrum according to the expression:

$$\tilde{p}(k_x,k_y)=\frac{\tilde{S}^*(k_x,k_y)}{\left|\tilde{S}(k_x,k_y)\right|^2+\alpha}, \qquad (2)$$

где $\tilde{p}(k_x,k_y)=\frac{1}{4\pi^2}\int\limits_{-\infty}^{\infty}\int\limits_{-\infty}^{\infty}p(x,y)\exp(-ixk_x-iyk_y)dxdy$;

$\tilde{S}(k_x,k_y)=\frac{1}{4\pi^2}\int\limits_{-\infty}^{\infty}\int\limits_{-\infty}^{\infty}S(x,y)\exp(-ixk_x-iyk_y)dxdy$; α – parameter of regularization that is chosen to be sufficiently small to achieve a concentration of the field, but large enough to avoid the division into small quantities and cause an increase in the level of artifacts.

The seeking distribution of elementary eddy current is calculated using the Fourier transform of the calculated spatial spectrum:

$$p(x,y)=\int\limits_{-\infty}^{\infty}\int\limits_{-\infty}^{\infty}\tilde{p}(k_x,k_y)\exp(ixk_x+iyk_y)dxdy. \qquad (3)$$

By means of formula (3) we have calculated the current distribution on the plane (Figure 3). In this case the current distribution is the convolution integral of $p(x,y)$ with elementary vortex function.

It can be seen that the resulting current distribution is axially symmetrical and looks like a set of ring currents which change direction on different diameters. Thus, currents at different diameters compensate each other field at low spatial frequencies. Due to the balance which was achieved there are only high spatial frequencies. This allows concentrating the field around a center point.

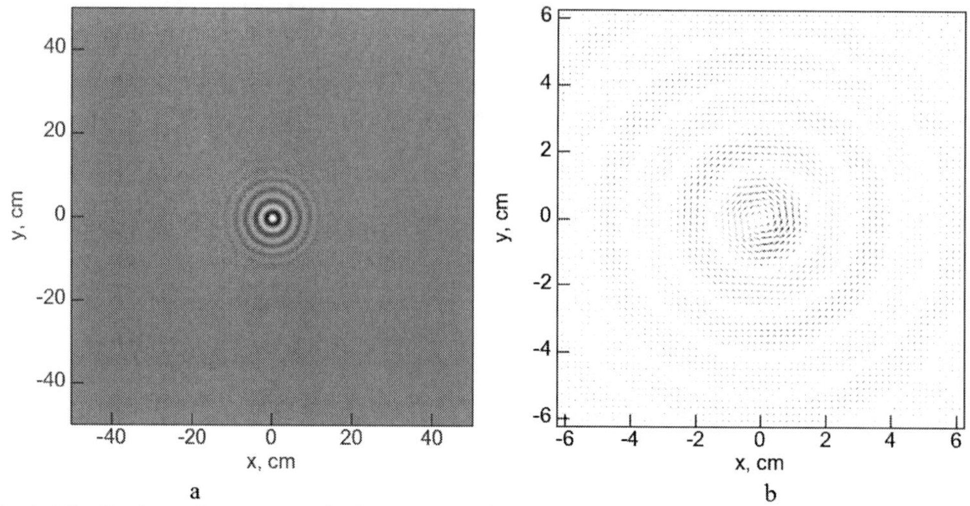

Fig.3. Distribution of current to obtain concentrated magnetic field: a – distribution of elementary eddy currents; b – distribution of place current density vector.

The Figure 4 shows the resulting field distribution which was calculated by using the formula (1) and (3). It can be seen that the field concentration reached 7 times stronger than the field concentration of elementary eddy current.

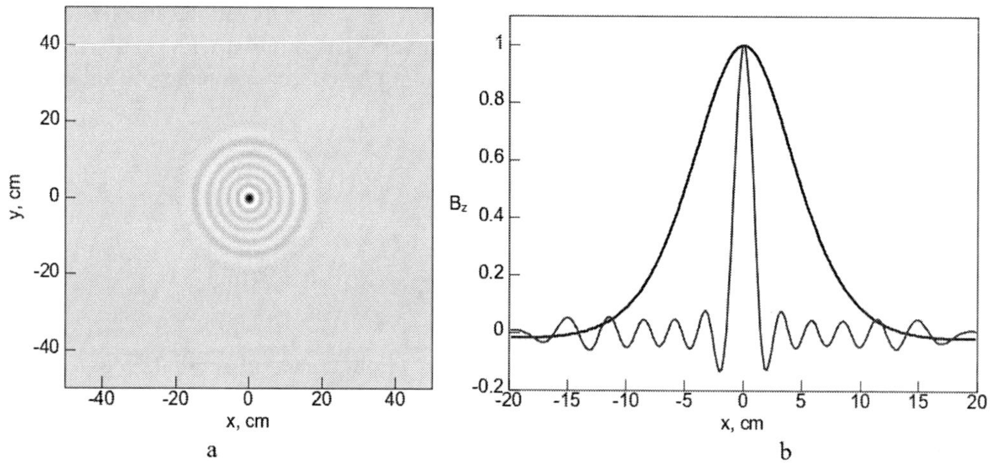

Fig.4. Distribution of z-component of magnetic induction field at a distance of 10 cm created by special current distribution: a – plane image of field; b – section of field (green) along x axis compared with the field of elementary eddy current (blue).

Should be noted that any distortion in the current distribution cause the disturbance of the balance and a sharp increase in low spatial frequencies in the magnetic field. This significantly impairs the concentration. This creates difficulties for the practical creation of such a power supply. It is known that if the wave field was focused that occurs multiple amplification of the field at the point of in-phase addition of the waves and at the concentration field. In this case, the concentration of the magnetic field is achieved through the suppression of low spatial frequencies, which significantly reduces the amplitude of the resulting field. However, there is a way to solve this problem. We can increase the amplitude of the current in the source.

The achieved effect is different from the focus, because it does not increase the amplitude of the field but rather reduces it repeatedly. The numerically calculated current

distribution is extremely difficult to create physically. It is proposed to simplify the problem by considering only 3 square coils located on a plane. For each coil, current values will be selected to ensure the formation of a concentrated field.
We calculated the magnetic field of a system of three coils with a current. Numerical values were obtained for three square coils of different sides: 8 cm, (current 1 A), 11 cm (current - 1.021 A), 14 cm (current 0.306 A). The result of modeling the field at a distance of 10 cm is presented in Figure 5.

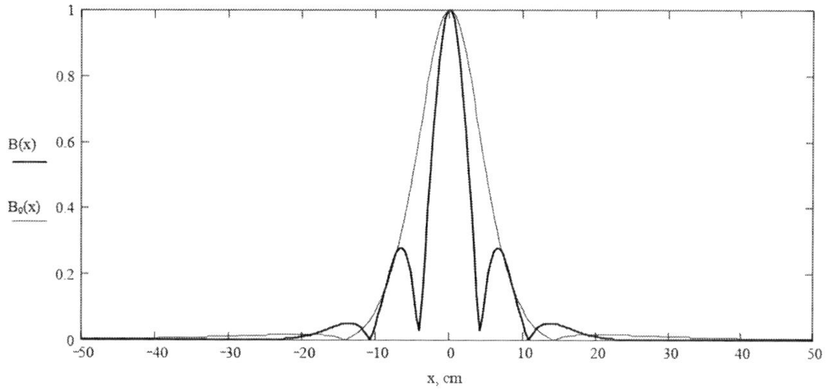

Fig.5. Magnetic field distribution at 10 cm distance.

It can be seen that the field concentration has broadened than for a continuous current distribution. In this case it is possible to reach the field concentration twice as much as for an elementary source. However, a similar configuration of the coils can be made practically.

4 Conclusion

Here is proposed a method of synthesis of current distribution which provides a concentration of the magnetic field at a distance. The concentration is achieved by reducing the level of low spatial frequencies of the field, while maintaining high spatial frequencies. The sources of a similar type field will be used in magnetic induction tomography of high resolution.

Acknowledgment

The research is supported by Russian Science Foundation project No 16-19-10272.

References

[1] A.A. Kolyshkin, R. Vaillancourt, Journal of Applied Physics. **77**, 4909 (1995).
[2] S. Li, S. Huang, W. Zhao, Sensors and Actuators A. **135**, 675 (2007)
[3] S. Li, S. Huang, W. Zhao, P. Yu. Sensors and Actuators A. **141,** 13 (2008)
[4] X. Peng, H. Songling, Z. Wei, NDT&E International. **44**, 339 (2011)
[5] R.J. Ditchburn, S.K. Burke, NDT&E International. **38**, 690 (2005)
[6] J.O. Fava, M.C. Ruch, NDT&E International **39**, 414 (2006)
[7] T.P. Theodoulidis, E.E. Kriezis, NDT&E International **35**, 407 (2002)
[8] A. Korjenevsky, V. Cherepenin, S. Sapetsky, Physiol. Meas. **21(1)**, 89 (2000)

RADIO-WAVE TOMOGRAPHY OF INHOMOGENEITIES IN BIOLOGICAL MEDIA WITH MULTI-FREQUENCY SOUNDING IN THE RANGE 2-8 GHZ

Sergey Shipilov[1,1] *Aleksandr* Eremeev[1], *Evgeny* Balzovsky[1], *Elena* Yurchenko[2]

[1]National Research Tomsk State University, the Department of Radiophysics, 634050 Tomsk, Russia
[2]National Research Tomsk Polytechnic University, 634050, Tomsk, Russia

Abstract. In this paper, a method for detecting and mapping inhomogeneities in biological tissues using the radio-wave tomosynthesis method is presented. The proposed method of radio-wave tomosynthesis allows us to calculate the three-dimensional distribution of the permittivity of the space under study and, thereby, to detect tissue inhomogeneities and to determine their location and size. Due to their harmlessness for humans, these methods are suitable for dynamic observation of changes in the size of formation, in contrast to x-ray methods, for which regular doses of ionizing radiation are contraindicated. Therefore, the development of non-invasive methods for the search for inhomogeneities in biological media based on radio-wave sounding, which makes it possible to identify pathological formations, is now very relevant.

1 Introduction

To date, due to the development of information sciences and computer technology, it has become possible to use the methods of radio-wave tomography as the means of remote non-destructive testing and diagnostics of the internal structure, semitransparent media for radio emission and restoration of the form of opaque objects. There is a great need for contactless and automated medical diagnostics. The best means for such tasks is radio wave tomography.

At present, ultrasonic scanners and x-ray tomographs are used to detect tumors in biological tissues, but it is proposed to use a radio-wave tomograph for earlier and safe control of the appearance of neoplasms. In pathological and physiological processes in biological tissues, as a rule, there are changes in their electrophysical properties, and measurement task of which is of considerable interest in various medical applications. This is important for the detection of cancer. When neoplasms appear, they accumulate liquid and blood, that leads to a change in the permittivity of the given formation [1, 2]. A change in the dielectric constant can be detected using methods of radio wave tomography [3, 6–8]. Due to their harmlessness to humans, these methods are suitable for constant monitoring of changes in the size of formation, in contrast to x-ray methods, for which regular doses of ionizing radiation are

[1] Corresponding author: s.shipilov@gmail.com

contraindicated. Therefore, the development of non-invasive methods for the search for inhomogeneities in biological media based on radio-wave sounding, which makes it possible to identify pathological formations, is now very relevant.

2 Measurement of the permittivity of a biological medium

In this paper, to solve the problem of tomography of inhomogeneities in biological medium is used the known method of radio-wave tomosynthesis [3]. For its application it is necessary to know the permittivity of the medium. With the help of methods and a coaxial measuring cell developed at the department of radiophysics of Tomsk State University, it is possible to calculate the dielectric constant of a biological material as measured in a wide frequency range.

Figure 1 shows a coaxial measuring cell.

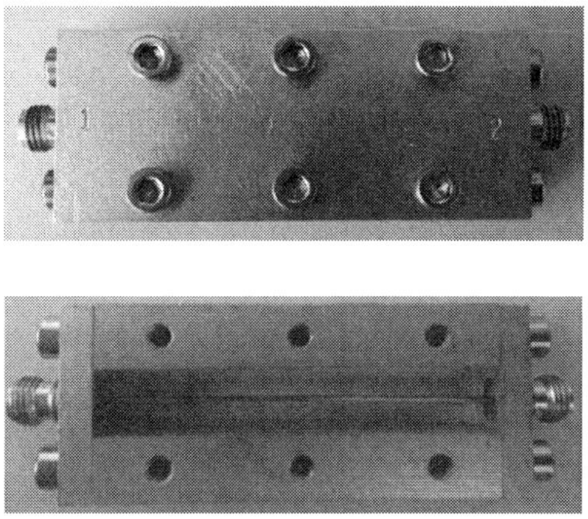

Fig. 1. Coaxial measuring cell.

To measure the complex dielectric constant in the 10 MHz-8 GHz frequency range, a coaxial line measurement method was used which was successfully used to measure the dielectric properties of liquids and bulk materials [4, 5]. This method is based on measuring the transmission and reflection coefficients of a section of a coaxial line. This method is characterized by a large broadband and comparative simplicity of processing the results of measurements. It makes it possible to measure the complex dielectric permittivity of samples, with both low and high dielectric permittivity and dielectric losses. Figure 2 shows the measurements. The investigated medium was placed in the cell, which is the section of the coaxial line with the wave resistance of 50 Ohm and the length of 5 cm. Electromagnetic waves of the given frequency were passed through the cell with matter. Pork melted fat was used as a biological medium.

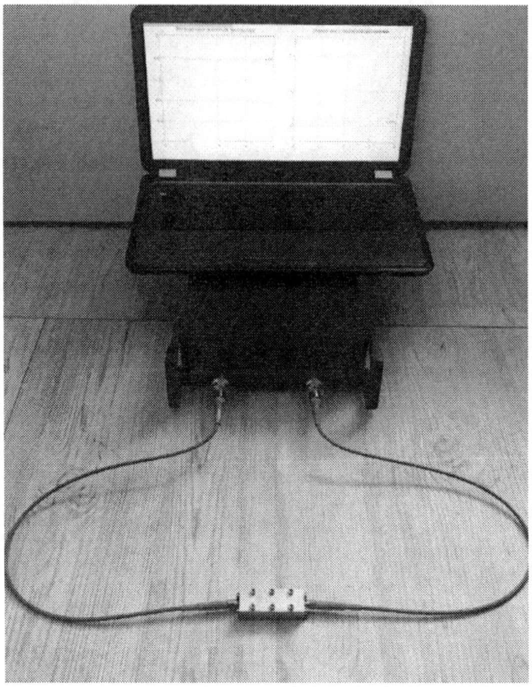

Fig. 2. Measurement of the permittivity of a biological medium.

The components of the scattering matrix of the coaxial measuring cell - complex reflection (parameters S11, S22) and transmission (parameters S12, S21) coefficients – were measured using "Planar" vector of network analyzer C1209.

Figure 3 shows the results of measurements - the dependence of the real and imaginary parts of the refractive index of pork fat on frequency at room temperature.

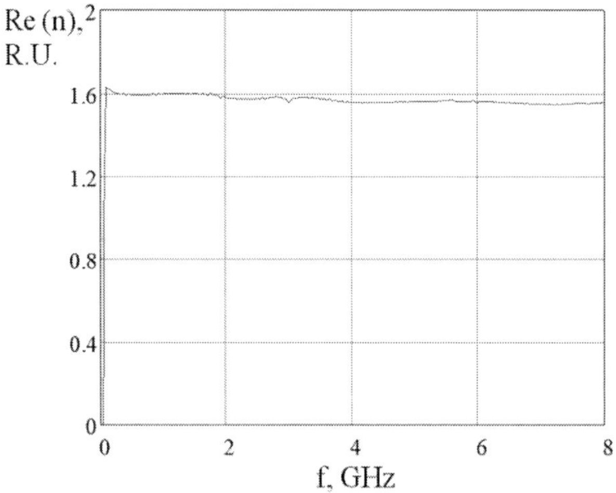

a)

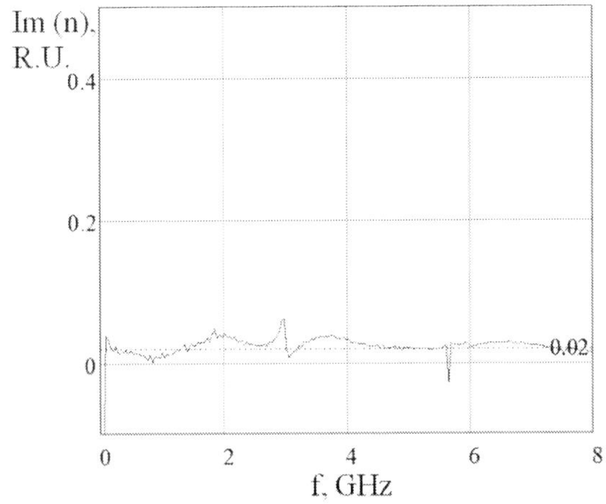

b)

Fig. 3. Dependence of the refractive index of pork fat on frequency: a–real part, b–imaginary part.

A decrease in the real part of the complex dielectric constant with increasing frequency was observed. The results obtained are in good agreement with the results known from the literature.

3 Experiment on the detection of inhomogeneities in the biological medium

The measurements were carried out in the frequency range 2–8 GHz. To carry out these experiments to search for inhomogeneities in a biological medium, a measuring device was used. This device consisted of the vector reflectometer "Caban R140" with ultra-wideband (UWB) antenna (operating frequency range 1.5–8.5 GHz) [9]. Figure 4 shows the scheme and the photo of the experiment.

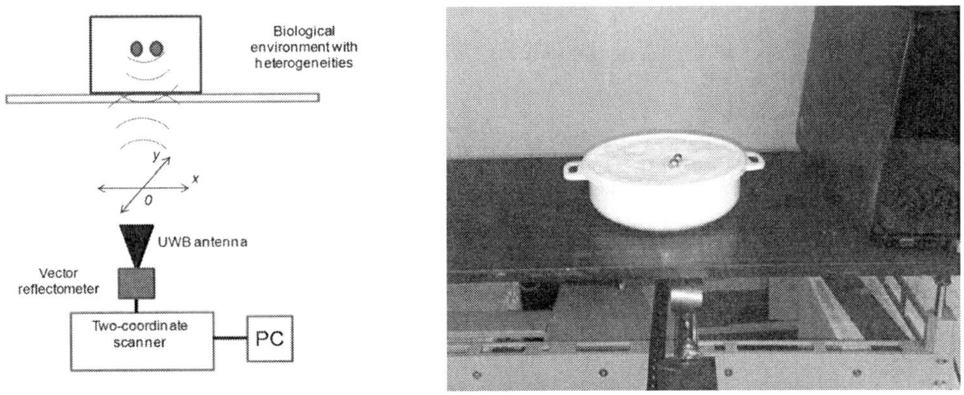

Fig. 4 Scheme of the experiment on the detection of inhomogeneities in the biological medium and its photograph.

The measuring device was fixed on a moving carriage of a two-coordinate scanner for obtaining multi-angle projections of a probable biological medium with heterogeneities hidden in it. Heterogeneities were test objects in the form of metal and plastic balls filled with saline solution. The diameter of the balls was 1.3 cm. The distances between the centers of the balls were 3 cm. Pork fat which is close to the human in its electrodynamic properties was used as a background biological medium.

The images of artificial heterogeneities in the biological medium were obtained after the experiment. Figure 5 shows the results of the experiments.

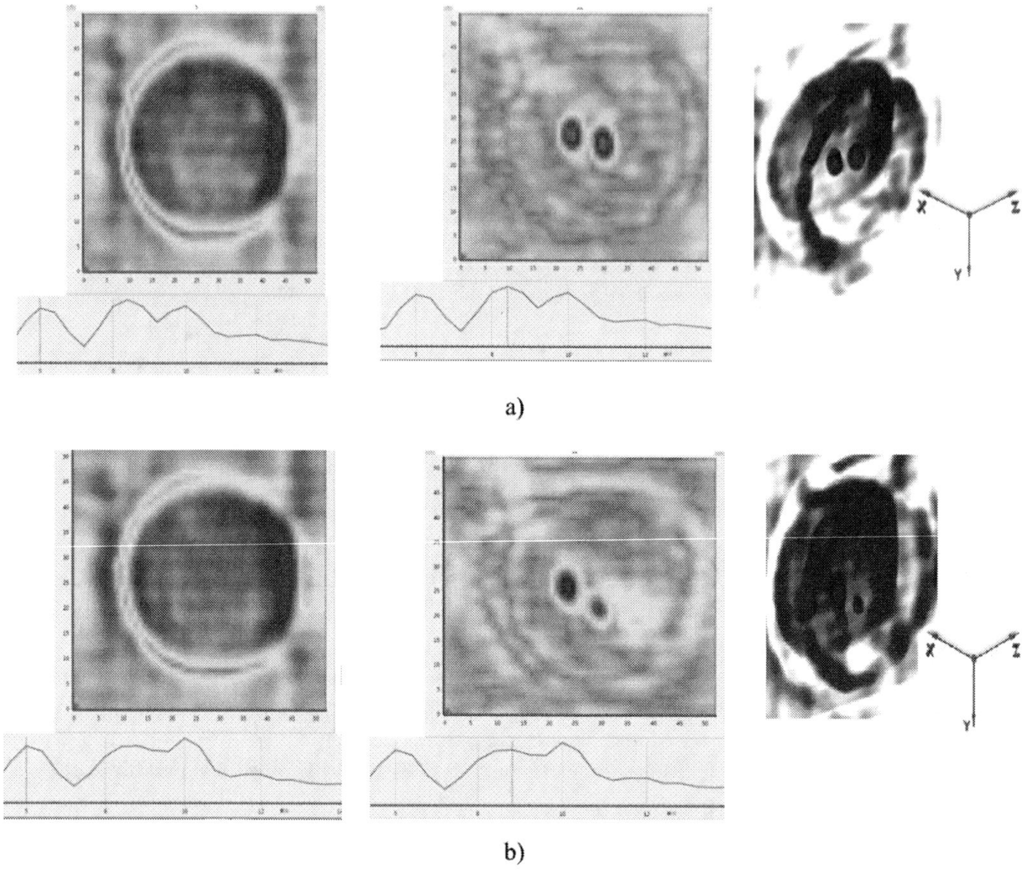

Fig. 5 The result of tomography of the biological environment: a – metal balls, b – plastic balls filled with saline solution.

Thus, the results obtained make it possible to construct a three-dimensional tomogram of the distribution of inhomogeneities in a biological medium.

4 Conclusion

In this paper, the method for detecting and mapping inhomogeneities in biological tissues using the radio-wave tomosynthesis method is presented. The proposed method of radio-wave tomosynthesis makes it possible to calculate the three-dimensional distribution of the permittivity of the space under study and, thereby, to detect tissue inhomogeneities and to determine their location and size.

Based on the results of the experiments, it is possible to estimate the resolving power not worse than 2 cm when using the frequency range of 2–8 GHz for probing.

Acknowledgements

The research was supported by the Russian Ministry of Education and Science as a part of the state order No 3.2068.2017/4.6.

References

[1] M. Lazebnik, D. Popovic, L. McCartney, C.B. Watkins, M.J. Lindstrom, J. Harter, S. Sewall, T. Ogilvie, A. Magliocco, T.M. Breslin, W. Temple, D. Mew, J.H. Booske, M. Okoniewski, S.C. Hagness, Phys. Med. Biol. **52**, 6093 (2007)
[2] M. Lazebnik, E.L. Madsen, G.R. Frank, S.C. Hagness, Phys. Med. Biol. **50**, 4245 (2005)
[3] S.E. Shipilov, R.N. Satarov, V.P. Yakubov, A.V. Yurchenko, O.V. Minin, I.V. Minin, Optical and Quantum Electronics, Optical and Quantum Electronics **49**, 339 (2017)
[4] A.G. Gorriti, E.C. Slob, IEEE TGRS **43**, 2051 (2005)
[5] J. Baker-Jarvis, E.J. Vanzura, W.A. Kissick, IEEE Trans. Microw. Theory Tech. **38**, 1096 (1990)
[6] A.V. Klokov, A.S. Zapasnoy, A.S. Miron'Chev, E.A. Dyukarev, S.V. Smirnov, S.S. Shipilova, IOP J. Phys: Conf. Series **881**, 012013 (2017)
[7] A.V. Klokov, A.S. Zapasnoy, A.S. Miron'Chev, V.P. Yakubov, S.S. Shipilova, IOP J. Phys: Conf. Series **671**, 012028 (2016)
[8] D. Sukhanov, K. Zavyalova, MATEC Web of Conferences **79**, 01040 (2016)
[9] E. Balzovsky, Yu. Buyanov, MATEC Web of Conferences **79**, 01036 (2016)

OPENNESS OF NATURAL INFORMATION PROCESSING SYSTEMS AND SHORT-TERM AND LONG-TERM MEMORY

Vladimir Shumilov*, *Vladimir* Syryamkin

[1]National Research Tomsk State University, 634050, Tomsk, Russia

Abstract. In this paper, various types of memory are considered. Development of the nervous system and the brain of organisms ensures expansion of species with more developed brain. Different characteristics of the brain zones have created different conditions for forming traces of events. The various traces of trace events formed in different parts of the brain provided a wider range of functions and brain responses, which contributed to the survival and expansion of species. The work describes the appearance of memory varieties and the causes of these varieties.

1 The need for openness of self-organizing information processing systems

Let us consider the philosophical aspect of the amount of information in the device being created (or in the person being brought up), which arises during the creation and education of the mind both artificial and natural. You can often hear statements as if reinforced by reasoning that no creator can create something more perfect than he himself is. This, ostensibly, follows from the statement that information does not arise from nothing. Therefore, a person or another creature cannot create something more perfect and more informed than a creator is. If a creator did not have information, where did it come from? This question arises when information is placed in the information storage medium of artificial intelligence (AI) directly during the process of its manufacturing.

The statement about impossibility would be legitimate if the person and the device being created or the creature being brought up (artificial intelligence or a child) were an isolated system. That is the person or the device being created or the creature being brought up could not use other sources of information, except for the information of the creator. In fact, the subject being created (or brought up) is comprehensively open, therefore, can receive and collect information from other sources than its creator or educator. Other teachers and educators, sources of concentrated information accumulated by other people, results from various means of information fixation and storage, and information carriers can act as third-party sources. They also include rock paintings, and books, various paper and electronic media, various artifacts created by other media, and various natural and social phenomena, and the primary sources of information (sources of relations, which are of

* Corresponding author: vnshumilov@rambler.ru

interest to the observer). Therefore, because of the openness to other sources of information, the creature (trainee) can accumulate more information than the creator possesses and become more perfect than the creator (teacher) is; we can notice this very often. There examples of outstanding scientists, philosophers, public Figures, military leaders, and athletes, as well as examples at the family level because children are often smarter than parents are.

Similarly, the created artificial intelligence in order to surpass its creator must be capable of perceiving information not only from its creator, but also from the surrounding world. Therefore, it is necessary to create not the AI, which is filled with necessary information and is ready for immediate functioning, but a device (system) capable of training and self-training, that is a device capable of high-speed receiving and fixing information, which had not been embedded in it during the creation. A device should be able to use this information for functioning, rather than for simple storage of this information, which cannot be controlled entirely by the creators because of the enormous speed of information flow. It is almost impossible and economically inefficient to control completely the information flow received by the intellect developed because of their volume. Therefore, the real (not demonstrational) intellect (both artificial and natural) should be supplemented with information continuously during functioning.

Developed intellect regardless of whether it is natural or artificial, operates with a huge amount of information. In [1], researchers give an estimation of the information capacity of the human brain (no less than 300 terabytes) and the resulting rate of filling this capacity with information (on average, no less than 30,000 connections between neurons, or 60 kilobytes per second). Most likely, information is perceived by the human brain even faster, because the perception channels, for example, visual information channel (eyes) can skip information faster than the mentioned 60 kilobytes per second. Indeed, the human eye contains several tens of millions of photosensitive cells that can trigger several times per second.

2 A mechanism for implementing the openness of information processing systems

How can it be possible to record a large amount of information in the created artificial or born up natural intelligence, as well as the observed speed of perception and full processing of a powerful information flow throughout the life (the entire time of functioning) of the subject (or device with AI)? Such a system should be open to the perception of information in a variety of forms. The natural system that processes this powerful flow is being self-organized throughout the life, that is during the period of formation, growth of the organism, and in the adulthood. This system is formed by a simple cell division starting with one original (maternal) cell. We have proposed a purely local mechanism for self-organization of the information processing system in the organism. Processing consists in the passage of signals coming from the receptors along various paths through the brain structures to the effector (manipulators). Because of the fact that receptors require the expenditure of always-limited resources of the organism, first, receptors reacting to environmental influences dangerous to the organism were formed. Signals are generated only in receptors when the environment exposes them. When effectors receive signals that have passed through the structures of the brain, they perform some actions and carry out the organism's reaction to the irritating effect of the environment. If (successful) actions of the effectors result in stoppage of the irritating effect of the medium on the receptors, then the brain ceases to receive signals, that is the movement of signals through the brain and actions of the effectors cease. The system will go into hibernation until the next signal from the receptors arrives.

When passing through the structures of the brain signals form traces of events that had given rise to these signals, that is they form new connections between neurons. These connections represent the organism's memory. They are associative, because they are formed exclusively (only) between two simultaneously excited neurons. Their association consists in the fact that the events that gave rise to the signals occurred about the same time. The connections accumulate the organism's experience. Subsequently, newly formed links along with the old ones ensure the forecast (prediction) of the arrival of dangerous events from the chains of events when one of the events in this chain is detected. Connections represent pathways for propagation of signals that have entered the brain. So, natural systems of information processing are organized extremely simply. Therefore, their self-organization is possible beginning with a single initial (mother) cell to a highly developed organism and a brain in it. Self-organization of the brain proceeds without a centralized assessment of the situation and control of the actions of neurons from the outside. Neurons functioning proceeds completely independently. They have no means of interacting with a distant or intimate environment, except for their entrances and exits.

Therefore, formation of links is carried out purely by a local mechanism. The weights of the connections are constantly changing. A continuous decrease in weights causes the process of forgetting.

3 Short-term and long-term memory

A brief preliminary examination of the functioning of the natural system of information processing shows the parameters of the memory of the organism, that is the parameters of the inter-neural connections, are determined by the elements of the mechanisms of fixing information (bonding) and by the rate of reduction of the weights of these links.

It becomes obvious that the properties of memory (inter-neuronal connections) depend on the parameters of the environment, in which these links reside, and on the parameters of the active elements (neurons), which form the bonds between themselves. The dependence of the memory decay on the environment, in which neurons and the connections between them are located (not the organism's environment), is that the connections, whatever their nature is, reduce their weights depending on the biological activity of the environment. In any case, the reduction in the weights of the connections will occur as a result of processes such as diffusion (resorption and decrease in the density of structures that form connections). The rate of processes that reduce the weights of connections will depend on the gradients of potentials of various types (electrical, chemical and others) in a given medium.

Just like memory fading, formation of connections and their initial weight depend on the environment and on the properties of the neurons that form the connections. Formation of connections depends on the characteristics of the medium. For example, it depends on the electrical strength of the surface separating the excited neurons as well as on the magnitude of potentials at the excited neurons. It depends on whether the potential difference between the inputs and outputs of the excited neurons exceeds the formation potential (**FP**) of the connection, and how much. First, **FP** depends, on the environment in which the neurons are formed. The potentials at the outputs and at the inputs of the excited neurons depend, first of all, on the structure and parameters of the neurons. These potentials also depend on the loads (on the number of associative connections for a given neuron).

It is clear that both the characteristics of the medium and the characteristics of the neurons (especially the magnitude of the loads (the number of associative connections)) will differ significantly in different areas of the brain. And these zonal differences will also have external manifestations: the ease of memorization and the various rates of forgetting.

Manifestations of these differences are classified by psychologists as the existence of various types of memory [2].

We cannot control the area of our brain (memory), which participates in the reproduction of the image of an event. Therefore, based on statistics, a conclusion is made simply about the nature of memory: if it is a long-term or short-term, or some kind of intermediate. Psychologists identify almost dozens of memory types depending on the speed of forgetting the recorded events [2, 3].

Figure 1*a* shows the curve of forgetting of the events (word reproduction) obtained experimentally. The resulting curve differs significantly from the usual type of curves in similar processes. In all processes, the measured quantities relating to flow and accumulation vary approximately exponentially (capacitor charge, potential equalization, the number of atoms that have not yet disintegrate up, and the level of the liquid flowing out of the capacitance. This nature of the change in quantities is because the rate of change of quantities is proportional to the difference between the magnitude and the limit, to which this quantity tends. What does it mean? Most likely, the result (the ability of human memory structures to reproduce event traces) is because it appears as the sum of several parallel processes and several indicators corresponding to these processes.

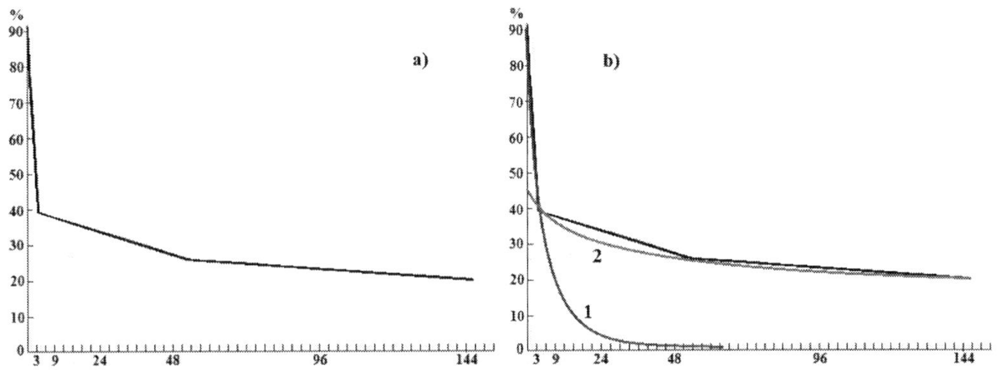

Fig. 1. The playback graph of the stored information.

Figure 1*b* shows the representation of the experimental curve through two exponents. One of them falls very quickly (curve 1), while curve 2 falls relatively slowly. Therefore, the concepts of short-term and long-term memory were introduced into consideration. What is the reason for the presence of two types of memory? What is the mechanism of this phenomenon?

Here is a simplified scheme of the simplest natural information processing system (Figure 3*a*) and a slightly more complicated one (Figure 3*b*), in which connections were formed responsible for the reproduction of images of events that left traces (connections).

In the upper part a) of Figure 2 is a diagram of the simplest nervous information processing system consisting of one receptor, one effector, and the connection between them.

Figure 2*b* shows a slightly more complicated system, which has the following distinctive features: 1) it reacts not to one, but to 3 stimuli, 2) it has 3 receptors and 3 effectors, and simple and direct (direct, without branching) connections between them. As a result of the increase in the number of connections in a limited volume, they begin to touch each other. The possibility of mutual influence between neighboring connections occurs because of formation of connections between the paths of signal transmission.

In the course of evolution, the nervous system is developing. An increasing number of input sensors (receptors) appears. And, correspondingly, the number of ways to promote the input signals from the receptors to the effectors is increasing. These paths necessarily influence each other, because they are in close proximity. Moreover, the number of effectors becomes much less than the number of receptors. Thus, the number of photosensitive elements (receptors) in the eye is estimated in tens of millions of elements. It is clear that the organism cannot have the same number of effectors reacting to the incoming stimuli of these receptors.

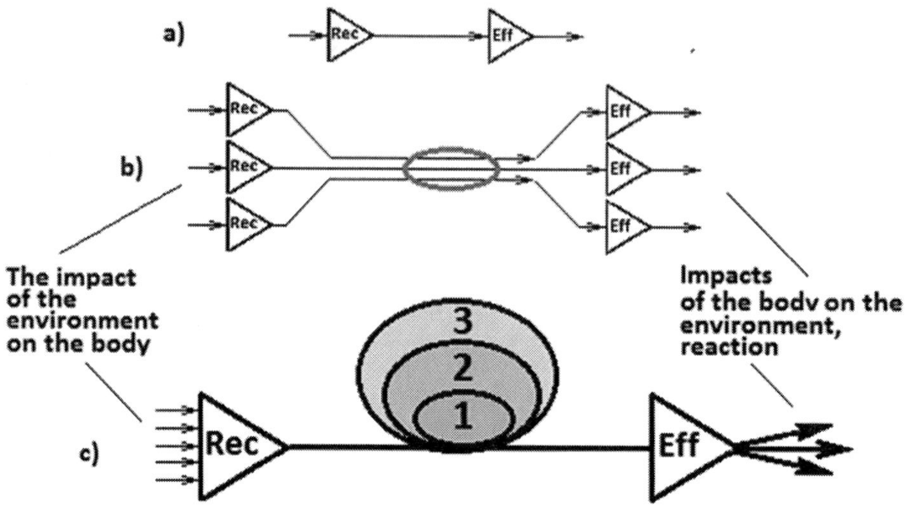

Fig. 2. Simplified diagram of the evolution of natural information processing systems.

So the signals from all the receptors are sent to the zone of their processing. These signals (afferent) after their processing (already efferent) pass through the existing paths to the corresponding effector. The actuation of the effectors carries out the organism's reaction to stimuli signals that had entered the brain through detectors. When a group of receptors and effectors are located remotely from the brain, afferent and efferent signals (centripetal and centrifugal) pass in close proximity to each other (in common channels) along the path from the place of generation to the place of effective response, where they can also interact. However, it is not so important for our problem.

Because the central part of the nervous system (the brain) was formed in the course of evolution for many generations, then its structures grew not at once but gradually. We can observe the consequences of this development in the brain structures of more developed animals (for example, fish, amphibians, birds, and mammals) [4, 5].

We need this observation to show that the signal pathways from the receptors to the effectors pass through the zones of the brain with different degrees of deepening into the brain and into its central zones of the brain, which have larger signal processing capabilities.

Inter-neural connections that conduct signals through different areas of the brain, of course, differ in their characteristics. In this case, we are interested in the ease of formation of inter-neural connections and the rate of their destruction and extinction (reducing their weights), and the speed of forgetting the information that gave rise to these connections.

Let us consider how the properties of memory can depend on the area of the brain, in which the connections were formed by the signals generated by the event of interest. For this, we refer to the mechanism of formation of connections between neurons [1]. Formation of connections between neurons occurs when neurons are in contact, and are both excited. If the potential difference between their points becomes higher than the potential of forming of the connection (PF), a new connection is formed.

Subsequently, the formed links (traces of events) will conduct up signals through themselves based on this connection, thus forecasting the arrival of subsequent events. This ensures the possibility of evading the organism from the impact of predicted events.

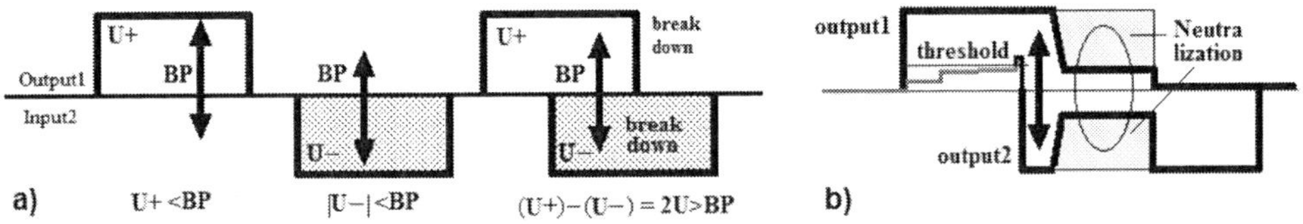

Fig. 3. Conditions of formation of communication between neurons and compensation of potentials of excited neurons.

This Figure clearly shows how the parameters of neurons and their loads (the number of connections in neurons) affect formation of connections. Figure 3*a* shows schematically the moment of formation of the connection between the output of one neuron and the input of another one. We can see that formation of a new connection depends on the magnitude of the potential generated by the neuron and on the electrical strength of the surface, which separates the neighboring excited neurons. Figure 3*b* shows the compensation of the different polar potentials of the outputs and inputs of neurons with galvanic contacts (connections). The process of formation of connections between neurons is described in more detail in [1].

We note that the processes of formation of traces of events (formation of new links between excited neurons) and new signal paths (accumulation of experience) take place completely locally regardless of any external influences. This ensures parallelism and high speed of information processing in the brain. Information is fixed in the brain not in the storage mode (as simple placing of information for storage), but by means of changing the brain structures, which changes the possible paths of signals through the brain structures and the brain's reaction to future events. Continuous training and retraining of the brain in accordance with the environment occurs. These properties of the brain ensure its openness to the assimilation of information from a variety of sources throughout life.

According to the same principle, connections can be formed not only between natural biological neurons, but also between artificial neural-like elements in engineering devices created by engineers [6]. The efficiency of this principle of recording events in memory has been confirmed on the electronic model of the simplest nervous system created at Tomsk State University (TSU) [7].

In the course of evolution, the human brain has increased (through mutations of individual organisms during the process of generating new generations, survival and expansion of more viable organisms), because a larger brain contributed to a more successful survival and expansion of species. This growth of the brain most easily could occur because of the growth of new structures of the brain with the preservation of the already existing ones. Because formation of new areas of the brain occurred under slightly different conditions than of the earlier structures, the parameters of new neuronal structures differed somehow from the parameters of the earlier (ancient) structures. That is, new thresholds of neuron excitation, different potentials for formation of new connections, and other extinction rates of information links could have formed on new sites. It means that when the brain and cortex volume increased, regions with slightly different parameters of active and passive memory elements (neurons and connections between them) were formed in the new regions of the brain. These changes turned out to be useful and, therefore, became constant.

Let us demonstrate the usefulness of forgetting in general, and different rates of forgetting information, which has different significance for survival of the organism. Forgetting gives an organism an opportunity to learn some reaction and do not loop on it. Without forgetting, the organism could not change its behavior in conditions of changes in environment, which occur constantly. Some changes are insignificant and information about these events does not penetrate deep into the brain, but some are very important and they cause with many associations. Therefore, they are fixed in different regions of the brain with different characteristics. For example, we are walking along a road covered with leaves; we see them and perceive them. However, there is no sense to keep in memory all the information about each of the leaf seen. Therefore, according to experimental data [8], information of this kind is stored for 20 seconds on the average. Similarly, when traveling along this road, we see all the defects on the road surface and even remember them all for a while. However, tomorrow we will be able to remember only the most dangerous and unusual defects and we will successfully forget all minor defects. Storage of such information in our memory does not make sense for our behavior.

Our brain acts rationally, it remembers important information and forgets the secondary one. In fact, the brain remembers everything. However, the rate of forgetting different types of information (the extinction of ties) is different. Therefore, psychologists talk about long-term and short-term memory; they experimentally measure the rates of forgetting [9]. The information values (remembered by subjects of an experiment) represented on graphs of psychologists cannot be represented as simple dependences of the type of declining exponents. However, they can be represented as a sum of simple dependencies. This indicates the placement of the same information in different zones, where the conditions for fixing information in the form of inter-neural connections and the rate of extinction of connections (connections weights reduction) differ.

How does the brain distinguish (with the declared locality of the mechanisms of memory), which traces of events should be placed in a more operative “short-term” memory, and what should be placed in a long-term memory? It turns out that there is no need to sort information when it arrives, it can be fixed simultaneously in many places. However, the more important is information, the more associative links it has. Therefore, a greater number of connections is formed for the same reason. In addition, new connections that are more persistent can be formed where the excitation of neurons (and connection formation) requires a larger signal provided by a large number of associative connections. In such zones of the brain, the formation of connections is heavier in comparison with the zones where connections with a high rate of extinction are formed, but the rate of extinction of the connections (information) in these zones is lower than in those zones where the connections are formed easily. If the information could have been fixed only in a zone with a high rate of forgetting (with a small number of associations), it effaced from the memory quickly. Perhaps this resource can be reused.

The shortest (instantaneous) memory is conditioned by the inertia of sensations received by the brain when the environment influence on the receptors. For example a bright flash of light remains in sensations up to a few seconds.

Next comes a little more long-term memory (perception), which is necessary for primary processing in the brain (to have time for perception). Not only sensations from the receptors, but also a certain number of connections from the information processing zones act as inputs. Perception allows you to embody sensations in images that are acceptable to the brain, it means that illuminated spots and points on the retina turn into signals indicating lines and certain silhouettes because of connections between neurons formed in the brain that has just started functioning. The brain is trained to act like this at the very beginning of its activity. In addition, there are also connections between things (events) felt by the young organism by various organs (receptors), what it sees with the eyes and feels with hands and a tongue. The results of perception move further along the brain as already formed images of a specific brain and can make traces not of signals from individual receptors but signals of image activation in subsequent parts of the brain.

Even further, (logically) from the signal input to the brain, a slightly more long-term (operational) memory is located; it has a large number of associative links. The signals in it remain active for a while, which helps to identify links between events separated by a certain period. However, connections formed in it quickly dissolve, and it is a useful property for the organism. The question that has not yet received an answer is very interesting for it: "Why the Miller's magic number is 7 (± 2) (the number of elements held in this memory)? What is the mechanism for regulating the number of elements held?" [10].

The existence of images in this memory may be because traces of events in this memory fade very quickly. On the other hand, from the point of view of a possible biological realization Miller's number may be because the images of a certain number of events in a certain area of the brain (near the entrance) circulate until attenuation. This number cannot be more than a certain limit because of mutual suppression of signal flows through brain structures. From the point of view of functioning such circulation increases the possibility of formation of connections between neurons. Because the time intervals for the excitation of neurons are small, it is almost impossible to form associations between events separated by time intervals, more time is required to find the neuron in the excited state. While the time-extended activity of neuron chains contributes to formation of connections.

Further, specialized in perception and association areas of long-term memory come (motor, visual, auditory, mathematical, and others).

As a signal generated in the receptors moves, it meets different types of memory (including the operational one). This division has no clear boundaries. Such a progressively developed structure of the signal propagation pathways ensured detection and fixation of cause-effect relationships between events separated by sufficiently large (all `large) time intervals, much greater than the time of the complete cycle of excitation of one neuron. This gave the organisms greater chances of survival and expansion than detection of an associative link between events occurring simultaneously. Because of the more successful survival and expansion of the species with such brain structure, not only the population of the species but also the volumes of the corresponding parts of the brain increased.

The structure of the brain has been formed in the course of evolution. Evolution of the species occurs as a result of mutations during the process of appearance, better survival and expansion of individual species organisms because of their competitiveness. That is, the evolution of the species occurs through small discrete changes at the birth of each new organism and fixation of the features by means of a faster expansion of well-formed organisms that have won competition in the existing environment. The same properties can be either winning or losing under different conditions. For example, the presence of a warm cover gives advantages to animals living in cold regions of the Earth. At the same time, such animals cannot exist in conditions of a warm climate.

4 Conclusion

Analysis of functioning of the natural memory makes clear that there is no special mechanism, which analyzes the events occurred and recorded in a short-term memory, and transfers the most important of these events to a long-term memory. All this would be too difficult even in the case (with a low probability of achieving), if scientists and engineers could have created such a mechanism, whereas natural information processing systems are formed spontaneously in each organism as a result of the self-organization of elements (neurons and connections between them) that form the structure of the brain.

Our analysis shows that the paths of signal (information) propagation in the brain parallelize and fall into different areas of the brain, where they are fixed. In these zones, as a rule, information is recorded that relates primarily to a certain group of receptors and to a certain activity. Here, the word "primarily" means "mostly", but not "exclusively".

Therefore, associations between events of various kinds can be fixed, and reliability is ensured; if a part of the brain is damaged information is usually lost not completely because of the dispersal of the places of fixation of information, although reproduction can be difficult.

Since traces of events along the way are fixed in a variety of places including zones with different rates of forgetting, then after a while short-term memory fades so much that the information recorded in it can be considered lost irrevocably. As a result, when you try to remember some episodes, information about events recorded in areas with a lower forgetting rate comes to the fore, but already with several other associations than information in a short-term memory. Therefore, researchers at different levels have the impression that the processed information is moving into a long-term memory. It seems to be recycled (undergoing processing) because the information fixed in a long-term memory comes to the fore. Information in a long-term memory (with a lower rate of extinction) is connected with associations that differ from associations that are inherent in information in a short-term memory.

Relocation of information from a short-term memory to a long-term memory is absent. Just a short-term memory about the event (with its associations) is thrown into the background.

If nature, as a result of evolution, has found it expedient to have zones in the brain with different rate of memorization and memory fading, then it is probably expedient to create such zones in the memory of artificial devices of the intelligence carriers. It is clear that nature does not have any goals (processes in it simply proceed according to its laws), so we should talk not about expediency, but about mutual consistency of processes. Forgetting is useful, it provides variability of the organism's response.

Acknowledgements

The research presented in this paper was supported by the Russian Foundation for Basic Research (grant No 16-29-04388/17). The authors are grateful to Tatiana B. Rumyantseva from Tomsk State University for English language editing.

References

[1] V. N. Shumilov *Principles of the Brain Functioning. Engineer's look.* (without PH Ekaterinburg, 2008)
[2] Psychology. Features of Memory as a Psychological Process URL: http://psixologiya.org/obshhaya/pamyat/1647-osobennosti-pamyati-kak-psixologicheskogo-proczessa-diplom.html
[3] V. N. Lange About the Rate of Forgetting URL: http://www.voppsy.ru/issues/1983/834/834142.htm
[4] Evolution of the Animals' Brain http://biofile.ru/bio/19130.html
[5] Biology and medicine. Evolution of the Brain in Vertebrates URL: http://medbiol.ru/medbiol/antrop/00079dc7.htm
[6] V. N. Shumilov, I. V. Shumilov, V. I. Syryamkin, Neurocomputers: Development and Application **4**, 94(2015)

[7] V. N. Voronin, V. N. Shumilov, Yu. A. Shumilov, A. R. Tsoy, A.V. Kiselev, S. M. Sharshov, Neurocomputers: Development and Application **4**, 26 (2015)
[8] Short-term Memory URL: https://ru.wikipedia.org/wiki/%D0%9A%D1%80%D0%B0%D1%82%D0%BA%D0%BE%D0%B2%D1%80%D0%B5%D0%BC%D0%B5%D0%BD%D0%BD%D0%B0%D1%8F_%D0%BF%D0%B0%D0%BC%D1%8F%D1%82%D1%8C
[9] G. Ebbingauz "*Psychology*" (1907)
[10] G. A. Miller. The Psychological Review **63**, 81 (1956)

EVOLUTION OF MOBILE APPLICATIONS

Anachack Phongtraychack[1, 2*], *Darya* Dolgaya[1]

[1]National Research Tomsk State University,634050, Tomsk, Russia
[2]National University of Lao PDR. 7322, Dongdok, Vientiane, Lao PDR

Abstract. Currently, we can see the rapid evolution of mobile technology, which involves mobile communication, mobile hardware, and mobile software. Features of mobile phones largely depend on software. In contemporary information and communication age [1–4], mobile application is one of the most concerned and rapidly developing areas. At the same time, the development of mobile application undergoes great changes with the introduction of new software, service platforms and software development kits (SDK). These changes lead to appearance of many new service platforms such as Google with Android and Apple with iOS. This article presents the information about the evolution of mobile application, gives some statistical data on the past and present situation, demonstrates how individual users of mobile devices can benefit, and shows how mobile applications affect society from the ethical perspective.

1 What is a mobile application?

Mobile application is a software application designed to run on mobile devices such as smartphones and tablet computers. It is a result of recent technological innovations. Mobile applications have appeared because of the convergence of media, information technology, Internet and advanced technologies. In addition, for many years, mobile telecommunications have been under investigation by mobile device manufactures, mobile service providers, application developers, and many researchers in the sphere of information technology (IT) and information systems (IS). However, the most interesting area for research is mobile application evolution [5].

1.1 Transactions on mobile applications

Mobile devices and smartphones can be still considered as new ideas in the technology community. Compared to personal computers and laptops, the common mobile smartphones have only been around for a couple of years now. This leads to an interesting area of discussion on mobile payments online. In this increasingly digital world, it is not surprising that money will follow suit as well. Recent trends show that digital money kept in mobile wallets will soon replace physical cash and even credit cards. We can buy most products and

* Corresponding author: idea.phongtraychack@gmail.com

services online these days via our laptops and mobile devices. It is easy to link this process to our bank accounts and online payment processors like PayPal.

SMS payments are currently one of the most popular methods of using mobile phones to pay for goods, services, or even to make person-to-person payments. In order to make SMS payments, a user needs only a phone with SMS capability. That is why many experts believe that because of the simple nature of SMS payments, they will continue to be a prospective area in mobile payments [6].

1.2 Specific types of services by means of mobile application

There are four specific types of services, which can be made of mobile application.

- **Browser Access:** The applications, which we use through native browser. For example, m.yahoo.com, www.google.com, m.redbus.in, etc.
- **Hybrid Apps – Web:** You need to install an application in your device; function of the particular application requires Internet. For example, Social Networking Apps (Facebook, Twitter), Instant Messenger (Skype), E-commerce (Flipkart), Internet Speed Testing (Speedtest), etc.
- **Hybrid Apps – Mixed:** You need to install the application in your device and function of the application may require Internet. For example, there are some games to play alone or to go online for playing with different players (multi player). This category also includes medical apps where you can keep a record of your health in order to share with your friends or doctor via internet.
- **Native Apps:** The applications, which are installed in the device. For example, Reminders, Few Games, etc [7].

1.3 Types of software for building mobile application

With a small investment, you can create and manage your mobile site or application using one of the platforms listed below:

- Appery.io.
- Mobile Roadie.
- TheAppbuilder.
- Good Barber.
- Appy Pie.
- AppMachine.
- GameSalad.
- BiznessApps.
- AppMakr.
- ShoutEm [8].

2 The goals of mobile application

The goal for developing mobile application is to provide conditions when maximum number of users use applications for maximum time. This means that the app developed should have the appropriate quality and purpose, so that maximum users across the globe like it. In this regard, applications should meet the following requirements:

- Conformance to the purposes of users.
- Capability to reach the majority of users.

- Capability to be secure.
- Being user-friendly.
- Supporting continuous improvement and engagement [9].

3 Mobile applications and mobile technology development

3.1 How do applications facilitate the development of the society?

Mobile applications can facilitate the development of the society by contributing to resolution of the following problems:

- Quick search of job vacancy.
- Quick communication.
- Considerable cost saving.
- Saving time and increasing productivity.
- Less power consumption because of less computer use.
- Improvement of IT infrastructure in developing countries.
- Entertainment.

3.2 Generations of mobile applications

The handset manufacturers designed and developed mobile phones of the first-generation. The competition was fierce and trade secrets were guarded. They did not want to expose the secrets of their handsets, so they developed the phone software in-house. During this period, the first "time-waster" games started to appear. Nokia was famous for putting the video game Snake on some of its earliest phones (1970-ies). Because of such things, people started to change their attitude to communication. When prices for mobile phones have dropped and batteries were improved, more people began carrying handy devices.

3.3 Why are mobile applications necessary?

Currently, almost 80% of people are online through mobile devices. Most of these people prefer mobile applications because they are easy in use and perform tasks instantly.

Users become more dependable on mobile application when solving easy tasks like booking movie tickets, checking sport's scores, buying and selling, or many other similar routine activities. These tasks can be solved by a single click. A mobile app available on apps store has more chances to convert visits into business. That assists to increase rate of sales [10].

4 Advantages, disadvantages and challenges of mobile application evolution

4.1 Advantages of mobile applications

Mobile applications help users by connecting them to Internet services, which are more commonly accessed by means of notebook computers. People, especially young, are constantly using their mobile devices to browse the World Wide Web and if you are not there, you are missing the opportunities. There are three main advantages of mobile application:

- Apps are a constant reminder of your business.
  Mobile apps reinforce your brand by increasing your visibility. An app gives a business more presence on a phone than a browser bookmark does because it is always visible on the screen of a phone. This helps building loyalty with customers because your business is around them all the time.
- Apps increase customer engagement.
  Customers are calling out for mobile apps because they quickly connect them to businesses they most commonly want or need. Businesses are using apps to improve their processes and increase the level of accessibility by customers. The point of mobile apps is to smoothly connect and interact with customers, making it a valuable tool for the modern business.
- Apps reduce costs.
  Apps reducc costs of SMS messages and paper newsletters. They simplify communications by securely, instantly and directly messaging customers. Apps reduce staff workload by information requests and phone calls [11].

4.2 Disadvantages of mobile applications

Mobile applications may also have some bad effect on the society. It includes the following aspects:

- When many Internet based mobile applications are available to teenagers, they are wasting time by using Facebook, Skype, YouTube etc. The young people are in risk when using Internet games or not appropriate applications.
- Most of people use mobile devices everywhere including buses, trains, offices, colleges, universities. It may disturb other people.
- Frequent use of mobile apps is bad for human health.

4.3 The challenges of mobile applications

One of the greatest challenges is to make your app visible so that people want to download it. If an app is not among the top 50 – 100 across a category in an app store, people will hardly ever download it.

One of the biggest challenges of mobile applications is their platform capability and limitation. In addition to interesting usability of mobile applications, they have some problems connected with platform and limitations. Let us consider a list of the main problems:

- Small screen size. On a mobile platform, it is difficult or impossible to view text and graphics like on a desktop computer screen.
- Lack of windows. We can see many windows at a time on a desktop. However, it can hardly be realized on a mobile platform.
- Navigation. Most mobile devices do not have mouse like pointer, so it has limited flexibility in navigation.
- Types of pages accessible. As a rule, mobile platforms do not support all types of file formats.
- Speed. The speed of processing and speed of connectivity of mobile platforms is slow.

- Size of messages or email. Many devices support limited number of characters in message or email.
- Cost. The cost for cellphones, mobile applications and Internet connection is high [12].

5 An overview of the famous mobile applications (2017)

According to comScore's annual U.S. mobile apps report, Americans spend more than half (57%) of their digital media consumption time in apps, which is about the same as a year ago. It means that the dramatic shift to mobile apps has now leveled out in the U.S. Table 1 shows the apps-leaders, according to comScore, as measured by their penetration of the U.S. mobile app audience [13].

Table 1. Top 10 Mobile Apps by Penetration of App Audience in the USA, June 2017

Rank	Applications	Percentage
1	Facebook	81%
2	Youtube	71%
3	Messenger	68%
4	Google Search	61%
5	Google Maps	57%
6	Instragram	50%
7	Snapchat	50%
8	Google Play	47%
9	Gmail	44%
10	Pandora	41%

5.1 Case study of mobile applications of China

China is the biggest mobile phone market in the world. The number of mobile phone users is more than 850 million. Among them, 277 million people use mobile net citizens, which is becoming the main driving force for the rise of general net citizen scale in China.

Teenagers contribute a considerable part to mobile communication market. A large proportion of teenagers in Korea (80.6%) and Japan (77.3%) own mobile phones. In China, 48.9% of teenagers aged between 12 and 18 are mobile users. There is a big market potential to attract the other half of teenagers. Chinese mobile booming market and the success of applications on smartphones are well-known facts. It is a huge booming market where all the companies are trying to be the best. However, it is not such an easy task, since there are many things to know about this market.

5.1.1 Understanding of Chinese mobile market

There is a huge potential of Chinese customers with 850 million users of smartphones. Let us go through several points here to understand the overall Chinese mobile market.

First, OS vs iOS and Android are widely dominating the market owning most of the market share; however, even if the non-friendly relation with the big Apple, iOS is the second one owning the market share. We will go deeper on this point in further analysis.

Second, Chinese people are not willing to pay for mobile applications. They consider the new wave of technology as a way to make their everyday life easier. However, most of applications revenue in China are made inside the application, which includes all the purchases inside the app, for example, additional features for a game, additional functions for dating app, etc.

Third, Chinese people are selective about the choice and the selection of their applications. They need to have a wide range of options among applications. Most of downloaded applications are still mobile games.

Fourth, you have to choose your mobile application partner. There are more than 500 application stores, you might feel lost among this numerous number of applications stores, they all have different functions, some focus on games, while others tend to be more social [14].

6 Conclusion

Mobile phones have changed our world completely. People use them for leisure time, business, and education. Mobile phones are not just mobile phones anymore. People rely on them for video chatting, voice calls, surfing internet, taking pictures, playing games, etc. Mobile phones have become a part of everyday life and we cannot imagine our life without a mobile phone. In addition, much of the capabilities of smartphones revolve around the applications they offer. This is very important in determining the future of smartphones. Applications give smartphones an extra edge that ordinary phones do not have. Such apps include advanced photo sharing, gaming or internet service. The opportunities for the future of mobile application are endless. It will only take some time to see, which of these innovations will be a part of our everyday life.

Acknowledgements

The research presented in this paper was supported by the Russian Foundation for Basic Research (grant No 16-29-04388/17). The authors are grateful to Tatiana B. Rumyantseva from Tomsk State University for her assistance in preparing the article.

References

[1] V. Syryamkin, V. Shidlovsky. *Correlation-extremal direction-finding systems* (Tomsk State University publishing house, 2010)
[2] S.V. Shidlovskii, J. Comp. Sys. Sci. Int **45**, 282 (2006)
[3] D.V. Shashev, MATEC Web Conf. **79**, (2016)
[4] D.V. Shashev, S.V. Shidlovskiy, J. Phy: Conf. Ser. **881,** (2017)
[5] Mobile app. [Electronic resource] URL: https://en.wikipedia.org/wiki/Mobile_app
[6] J. Rocheleau. 6 Safety Steps To Making Secure Mobile Transactions. URL: http://www.hongkiat.com/blog/secure-mobile-payment-tips/
[7] Types of mobile application. URL: http://softwaretestinggarbage.blogspot.ru/2012/10/types-of-mobile-application.html
[8] Gr. Smith. 10 Excellent Platforms for Building Mobile Apps. URL: http:// http://mashable.com/2013/12/03/build-mobile-apps/#XycLK6j7ZkqA
[9] Yo. De. What is the goal of the mobile app development? URL: https://www.quora.com/What-is-the-goal-of-the-mobile-app-development

[10] Ch. Siegl. Is it necessary to have a mobile app when you already have a website? URL: https://www.quora.com/Is-it-necessary-to-have-a-mobile-app-when-you-already-have-a-website
[11] Benefits of mobile apps. URL: http://vividus.com.au/insights/benefits-of-mobile-apps/
[12] J. Shore. Challenges of developing applications for mobile platforms. URL: http://searchcloudapplications.techtarget.com/podcast/Challenges-of-developing-applications-for-mobile-platforms
[13] Top 10 most popular apps of 2017. URL: https://www.recode.net/2017/8/24/16197218/top-10-mobile-apps-2017-comscore-chart-facebook-google
[14] Solutions to Market your mobile application in China. URL: http://marketingtochina.com/solutions-to-market-your-mobile-application-in-china/

COMPARATIVE ANALYSIS OF METHODS FOR ASSESSING THE EFFECTIVENESS OF GEOLOGICAL AND ENGINEERING OPERATIONS

Boris Ivanenko and *Alexander* Petelin[1]

National Research Tomsk State University, 634050, Tomsk, Russian Federation

Abstract. The paper presents a comparison of the methods for assessing the baseline oil production level based on the use of nonparametric statistics, integral and differential models, and neural network algorithms.

1 Introduction

As is known, geological and engineering operations (GEO) are directed methods of influencing the development of a field for the purpose of increasing the oil production. Depending on the action mechanism and the action object, the diversity of GEO can be divided into several groups: methods of oil production intensification, physicochemical methods, hydrodynamic methods, and gas methods.

GEOs are considered to be the most effective with the smallest discrepancy between the actual and basic (calculated) indicators that would be characteristic for the baseline level of the object development, i.e. the method used before the performance of GEO. Construction and further support of a special hydrodynamic model of the field can be used to calculate the baseline oil production level and assess the effectiveness of geological and engineering operations. However, this approach, in addition to substantial economic costs, requires special training of specialists in the area of physical process modeling. In other words, it is practically not suitable in production conditions when solving operational problems of optimization of the field development system.

2 Methods for forecasting the baseline oil production

The following methods for forecasting the baseline oil production level are common in modern practical oilfield geology [1, 2]:

- displacement characteristics (integrated models);
- decline curves (differential models).

Both methods are based on the regression analysis and are designed to forecast operational characteristics of oil production.

[1] Corresponding author: aepetelin@gmail.com

Displacement characteristics are used in the event if the development of oil wells is carried out by means of waterflooding. A reliable and a long-term forecast of oil production characteristics can be implemented only when the rate of change in the water cutting of oil wells is stabilized. It is believed that stabilization occurs at water cutting of more than 70%. At water cutting values from 50% to 70%, only a short-term forecast is possible. For values below 50%, the forecast cannot be consistent.

Decline curves are used when the depletion mode is the main method of developing oil wells, allowing to characterize the dynamics of decline in oil production rates.

Today, more than 70 kinds of displacement characteristics and decline curves are known. The forecast model of the baseline oil production level is selected based on the field description and the modeling purpose. The type of dependence is determined empirically [1].

3 Artificial neural networks

Methods of information simulation or mathematical modeling of processes and phenomena become relevant when assessing the baseline oil production level and the efficiency of geological and engineering operations. Artificial neural networks (ANN) is a convenient and a natural way of recording information models.

Information models can also be built on the basis of nonparametric statistics methods [2]. However, in practice neural network models are more preferable due to the low cost, the impossibility of obtaining a sufficient amount of experimental data, often high noise, and inconsistency of experimental methods.

An ANN can be quite formally defined as a set of neurons (simple processor elements with local functioning) and synapses (unidirectional links). An INS receives an input signal from the outside world. The signal passes through a network with transformations in each processor element. The signal is converted into certain output data when it passes through network connections. As is known, the neural network formulation of the problem consists of several stages:

1. Statement of the problem.
2. Selection of the algorithm for training the network and its architecture.
3. Construction of training samples. The following is used to create training samples: 1) information from databases of the field development history, physical and chemical properties of fluids, geological and hydrodynamic characteristics of formations, etc.; 2) the results of numerical modeling of processes and phenomena occurring in reservoirs.
4. Training and testing of the network.
5. Execution of the algorithm, obtaining of the result, and the analysis of the obtained data.

To determine the structure and the composition of the training sample fields for various input parameters (signals) of a neural network with their further detailing, the input signals are classified into the following groups:

- time and frequency of information selection (selected training period and dates of measurement taking);
- characteristics of injection wells (coordinates, operating time and number of wells, intake, injection volume, etc.);
- characteristics of production wells (well production volume, fluid extraction volume, well production rate, well operation time, location of producing wells, etc.);
- equipment parameters (pump capacity, pump immersion depth, etc.);
- geological and hydrodynamic characteristics of the formation (formation pressure, permeability, bottomhole pressure, dynamic level, porosity, etc.).

It is obvious that building of training samples is essential when solving problems of oilfield geology: determination of their composition and structure and division into subsets –

training and verification. In [3] it is proposed to conditionally divide them into two types – samples with a vertical and a horizontal structure (depending on the type of tasks being solved).

Before proceeding to the discussion of practical results, we shall note that the framework of this paper will not include the issue on the selection of a network architecture and algorithms for its optimization and training. It shall be noted that on the basis of our earlier studies [3], this paper will include only universal, multi-layer neural networks trained using the method of "back propagation" and linear neural networks.

The task of assessing the baseline production level in a neural network setting is the classical forecast problem, therefore, it cannot be separated into a separate class of problems. The difference is only in the method of specifying the appropriate parameters of the neural network model in the forecast period and the selection of the training period for the network, or the base interval. In practice, it is recommended to resort to two approaches of specifying forecast parameters of the model.

In the first case, the input of the trained neural network for the forecast period consists of some average characteristics of oil production obtained from the history of the formation development for the baseline interval. This approach is justified for short-term forecasts (in which the extraction dynamics is determined only by the rate of fluid extraction and the injection modes and, above all, the operation time of wells). In this case, it is enough to limit oneself to setting some constants for the forecast period, i.e. the average time of the well operation, the average fluid extraction, etc.

In the second case, when we are faced with a pronounced tendency to a decline or to an increase in the characteristics of oil production, it is recommended during the training period to construct regression curves for the corresponding engineering characteristics of each well (production rate, injectivity...), to prolong them for the forecast period and to submit to the input of the existing neural network that has been trained. These two methods are particularly effective in assessing the cumulative baseline production level simultaneously for several wells or for the entire field.

3 Analysis of the methods for assessing the effectiveness of geological and engineering operations

It is possible to do the same when solving the tasks of assessing the baseline oil production level for one well.

We shall consider two selected cells 2.2 and 3.2 of the formation U32 of the Severo-Vakhskiy deposit as an object of study (Figure 1). The total number of wells for two selected cells is 31. 18 of them are producing and 13 are injection wells. The data were collected from 01.01.1997 to 01.10.2000, i.e. over 58 months. Information content: current time, average monthly production and injection volumes, operating time and downtime, penetration depth of pumps.

GEOs, such as pump penetration and formation hydrofracturing (FHF) with simultaneous almost threefold increase in injection volumes, which significantly complicates the construction of the neural network model, were carried out in the period under the investigation in wells 1021, 1022, 1030 of the cell 2.2 (hereinafter referred to as intensified wells). Moreover, the complexity of building a neural network model is conditioned upon the presence in the database of a significant number of gaps and the lack of accurate data on the geological structure of the oil formation.

The calculations were carried out using a PC SNN and NeuroPro [4–6], developed using object oriented programming similar to [7]. Operating hours of production wells, time of information selection, well coordinates, submergence depth of the pumps, and liquid withdrawal were taken as training characteristics. The period of ANN training was chosen from January 1, 1997 to September 1, 2000, i.e. before the intensification of the wells. The forecast period was from November 1, 2000 to September 1, 2001. After the training of the neural network, the average volume of fluid drainage and the average operating time of the well were supplied for the forecast period.

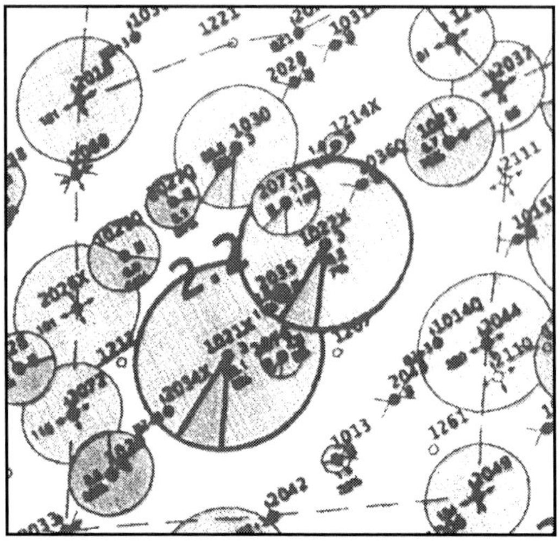

Fig. 1. Diagram of well location of the cell 2.2. Wells 2018, 2026, 2072, 2049, 2044, 2037 are injection wells.

Figure 2 presents the results of the assessment of the cumulative baseline oil production level for three intensified wells 1021.1022 and 1030 of the cell 2.2 of the formation U32 of the Severo-Vakhskiy field.

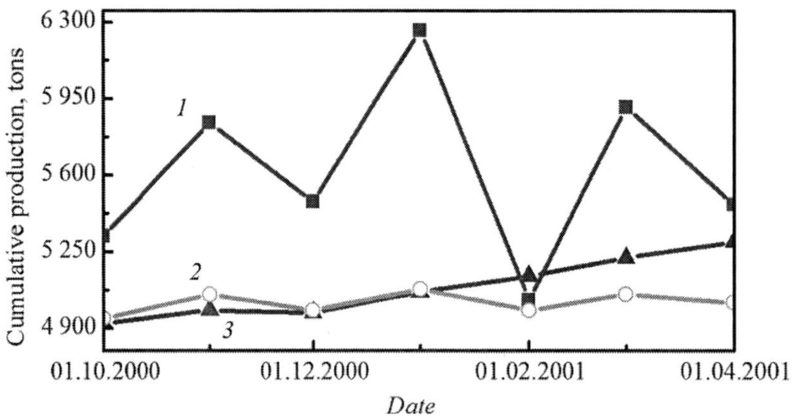

Fig. 2. Example of the calculating of the baseline level of cumulative oil production using different methods, without taking into account the injection: 1 (red line) – cumulative production; 2 (green line) – neural network forecast without injection; 3 (blue line) – standard forecast obtained using the method approved by the Yukos Methodical Committee.

Based on the analysis of the graphical data presented in Figure 2, it is possible to make an obvious conclusion that neural network methods, even in the simplest form (with a minimal training sample composition), offer results more adequate to the reality than conventional methods. It is really easy to see that the forecast curve (the baseline production level), calculated from the displacement characteristics, has a clearly pronounced tendency to increase, which in fact should not be the case.

Figure 3 presents the comparison results of neural network and non-parametric methods for estimating the baseline production level (taking into account the injection). The data on the injectivity of injection wells was specified in the training sample. It is easy to see that, in general, the results coincide quite well. Moreover, unlike conventional methods and methods of non-parametric statistics, the neural network methods have higher resolution capabilities and are capable to quickly solve the task of forecasting the baseline production level not only for several wells at the same time, but also for specific wells for any degree of flooding with a minimal set of input data.

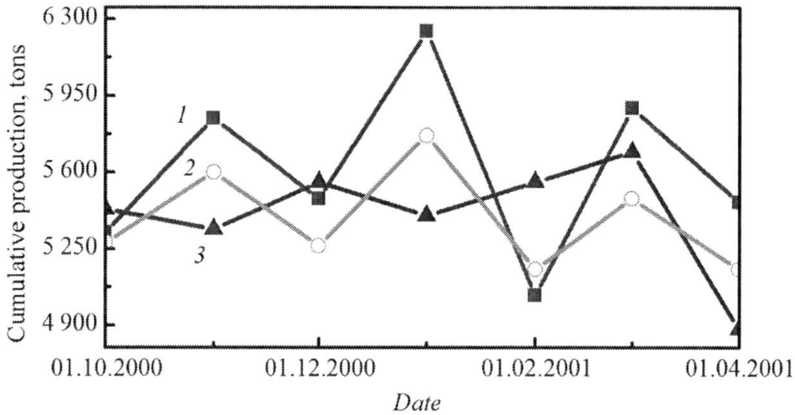

Fig. 3. Example of calculating the baseline level of the total oil production using various methods, taking into the account the injection: 1 (red line) – cumulative production; 2 (green line) – neural network forecast with injection; 3 (blue line) – forecast using the nonparametric statistics method.

The authors are grateful to V.L. Sergeyev, Dr. Eng. Sc., for the provided calculation results of the baseline production level using nonparametric statistics methods.

References

[1] *Assessment of the technological efficiency of geological and engineering operations. Methodical instructions. Version 1.0.* (Moscow: NK Yukos, 2001).
[2] V. Sergeev, *Training algorithms in control systems and information processing* (Novosibirsk: Nauka, 1978).
[3] E. Gromakov, T. Aleksandrova and B. Ivanenko, IOP Conf. Series: Mat. Sc. and Eng. **124**, 012015 (2016)
[4] B. Ivanenko, S. Kostyuchenko, A. Parfenov, E. Muslimov and V. Yampolsky, Proc. – KORUS 2000: 4th Korea-Russia Inter. Symp. on Sci. and Tech. **XP002908186**, 17 (2000)
[5] *Neural networks. Statistica Neural Networks* (Moscow: Goryachaya liniya – Telecom, 2000)
[6] A. Gorban and D. Rossiev, *Neural networks on a personal computer* (Novosibirsk: Nauka, 1996)
[7] A. Petelin and A. Eliseev, IOP Conf. Series: Mat. Sc. and Eng. **225,** 012085 (2017)

MESOSCALE ACOUSTICAL CYLINDRICAL SUPERLENS

Igor Minin[a], *Oleg* Minin

National Research Tomsk State University, 634050, Tomsk, Russia

Abstract. We demonstrate experimentally for the first time the *acoustojet* (acoustic jets) formed from acoustic plane wave scattering by a penetrable cylindrical particle with dimensions of several wavelengths. It acts as a superlens with subwavelength localization of acoustical wave. During the scattering by elastic solid particles, additional internal shear waves are excited due to modes conversion. This mechanism allows achieving sharp focusing in the near-field zone. Such mesoscale single particle cylindrical lens may be considered as acoustic metamaterials free superlenses with resolution beyond the diffraction limit.

1 Introduction

Acoustic lenses find applications in various areas ranging from nondestructive detection to biomedical imaging and even noninvasive cancer treatment. One of the leading trends in the development of modern devices is miniaturization and integration. It is therefore of paramount significance to realize an acoustic lens that has a compact (mesoscale) size, a subwavelength resolution and a high transmission efficiency and thus should have deep implication for the acoustic field. However, its realization remains challenging due to the tradeoff between high refractive-index and impedance-mismatch.

The first acoustical lens has been probably described in [1, 2]. Last century several acoustic lens designs were considered: planar acoustic lens structures based on Fresnel diffraction [3], sonic-crystal slab composed of a two-dimensional square-array of aluminum rods in air [4], 3D phononic crystal [5, 6], metamaterial based acoustic gradient index (GRIN) lenses [7, 8], hyperbolic GRIN lens [9, 10], etc. It could be noted that GRIN lens are far less prevalent in practice due to the difficulty in finding real materials with desired parameters, since natural materials are generally unable to provide slow sound velocity and high mass density simultaneous In 1991 for the Soret geometry of zone plate shock diffraction and focusing has been simulated by referencing a characteristic wavelength of an attenuated shock pulse [11]. Strong focusing of a plane ultrasound wave by a plano-concave lens was investigated in [12]. Many of modern approaches use Helmholtz resonators or metamaterial based split-ring type resonators [13]. However, these metamaterials are all bulk materials and much larger than the working wavelength, which is to the disadvantage of the miniaturization and integration of acoustic devices. Brief review of acoustical artificial lenses was given in Ref. [14]. However, existing methods of subwavelength focusing in

[a] Corresponding author: prof.minin@gmail.com

acoustics are usually based on new artificial materials, or require complex signal processing methods.

It is well known, that due to diffraction theory, the resolution of ultrasound imaging is limited at best to a half-wavelength and features below this value cannot be resolved. Shorter wavelength can be used, but to the cost of a higher attenuation and lower penetration depth, which rapidly reduces the scope of the imaging technique. To date, this "resolution versus penetration" paradigm annihilates the quest for ultrasound imaging deep into target.

In the optical frequency range, optical jet-like structures for small dielectric particles were investigated at least since 1987 [15]. The well-known by now so-called photonic nanojets (PNJ) allow to form a localized hot-spot in subwavelength area in light scattering by low loss dielectric particles [16].

On the other hand, in the linear mode it should be expected that methods of sub-wave focusing, based on photonic jet phenomenon, can be successfully applied to the acoustic range as well. Formally this can be observed on the analogy between the wave equations, specifying acoustic and electromagnetic scattering processes [17]. Thus, PNJ effect have motivated us to propose the concept of acoustic jets, called "acoustojets" [18, 19]. It has been theoretically predicted for the first time [18, 19] that the existence of acoustic analogue of PNJ phenomenon is possible, providing an opportunity for subwavelength localization of acoustic field in shadow area of arbitrary 3D penetrable mesoscale particle. It is important to note that there is a principle difference between optical and acoustical materials such as existence of a shear speed of sound, i.e. acoustic materials are anisotropic due to the two speeds of sound [20, 21]. Recently, the phenomenon of acoustojet formation was experimentally verified for spherical Rexolite particle, immersed in water [22].

2 Cylindrical acoustojet

In this report, for the first time we experimentally demonstrate that highly intense acoustic beam with subwavelength waist is produced in the scattering by a mesoscale penetrable cylindrical particle. It is well known that metallic spherical particles have low acoustic absorption. In the case of the rigid particle pressure waves do not penetrate inside it, while in the case of the penetrable particle both cubical and spherical shape pressure waves may travel through it and can excite internal modes [23]. From our previous simulation results [18–20] it follows that metallic spherical particles (made of lead, stainless steel, brass, etc.) can produce acoustojets, but due to an acoustic impedance mismatch, the intensity is relatively low. Due to this reason for further experimental investigations we selected polymer mesoscale particles, like ABS plastic (black), which has perfect acoustic impedance matching with water (see the Table 1 [24]).

Table 1. Parameters of the used materials.

Material	Sound velocity, m/s	Density, g/sm^3	Acoustic impedance
Water	1490	0.998	1.487
Rexolite	2337	1.04	2.430
Plexiglas	2670	1.18	3.150
ABS Plastic (black)	2250	1.05	2.362

3 Experimental equipment

Let's briefly consider an experimental setup for cylindrical acoustojets measurements. The basic equipment was similar to that used in Ref. [25] – see Figure 1. A signal generator GZ-112 was used to excitation of transducer. The transducer assembly with the nominal thickness-resonance of ceramic emitter 1.092 MHz was fixed centrally into one of the square faces of a rectangular glass-walled tank, dimensions 20×20×30 cm, with the transducer axis horizontal and parallel to the length of the tank.

To measure the acoustic parameters in the focusing area we used a calibrated [26, 27] broad band needle-like hydrophone GUZ-2 [28]. A 3D printed black ABS cylinder with diameter of 4 wavelength was suspended by a thin (0.08 mm) fishing line and placed along the central part of transducer. The position of the hydrophone was determined by a three-coordinate system of positioning, controlled by a computer, similar to Ref. [29]. Such measurements were carried out with a spatial step of 0.5 mm in an area of 60×60 mm. The transverse and longitudinal distributions of the acoustic pressure were measured.

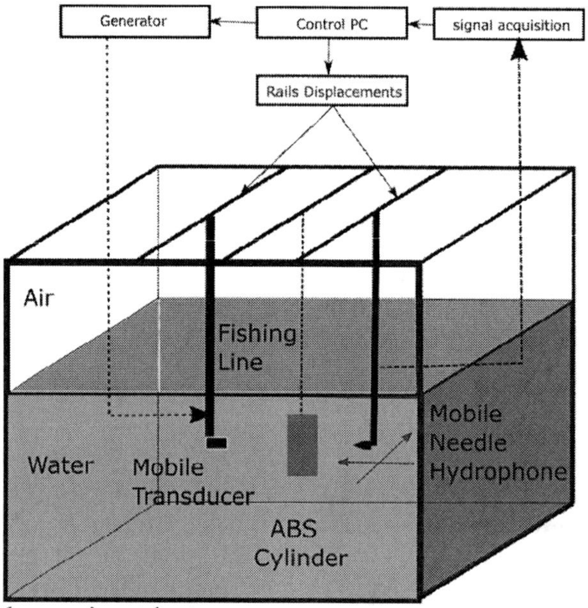

Fig. 1. Schematic of the experimental setup.

4 Simulations

A simulation model which mimics an experimental physical phenomenon could be a primary model to start different designs for improvement. We first simulated the cylindrical acoustic particle-lens using Finite Element Method (FEM) software Comsol Multiphysics. Boundary conditions near the surface of glass-walled tank were set as water acoustic impedance in order to avoid reflections of the propagation wave. We used a free tetrahedral mesh with maximum element size $\lambda/5$ for the free-space regions, and $\lambda/5/1.5$ for the particle. Additionally, we used a free triangular mesh with maximum element size $\lambda/20$ for the z-plane to obtain Figures of better resolution – the need for very fine meshes in acoustic jet numerical simulations is a known problem in optics, and an even finer mesh is required for obtaining the field's spatial derivatives to obtain the forces. The properties of the material were set through the longitudinal and shear velocities of sound.

Modeling the interaction of ultrasound with such a structure is a complex problem, and further modeling the resultant wave behavior after the interaction is also challenging. The complexity of these problems arise from parameters such as the number and type of boundaries involved, the number of reflections accounted for, the number and type of wave modes interacting, and how the focal spot is defined [30].

5 Experimental results

Similar to optics [16, 31], the analysis of acoustojet is characterized by such main parameters of ultrasound beams as peak intensity, its position along Z axis, full width at half maximum (FWHM), full length at half maximum (FLHM) at -3 dB [22, 32]. The experimental measurement of FWHM and FLHM are shown in the Figure 2. The pressure distribution along the wave propagation direction was measured from a distance of 0.5λ from the shadow surface of the cylinder to eliminate the influence of the probe. The enhancement of the measured acoustojet is 1.8 with sidelobes under 1.2. The similar results (with lower gain) were observed for cylindrical particles illuminated from a flat surface.

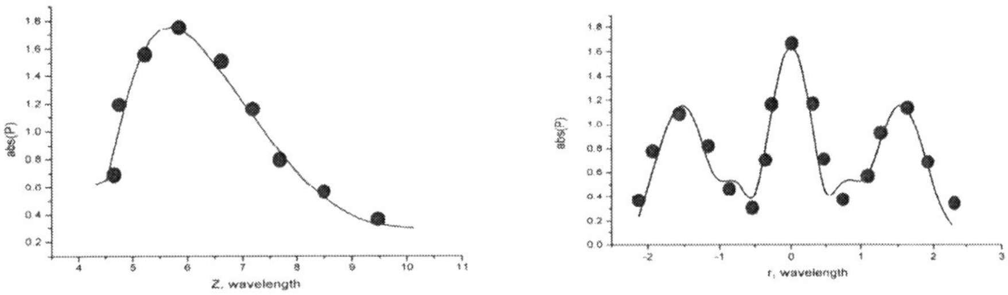

a) b)

Fig.2. The experimental result for the acoustojet relative pressure intensity in the xz (a) and xy (b) plane. The solid lines present the results of numerical simulations in Comsol Multiphysics software. The derived length of acoustojet FLHM is about 2.5λ and FWHM is less than 0.62λ.

6 Conclusion

In conclusion, we have demonstrated for the first time that *acoustojets* (acoustic jets) formed by acoustic plane wave scattered by penetrable cylindrical particle lens with the diameter of four wavelengths may be used for subwavelength localization of acoustical waves. This work presents a quantitative comparison between experimental and simulated acoustic fields of acoustical cylindrical particle lens. In the acoustic wave scattering by elastic solid mesoscale particles shear waves are excited due to mode conversion leading to a stronger localization effect and this is a fundamental difference from optical nanojet phenomenon.

Such mesoscale particles can be considered as an acoustic superlenses with resolution beyond the diffraction limit and, thus, can be used in acoustical tweezers for particle manipulation, medical diagnostics, non-destructive testing, sensors, acoustical microscopy etc, due to subwavelength resolution and high energy gradient.

Our results can be further extended to the particles made of different materials and of various shapes with the attention to the feasibility of acoustical subwavelength focusing with a single particle lens (both spherical and cuboid) from artificial periodic structures, core-shell particle with a liquid core (including blood) and a coating layer, metamaterials, particles with pupil mask and the whole other arsenal of methods previously developed in the optical range [26].

Acknowledgments

I.V.M. acknowledges the support of the Russian Scientific Foundation, grant 17-79-20051 (test object design and characterization), O.V.M acknowledges the support of the Mendeleev scientific fund of Tomsk State University (experimental setup). O.V.Minin and I.V.Minin contributed equally to this work.

References

[1] W.E.Kock, F.K. Harvey, Jour. Acous. Soc. Amer. **21,** 471 (1949)
[2] G.L. Augspurger, Electronics World **52** (1962)
[3] B. Hadimioglu, E.G. Rawson, R. Lujan, M. Lim, J.C. Zesch, B.T. Khuri Yakub, F.C., Quate Proc. IEEE Ultrasonics Symp. **1**, 579 (1993)
[4] T. Miyashita, R.Taniguchi, H. Sakamoto, Proc. WCU, Paris, Sep. 7-10, 911 (2003)
[5] J.T. Welter, S. Sathish, D.E. Christensen, P.G. Brodrick, J.B. Heebl, M.R. Cherry, J. Acoust. Soc. Am. **130**, 5 (2010)
[6] S. Alagoz, Kaya O. Adem, B.B. Alagoz, Applied Acoustics. **70**, 1400 (2009)
[7] V. Romero-Garcia, A. Cebrecos, R. Pico, V.J. Sanchez-Morcillo, L.M. Garcia-Raffi, a J.V. Sanchez-Perez, Applied Physics Letters. **103**, 264106 (2013)
[8] Li Y., Yu G., Liang B., Zou X., Li G., Su Cheng and Cheng J. Sci. Rep. **4**, 6830 (2014)
[9] W. Wang, Y. Xie, A. Konneker, B. Popa, S.A. Cummer. Applied Physics Letters. **105**, 101904 (2014)
[10] M. Moleron, M. Serra-Garcia, C. Daraio, Applied Physics Letters **105**, 114109 (2014)
[11] I.V. Minin, O.V. Minin, V.F. Minin, *Book of abstracts the 18th Int. Symp.on Shock waves, Ed by K.Takayama* (1991)
[12] M. Akiyama, T. Kamakura. Acoust. Sci. & Tech. **26**, 3 (2005)
[13] W.M. Robertson, US Patent 2016/0111080, A1 (2016)
[14] I.V. Minin, O.V.Minin, Proc. of APEIE, 2016 13th Int. Sci.-Tech. Conf. 136 (2016)
[15] D.S. Benincasa, P.W. Barber, J.Z. Zhang, W.F. Hsieh, R.K. Chang, Applied Optics **26**, 1348 (1987)
[16] B.S. Lukiyanchuk, R. Paniagua-Domínguez, I.V. Minin, O.V. Minin, W. Zengbo. Optical material express 7, 1820 (2017)
[17] L. Nicolas, M. Furstoss, M.A. Galland. The European Physical Journal Applied Physics **4**, 95 (1998)
[18] I.V. Minin, O.V. Minin, arXiv, 1604.08146 (2016)
[19] O.V. Minin, I.V. Minin, Optical and Quantum Electronics **49**, 54 (2017)
[20] J.H. Lopes, J.P. Leao-Neto, I.V. Minin, O.V. Minin, G.T. Silva, Proc. 2nd ICA2016, 1 (2016)
[21] B.D. Tartakovski, Akustitcheskij Zhurnal **7**, 349 (1961)
[22] J.H. Lopes, M.A.B. Andrade, J.P. Leao-Neto, J.C. Adamowski, I.V. Minin, G.T. Silva Physical Review Applied **8**, 024013 (2017)
[23] A. Maurel, J.-F. Mercier and Felix, S. J. Acoust. Soc. Am. **135**, 165 (2014)
[24] Selfridge A.R. IEEE Transactions on Sonics and Ultrasonics, **SU-32** , 381 (1985)
[25] Neppiras E.A. and Coakley W.T. Journal of Sound and Vibration **45**, 341 (1976)
[26] Papadakis P., Taroudakis M., Sturm F., Sanchez P., and Sessarego J-P. Acta Acustica united with Acustica **94**, 676 (2008)
[27] IEC 60565:2006, *Underwater acoustics - Hydrophones - Calibration in the frequency range 0,01 Hz to 1 MHz* (International Electrotechnical Commission, International Standard, 2006)

[28] A. M. Enyakov *Development of methods and tools for metrological support of hydroacoustic measurements in the frequency range from 0.5 to 15 MHz* (Degree of the Doctor of Technical Sciences Thesis (in Russian), Mendeleevo, 2007)
[29] A. Cebrecos, V. Romero-García, R. Picó, V.J. Sánchez-Morcillo, M. Botey, R. Herrero, Y.C. Cheng and K. Staliunas, J Phys D Appl Phys **48**, 025501 (2014)
[30] E. Twerdoski, M. Von Buttlar, A. Habib, R. Wannemacher, W. Grill, ECNDT 2006 - Fr.1.8.4 URL: http://www.ndt.net/article/ecndt2006/ doc/Fr.1.8.4.pdf
[31] I.V. Minin and O.V. Minin, *Diffractive Optics and Nanophotonics: Resolution Below the Diffraction Limit* (Springer, London, 2016)
[32] S. Yang, W. Qin, H. Guo, T. Jin, N. Huang, M. He, and L. Xi, Biomedical optics express **8**, 2756 (2017)

THE EFFICIENCY OF THE PUMPING OF THE LASERS BASED ON SELF-TERMINATING ATOMIC TRANSITIONS OPERATING IN THE ENERGY INPUT CUT-OFF MODE

Alexey Burdin, *Yury* Polunin, *Ivan* Razmahnin, *Anatoly* Soldatov, *Nikolay* Yudin, *Nikolay* Yudin*

National Research Tomsk State University, 634050, Tomsk, Russia

Abstract. The analysis of the electro-physical processes in the discharge circuit of the lasers based on the self-terminating transitions of metal atoms (LSTM) and the electrodes placed in the cold buffer zones of the gas discharge tube (GDT) is occurred. That design of the GDT can provide the efficient lasing at the reduction of the current flowing through the switch to zero after the charging of the capacitive components of the circuit from the storage capacitor. Under the circumstances the pumping of the active medium is determined by the energy input from the peaking capacitor and, consequently, the efficiency of the pumping can be increased by an order of magnitude, if (using a managed switch) the energy input into the active medium from the storage capacitor is "cut-off" after charging the capacitive components of the circuit. It was shown that the efficiency values of ~ 9-11 % and of ~ 5-6 % for the copper and gold vapor, lasers could be achieved.

1 Introduction

The lasers based on the self-terminating transitions of metal atoms (LSTM) are one of the most efficient coherent lasing sources in the visible spectrum among gas lasers and the predicted copper vapor laser efficiency is ~ 10 % [1], this is by an order of magnitude more than the achieved practical efficiency by now [2–4]. The highest value of the efficiency of the pumping of ~ 9 % of copper vapor laser operating in the energy input cut-off mode (the energy input into the active medium was excluded after a laser pulse) was achieved [5, 6]. It can be explained by the fact that sharp-front pulses, which duration is equal to the lifetime of the population inversion of the operating levels in the active medium [1], should pump the LSTM. Due to the fact, that the duration of a pump pulse is higher than the lifetime of the population inversion of the operating levels in the active medium [2–4], the energy input cut-off after a lasing pulse is the obvious way to increase the efficiency of the pumping of the LSTM.

* Corresponding author: yudin@tic.tsu.ru

In the experimental researches [5, 6] the managed switch consisting of the thyratron TGI1-270/12 and the tacitron TGU1-27/7 was used, what helped to realize the energy input cut-off mode at the circuit current of 27 A. According to the experimental researches [5, 6], the circuit current at the end of a laser pulse was higher than the mentioned value. Therefore, the tacitron should be close before the end of a laser pulse to realize the energy input cut-off mode. In this case, the duration of tacitron being open is equal to the thyratron jitter, that determines unstable operating, which was not observe during the experiments with the laser operating in the energy input cut-off mode. It is obvious that the modulation lamps GMI-29 [3] is the ideal switches for the lasers operating in the energy input cut-off mode. However, the efficiency of the pumping of the active medium of the LSTM with the modulation lamps that is equal to the value from [5, 6] was not realized. Thus, there are the several unanswered questions that limit using the energy input cut-off mode for the creating of the efficient sources of coherent light.

The purpose of the present work is the research of the electro-physical processes, in the discharge circuit of the laser, which determine the possibility of operating in the energy input cut-off mode.

2 The electro-physical processes in the discharge circuit

The population inversion of the LSTM is determined by the difference between the excitation rate of the upper and the lower operating laser levels by electron collisions during the non-equilibrium period of the plasma. The excitation rate constants of the laser levels of the LSTM are determined by the electron temperature, which depends on the voltage change on the active components of the active medium impedance [3]. Thus, the electro-physical processes in the discharge circuit of the LSTM determine the kinetic of the processes in the active medium. According to the electrical circuit theory, there are need to be created an equivalent circuit for the analysis of the electro-physical processes in the discharge circuit. The high pre-pulse concentration of electrons ($n_e \sim 10^{13}$ cm^{-3}) in the active medium of the LSTM is typical. Thus, as suggested the gas discharge in the active medium occurs without the electrical breakdown and the equivalent circuit is the simple oscillatory circuit [3]. However, this equivalent circuit cannot explain the observed experimental dependencies, particularly the realizing of the efficient pumping of the active medium of the LSTM operating in the energy input cut-off mode [5, 6].

The presented in [7] equivalent circuit of the lasers based on the electrodes placed in the cold buffer zones of the gas discharge tube is more suitable for the explanation of the electro-physical processes in the discharge circuit. The mentioned design of the GDT is traditional for research and development of the LSTM [2–4]. In [5, 6] was studied the similar laser operating in the energy input cut-off mode. The pumping of the active medium occurs in two stages [7, 8]. At the first (preparatory) stage, the charging of the capacitive components of the discharge circuit (peaking capacity C_0 and GDT's own capacity C_{GDT}) by the energy stored in the storage capacity C occurs. There are two modes of the charging of the capacitive components of the discharge circuit [7], which determine the population inversion, and these modes can be called — the modes of the "increased" and the "decreased" energy input. The existence of the two charge modes is determined by isolating the active medium, which geometric dimensions are govern by the heat-insulated discharge channel, from the electrodes by the cold buffer zones, in which there are no metal vapors. The charge neutralization during an inter-pulse period in the GDT, which has the mention design, occurs by the process of the three-particle recombination in the active medium and by the process of the dissociative recombination in the buffer zones [9]. Since the dissociative recombination rate is higher than the three-particle recombination rate [9], the charging of the capacitive components could occur at the total or partial recombination in the cold buffer zones. The mode change (from the "decreased" to the "increased" energy input) occurs with the increasing of the storage capacitor voltage [7] when the charging of the capacitive components (at the preparatory stage) of the discharge circuit from the storage capacitor at the partial recombination in the cold buffer zones occurs [7, 8]. Due to the increasing of the storage capacitor voltage results in the increasing of the energy output and the practical efficiency of the LSTM (relatively to the storage capacitor energy) [7], all research studied only the "increased" energy input mode and the "decreased" energy input mode was not studied.

The "decreased" energy input mode can be realized at the total recombination in the cold buffer zones and the decreasing of the current flowing through the switch to zero after the charging of the capacitive components from the storage capacitor, i.e. if the GTD voltage is equal to the breakdown voltage on the anode-side cold buffer zone of the GDT. After charging C_{GDT} during the breakdown process, according to [7, 8], the electric potential difference on the active medium occurs and (at the second stage) the pumping will be occurred during discharging C and C_0. The edge condition of the start of the pumping of the active medium is $I_1 = 0$ and $I_2 = 0$, where I_1 and I_2 — the circuit current related to C and C_0. Under these conditions the pumping of the active medium must be determined by the energy input from C_0, since the capacity value of the storage capacitor, ordinarily, is higher by an order of magnitude than the capacity value of C_0. That means that the efficiency of the pumping can be magnified by an order, if the switch like the tacitron excludes the energy input from the storage capacity into the active medium.

3The results of the experimental research and the discussion

The research results of the electro-physical processes in the discharge circuit of the copper and gold vapor lasers illustrate all of the above. The experimental equipment circuit is presented in Figure 1.

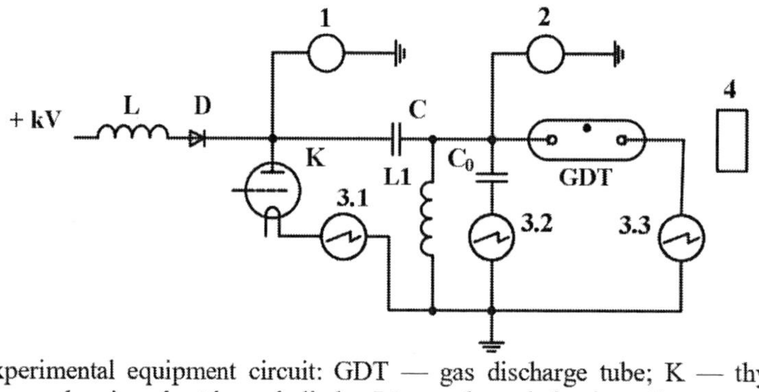

Fig. 1. The experimental equipment circuit: GDT — gas discharge tube; K — thyratron TGI1-1000/25; L, D — charging throttle and diode; L1 — shunt induction; C — storage capacitor; C — peaking capacity; 1 — static kilovoltmeter; 2 — voltage divider; 3 — current indicators; 4 — power meter.

The research was occurred with the heat-insulated GDT of the copper vapor laser at the pressure of the buffer gas (Ne) of ~ 80 Torr. The GDT discharge channel is made of the BeO-ceramics tube which the inner diameter of 8 mm and the length of 30 cm. The capacitors KVI-3 was used as the storage and the peaking capacitors (C = 2200 pF and C_0 = 330 pF). The oscilloscope "Tektronix DPO-4034B" controlled the current and the voltage pulses. The average laser power was measured by the power meter "OPHIR-NOVA". The parameters (Figure 2) at the pulse repetition rate (PRR) of 10 kHz, the power consumption from the high-voltage rectifier of ~ 350 W and the voltage U_r of ~ 2.5 kV, the average laser power of ~ 300 mW at the normal operating are presented. The pulses of the current flowing through the points (3.1), (3.2) and (3.3) (Figure 1) was measured by the same current sensor "Textronix" to exclude the measuring errors. The peak voltage on the storage capacitor of $U_C = 2U_r \sim 5$ kV under the said conditions of the pumping, i.e. at the matching mode of the electricity source with the electricity consumer, that did not allow to increase the voltage on the storage capacitor above 5 kV because work of the thyratron is unsustainable. The typical CVL oscillograms of the pulses of the current flowing through the GDT 1, the voltage on GDT 2 and the lasing 3, are presented. In addition the oscillograms of the current 4 flowing through the thyratron during the discharge process of the storage capacitor and the (inverted) current 5 of the charging (5.1) and discharging (5.2) of C_0 are given. The current 4 flowing through the thyratron consists of two parts 4.1 and 4.2, where the first is the current of the charging of C_0 at the preparatory stage, the second is the current of the discharging of the storage capacitor at the pumping of the active medium. The identity of the pulses of the current 4.1 and 5.1 is the confirmation of the aforesaid. The invert oscillograms of the current 5 are presented (Figure 2) to demonstrate that the pulses 4.1 and 5.1 are identical.

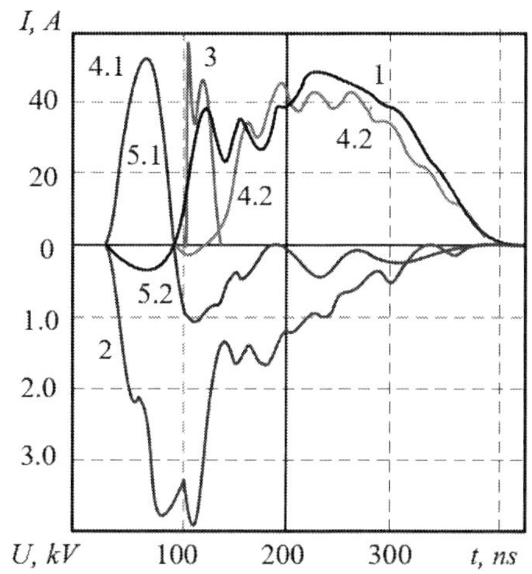

Fig. 2. The oscillograms of the pulses of the current (1) flowing through the GDT, the voltage (2) on GDT, the lasing (3), the current flowing through the thyratron (4) and the current (5) of the charging and the discharging of C_0. Where 4.1 is the current of the charging C_0 at the preparatory stage from the storage capacitor and 4.2 is the current of the discharging of the storage capacitor at the pumping, 5.1 is the current of the charging C_0 at the preparatory stage and 5.2 is the current of the discharging C_0 at the pumping.

The lasing (Figure 2) occurs during the discharging of C_0, and the energy input into the active medium from the storage capacitor occurs after the laser pulse and, accordingly, does not contribute in the population inversion generation. Thus at the excluding of the energy input into the active medium after the laser pulse in [5, 6] the changing of the position and the forms of the laser pulse did not occur. As follows from the presented results the excluding of the energy input into the active medium [5, 6] occurs before the laser pulse, i.e. after charging C_0, when the current flowing through the switch decreases to zero, that allows to close the tacitron.

The value of C_0, according to oscillograms (Figure 2), is ~ 500 pF. The similar value of $C_0 \sim 500$ pF can be calculated, according to the law of conservation of charge, under the mentioned conditions of the pumping, that less than the nominal value of $C_0 \sim 330$ pF on 170 pF, and is determined by the value of the capacity of the conductive bus, that connected in parallel to $C_0 \sim 330$ pF. The efficiency of the laser relatively to the using energy from the rectifier under the mentioned conditions of the pumping is $\eta_r \sim 0.09\%$; relatively to the energy stored in the storage, capacitor is $\eta_C \sim 0.11\%$ and relatively to the energy stored in

C_0 is η ~ 0.75%. These results confirm that the energy input cut-off mode results in the increasing of the efficiency of the pumping of the LSTM by an order of magnitude. Nevertheless, the obtained efficiency (η ~ 0.75%) is less by an order of magnitude than the mentioned (η ~ 9%) in [5]. The differences between the efficiency results are determined by the non-optimal value of the C_0 ~ 500 pF. To prove it purpose the GDT was connect to the storage capacitor and the shunt inductance (Figure 1) by the coaxial 2 meter in length cable. Under these conditions, the capacity of the coaxial cable act as C_0 and the conditions of the pumping can be optimized by changing of the length of the coaxial cable. The average laser power of the laser with the coaxial cable was ~ 350 mW under the mentioned conditions of the pumping, and relatively to the energy stored in C_0 η ~ 2.2 %. The oscillograms of the pulses of the GDT voltage 2 and the lasing 3, the current 4 flowing through the switch during the charging of the storage capacitor and (inverted) the charge (5.1) and discharge (5.2) current 5 with the coaxial 2 meter in length cable are presented on Figure 3.

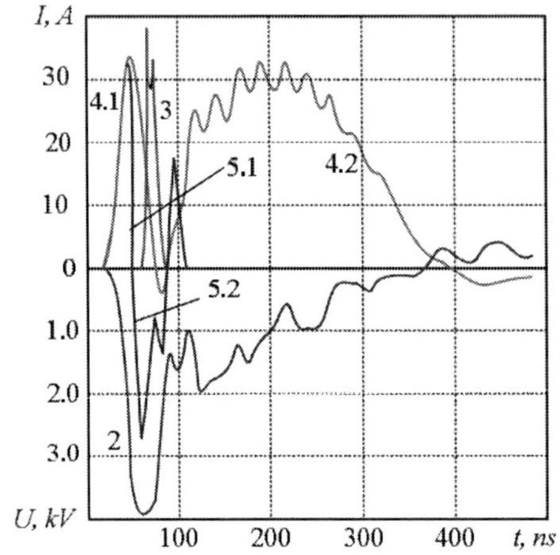

Fig. 3. The oscillograms of the pulses of the GTD voltage (2), the lasing (3), the current flowing through the switch (4) and the charge and discharge current (5) of C_0 (the coaxial 2 m in length cable). 4.1 – the charge current of C_0 from the storage capacitor at the preparatory stage and 4.2 – the discharge current of the storage capacitor at the pumping, 5.1 – the charge current of C_0 at the preparatory stage and 5.2 – the discharge current of C_0 at the pumping.

Under the similar conditions of the pumping using the coaxial ~ 1 m in length cable the efficiency of η ~ 2.8 % was occurred. This mode can be obtained at the less than 17 kHz PRR. The storage capacitor of ~ 1320 pF connected to the GDT through the coaxial 1 m in length cable was used for obtaining the self-heating mode at the 17 kHz PRR.

The further studies were carried out with the GDT of the gold vapor laser at the pressure of the buffer gas (Ne) of ~ 60 Torr. The discharge channel of the GDT is made of BeO-ceramic tube of 10 mm in inner diameter and 50 cm in length. The capacitor KVI-3 (C = 3300 pF) was used as the storage capacitor. The energy characteristics of lasing and the electro-physical parameters were measured at the PRR of the pumping of 12.5 kHz, the power consumption of the rectifier of ~ 1.2 kW and the voltage of U_r ~ 2.8 kV after the stabilization of the laser operating. The dependencies of the average laser power (λ = 627.7 nm) from the value of the peaking capacity and the efficiency of the pumping of the active medium from the energy stored in the peaking capacitor are presented on Figure 4. Furthermore the increasing of the average laser power (Figure 4, graph 2) after putting the induction L_d into the discharge circuit between the storage capacitor and the peaking capacitor (330 pF) was observed, due to there was the resonance charging of the peaking capacitor and the storage capacitor.

The energy parameters of the copper and gold vapor lasers was measured at the decreasing to zero of the current flowing through the switch after the charging of the capacitive components of the circuit (Figure 1 and Figure 2). However, the increasing of the peaking capacitor (C_0 ~ 220 pF) voltage of ~ 8 % (the energy stored in the peaking capacitor of ~ 17 %) results in the increasing of the average laser power by an factor of 2-3 times at the minimal value of the current flowing through the switch of ~ 20 A after the charging of the capacitive components of the circuit. It is obvious that these values of the current flowing through the switch allow the laser to operate in the energy input cut-off mode. Consequently, the efficiency of ~ 9–11 % and of ~ 5–6 % for the copper and gold vapor lasers operating in the energy input cut-off mode could be achieved.

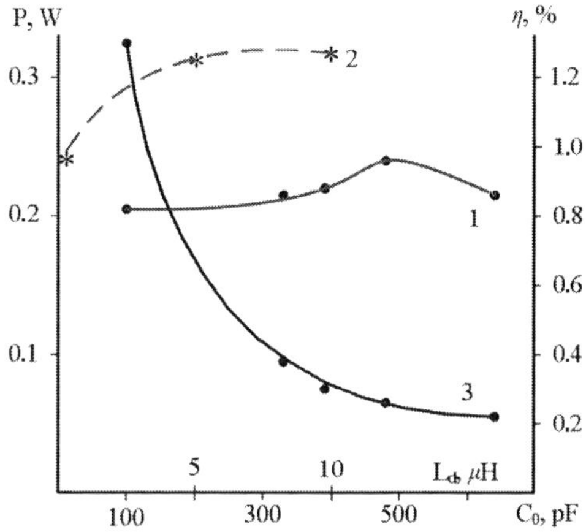

Fig. 4. The dependencies of the average laser power (λ = 627.7 nm) from the value of the peaking capacity – 1, the value of the inductance putted between the storage capacitor and the peaking capacitor – 2 and the efficiency of the pumping of the active medium from the energy stored in the peaking capacitor – 3.

There are need to be mentioned, that the energetic parameters of the copper and gold vapor lasers were measured at the source and load matching, that means the condition of the aperiodic discharge of the storage capacitor was fulfilled. Also the mismatch relatively to the energy input from the peaking capacitor was observed, that grows at the decreasing of the value of the peaking capacity in the copper and gold vapor lasers (Figure 2 (5.1)-(5.2) and Figure 3 (5.1)–(5.2), Figure 5 (1)). Higher mismatching was observed in gold vapor laser (Figure 5).

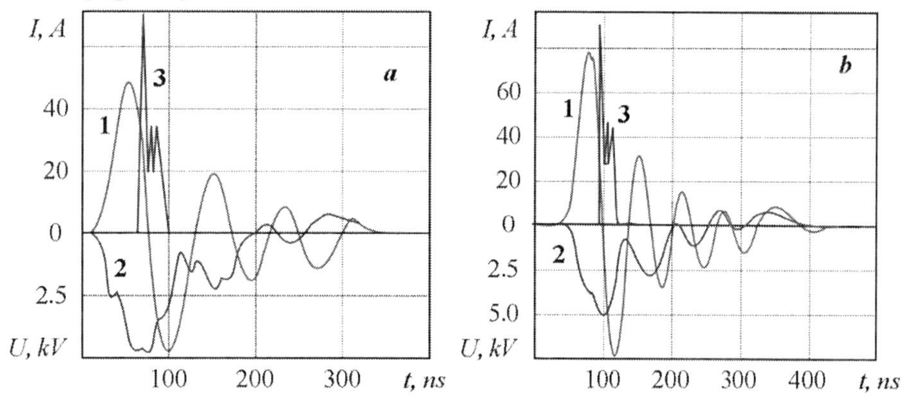

Fig. 5. The oscillograms of the pulses of the GDT voltage (2), the lasing (3) and the charge and discharge current (1) of the C_0 of the gold vapor laser: a) C_0 = 330 pF, b) C_0 = 220 pF.

The occurred researches confirm the possibility of the efficient pumping of the active medium of the LSTM and allow explain, according to the equivalent circuit [7], the causes of the decreasing of the efficient pumping at the growth of the storage capacitor voltage. This decreasing is explained by the fact that the capacitive components of the circuit from the storage capacitor (at the preparatory stage) are charged at the shunting of the capacitive components of the anode-side cold buffer zone of the GDT and the overrating of the voltage of the breakdown voltage U_{br} of the cold buffer zones. This determines the fact of the increasing of the amplitude of the GDT voltage U_a till the specified value U_{max} at the increasing of the storage capacitor voltage U_C. The specified value U_{max} is determined by the charge input rate υ_1 (the rate of the increasing of the current flowing through the switch) from the storage capacitor into the capacitive components of the circuit and the charge output rate υ_2 ($\upsilon_2 = f(U_a)$) to anode of the GDT. It is obvious that the increasing of the voltage amplitude U_a is observed at $\upsilon_1 > \upsilon_2$ till value U_{max} at $\upsilon_1 = \upsilon_2$ and determines the optimal storage capacitor voltage U_{opt} at the specified experimental conditions, that allow obtain the most efficient pumping relatively to the energy stored in the storage capacitor. The charging of the capacitive components of the circuit from the storage capacitor occurs till the capacitive components (C, C_0 and C_{GDT}) voltage equalization, when $U_C \sim U_{max}$. The further process is accompanied by the decreasing of the current flowing through the switch and the condition of $\upsilon_2 > \upsilon_1$ is fulfilled that determines the discharging of the GDT capacity. These processes are crucial for the population inversion generating in the active medium, due to during this period the features of the mentioned tube design better appear. The high-conductive ($n_e \sim 10^{13}$ см$^{-3}$) active medium, the discharge ceramic channel and the reverse current wire constructively are the capacitor C_{GDT}, one plate of which is the active medium. The feature of this is that the plasma, according to the definition, is quasineutral and an excess charge cannot accumulate in the plasma during the charging of the capacitive components. Consequently, an excess charge should be placed on the plasma surface, i.e. at the inner wall of the discharge ceramic channel. The potential difference between the plates of the capacitor occurs during C_{GDT} charging, but the plasma is still under the same potential, that prevents the pumping of the active medium during this period, that confirm the experimental research [10]. The charge transfer to the anode during the charging of the capacitive components determines the phantom current I_{ph}, which maximum value can be ~ 60-70 % relatively to the peak value of the circuit current. [7, 10]. The C_{GDT} charging occurs after the charging of the capacitive components from the storage capacitor discharging, the potential difference on the active medium occurs, that determines the pumping process. This process, called in [10] the "breakdown" and in [8] the "pseudo-breakdown", is determined by the fact that the discharging of C_{GDT} occurs not through the active medium and can be considered as the diffusion of electrons through the surface of the discharge ceramic channel to the anode through the cold buffer zone [11], that determines the sharp front of the pulse of the pumping. Wherein the significant share of the energy $E_L = L(I_{ph})^2/2$ accumulated in the inductance L almost does not contribute to the population inversion generation, due to the period of the discharging of the storage inductance into the active electrical consumer is ~ L/R and, consequently, the storage inductance is ineffective at the typical of LSTM the high plasma conduction.

The energy consumption in the mentioned processes was studied in [10]. The inductance of the discharge circuit of the laser, according to the oscillograms from [10], of ~ $1.1 \cdot 10^{-6}$ H. The GDT voltage amplitude is ~ 10 kV and the phantom current amplitude is ~ 600 A. The energy stored in the peaking capacitor at the moment of the breakdown under these conditions is E_{C0} ~ 0.16 J, in the circuit inductance is E_L ~ 0.19 J and in the storage capacitor is E_C^b ~ 0.35 J. Due to the fact that in [10] the full set of the necessary information for the energy consumption calculating was not presented the further calculations will be based on the indirect estimates. The estimation C_{GDT} of ~ 0.6 nF was based on the fact that the condition $E_{C0} + E_{GDT} = E_L$ should be fulfilled at the moment of the breakdown. The total energy stored in the reactive components of the circuit ($E_C^b + E_{C0} + E_{GDT} + E_L$) is ~ 60% relatively to the energy initially stored in the storage capacitor, what determines the significant energy consumption of the charge displacement operation during the charge process of the circuit capacitive components from the storage capacitor and determines the high initial energy losses in the switch [3]. Consequently, the consumed power by the laser should be ~ 8 kW at the PRR of the lasing of 6.5 kHz, and the efficiency of ~ 0.5 % at the average laser power of ~ 40 W or 3.3% relatively to the energy ($E_{C0} + E_{GDT}$), that determines the pumping of the active medium.

The mentioned above estimates explain the causes of the insignificant increasing of the efficiency of the pumping of the LSTM using the modulation lamps, like GMI-29, as the switch [3]. In this case, there is the decreasing of the energy losses (if the energy input is cut-off at the moment of the breakdown) by the value E_C^b or ~ 30% relatively to the energy stored in the storage capacitor. In fact, it is impossible to technically realize the excluding of the energy input into the active medium at the moment of the breakdown using the modern switches, because in this case there should be the voltage $U = -L(dI/dt)$ that is determined by the current break that significant exceeds the possible voltage value of the switch. For this reason, the pumping of the active medium of the LSTM occurs using the modulation lamps connected with the GDT from the storage capacitor through the 10 m in length cable acting as the long line and the peaking capacitor C_0. The charging of the long line occurs from the storage capacitor when the lamp opens for $t \leq 50$ ns. Then the charging of C_{GDT} and C_0 occurs and the further processes are identical to the mentioned above.

4 Conclusion

The carried out researches confirm the possibility of the gas discharge pumping of the LSTM with the predicted efficiency and allow evaluate the technical solutions operating in the "increased" energy input mode. The general approach to solving of this problem – is the resonance charging of the circuit capacitive components with the further discharging of them into the active medium, what can be done, at least, by the two ways. The first was studied in [12] where the discharging of the storage capacitor by the high-rate switch through the active medium of the GDT, which electrodes are placed in the hot zone of the discharge channel, was proposed. The second is suitable for the pumping of the active medium of the GDT, which electrodes are placed in the cold buffer zone. This approach of the pumping can be realized using the managed switch that should be closed after the resonance charging (by the putting of the additional inductance into the circuit between the storage capacitor and the peaking capacitor) C_0 and C_{GDT} from the storage capacitor. For realizing this condition there are need to be the additional switch putted in the anode-side zone of the circuit, which should be opened after charging C_0 and C_{GDT}, when the current flowing through the switch decreases to zero. This condition can be realized using the kiwotron [13, 14] as the additional switch.

The carried out researches demonstrate the possibility of the increasing of the efficiency of the pumping of the active medium of the LSTM due to the decreasing of the energy consumption of the population inversion generation. However, the energy consumption of the population inversion generation is not the main factor that determines the energy characteristics of the LSTM. Now it is well known, that the main cause of the limiting of the output characteristics of the LSTM is the high rate of the de-excitation of the laser levels by the stepwise ionization [15–17], that limits the population of the resonance levels [18] and, consequently, the energy output [16]. The population inversion value under these conditions is totally determined by the processes of the excitation the lower laser levels during the front pulse of the pumping and by the processes of their de-excitation during the inter-pulse period, and the increasing of the average laser power is possible due to the increasing of the pulse repetition rate (PRR) of the lasing. Consequently, the energy output is determined by the electro-physical process in the discharge circuit of the LSTM, which determines the need for further research. The interpretation difficulty of the research results of the electro-physical processes in the discharge circuit of the laser is determined by the fact that the electrical circuit includes plasma object, in which the processes were not studied until now, particularly, the processes in the cold buffer zones. That determines the need for the further detail research of the electro-physics and the physics of the pulse period gas charge. However, the carrying out this research is expediently if there is the potential possibility of the increasing of the frequency-energy characteristics of the LSTM. The mentioned above estimates, which demonstrate that the energy consumption of the population inversion generation in the laser [10] at the energy output of ~ 3–4 $\mu J/cm^{-3}$ can be decreased by a factor of 4 without changing the parameters, determine the continuation of the research. Thus, there is the potential possibility of the increasing of the PRR of the lasing to ~ 26 kHz and the average laser power to ~ 160 W, what confirms the research results [14]. Wherein the average laser power can be increased due to the optimization of the parameters of the pumping to ~ 500–800 W, because the energy output value can be ~ 20 $\mu J/cm^{-3}$ at the GDT voltage of 10 kV.

References

[1] G.G. Petrash, Sov. Physics - Uspekhi **14**, 747 (1972)

[2] A.N. Soldatov, V.I. Solomonov, *Gas-discharge lasers on self-limited transitions in metal vapor* (Science, Novosibirsk, 1985)

[3] V.M. Batenin, V.V. Buchanov, M.A. Kazaryan, I.I. Klimovsky, E.I. Molodykh, *Lasers on self-limited transitions of metal atoms* (Scientific book, Moscow, 1998)

[4] C.E. Litlle, *Metal Vapour Lasers. Physics, Engineering and Application* (John Wiley & Sons, New York, 1999)

[5] A.N. Soldatov, V.F. Fedorov, N.A. Yudin, Quantum Electronics **24**, 677 (1994)

[6] A.N. Soldatov, N.A. Yudin, J. of Russian Laser Research **16**, 128 (1995)

[7] N.A. Yudin, V.B. Sukhanov, F.A. Gubarev, G.S. Evtushenko, Quantum Electronics, **38**, 23 (2008)

[8] N.A. Yudin, N.N. Yudin, Russian Physics Journal **58**, 1782 (2016)

[9] K.I. Zemskov, A.A. Isaev, G.G.Petrash, Quantum Electronics **29**, 462 (1999)

[10] G.P. Hogan, C.E. Webb, Optics Communication **117**, 570 (1995)

[11] Yu. Akishev, G. Aponin, A. Balakirev, et al., J. Phys. D: Appl. Phys. **46**, 464014 (2013)

[12] N.A. Yudin, N.N. Yudin, Russian Physics Journal **59**, 809 (2016)

[13] P.A. Bokhan, P.P. Gugin, Dm. E. Zakrevsky, M.A. Lavrukhin, Technical physics letters **38**, 383 (2012)

[14] P.A Bokhan., P.P. Gugin, Dm. E. Zakrevsky, M.A. Lavrukhin, M.A. Kazaryan, N.A. Lyabin, Quantum Electronics **43**, 715 (2013)

[15] A.V. Eletsky, Yu. K. Zemtsov, A.B. Rodin, A.I. Starostin, Reports of the USSR Academy of Sciences **220**, 318 (1975)

[16] Yu. A. Piotrovsky, N.M. Reutova, Yu. A. Tolmachev, Optics and Spectroscopy **7**, 99 (1984)

[17] N.A. Yudin, V.M. Klimkin, V.E. Prokopyev, Quantum Electronics **29**, 828 (1999)

[18] A.A. Isaev, V.T. Mihkelsoo, G.G. Petrash, B.E. Peet, I.V. Ponamorev, A.B. Treshchalov, Sov. J. Quantum Electronics **18**, 1577 (1988)

DEVELOPMENT OF TECHNOLOGICAL SCHEMES FOR CONDUCTING MINE WORKINGS WITH THE USE OF AN ANCHORAGE WITH ALLOWANCE FOR THE STRESSED STATE OF THE ROCK MASSIF

Alexander Tomilov*, *Nadezhda* Tomilova, *Assiya* Batyrkhanova

Karaganda State Technical University, 100027, Karaganda, Kazakhstan

Abstract. The article presents a structural scheme for solving the problem of creating a highly effective technology for conducting and supporting preparatory workings, consisting in establishing the laws governing the development of geo mechanical processes for the development of rational technological schemes for conducting mine workings with anchorage. The order of development of technological schemes of carrying out of mine workings with application of an anchor fastening taking into account the intense condition of a file of rocks is generated.

1. Introduction

Anchor fasteners are used in all countries of the world with a developed mining industry, in France, England, Spain, Kazakhstan, etc. This support is used in a wide range of fastening of preparatory, opening and other excavations.

The main task of the anchorage system is to mobilize and preserve the strength inherent in this array, so that it becomes self-sustaining. The conceptual model of the classifier of an anchoring system for mine workings is shown in Figure 1.

The mine workings in which the anchor bolt is installed in integration with frame supports are less costly compared to the workings, which are fixed only by frames (Table 1) [1].

Table 1. Maintenance costs of mine workings.

Costs to maintain one kilometer of mine workings, people / shifts.	Mine, Karaganda	
Type of fasteners	Kostenko Mine	Baizhanov Mine
Metal	1962	1510
Prefabricated reinforced concrete	2411	13164
Anchor, in integration with the frame	1133	1261
Anchorage	37	–

* Corresponding author: tomilova@kstu.kz

The introduction of a rectangular section of the preparatory excavations with an anchor support guarantees the probability of installing a mechanized lava mount, namely, in the conveyor and ventilation workings, which actually allows to abandon the complicated circuits of movement of sections on the end sections of the lava and to perform these operations without downtime.

The use of anchor bolts in mines of states with a developed coal mining industry allowed to reduce the consumption of metal, concrete and timber 5…10 times; in 3…5 times to increase productivity of work at fastening of developments; 2…3 times increase the rate of penetration; to halve the cost of fastening and maintaining the support in working condition during operation.

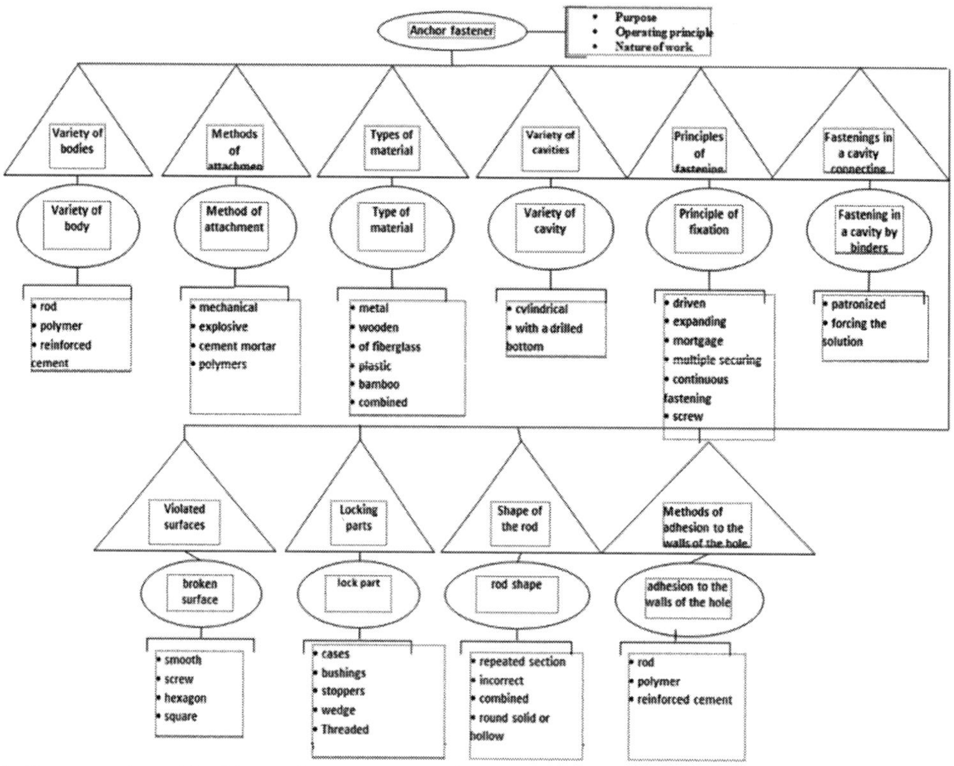

Fig. 1. Conceptual model of the anchorage system classifier.

2. Methods of solution

In connection with the complication of mining and geological conditions and increasing the depth of development in the creation of modern technological schemes for conducting mine workings with anchorage, a comprehensive approach combining the study of the stress-strain state of the rock massif and the synthesis of the optimal components of the subsystem "mining and preparatory work" was applied.

The experience of using anchor support in France, England, USA, Germany, Japan has shown that the main directions of improving the anchor support are: increasing the bearing capacity of the anchor, reducing the duration of the installation of anchors and reducing the cost of anchorage.

Not paying attention to the fact that the installation of the anchor support requires less time than the installation of the arch support, in itself, this operation is considered to be the longest in the run-through cycle, the utilization factor of the short-cutter takes 20-30% of the working time. To increase the productive time of the road headers, a systemic approach is required, in which the processes of coal and rock breaking in the face, fixing, ventilation, transportation, auxiliary operations, supplies of materials must be optimized in accordance with the productivity of the equipment used.

For this, a "tree" of variants of qualitative characteristics and quantitative parameters of mining preparatory work is formed on the basis of establishing ranges and steps of changing the dimensions of mine workings, their elements, number, length, speed of advancement of bottom faces, etc.

The presentation of the determining factors on the technological scheme of mining-preparatory works is possible in the following form:

$$\boldsymbol{T} = \boldsymbol{T1} \cup \boldsymbol{T2} \cup \boldsymbol{T3} \cup \boldsymbol{T4} \cup \boldsymbol{T5} \cup \boldsymbol{T6}, \quad (1)$$

where T1–T6 – respectively, the vectors of mining and geological (T1), mining conditions (T2), cost indicators (T3), socio-economic conditions (T4), conditions determining geomechanical features of mining (T5), and the possibility of actual reproduction of the front of clearing works (T6).

The set of initial mining and geological conditions T1 has the form (Table 2) [2]:

$$\boldsymbol{T1} = (W_n, m, \alpha, Y, P, \mathrm{H}_g, \mathrm{O_в}, q) \quad (2)$$

Table 2. Characteristics of mining and geological developments.

Mining-geological conditions	Value ranges	Initial values
Formation thickness (m), m	1.2–3.5	1.2; 1.5; 2; 3; 3.5
Angle of fall (α), degree	0–30	0; 3; 8; 10; 15; 25; 30
Stability of the roof rock layers (Y)	From unstable to sustainable	Unstable (ky = 3) Medium stability (ky = 2) Sustainable (ky = 1)
Natural gas content of the reservoir (q), m^3/t	0–15	0–15
Depth of mining (H_g), m	300–1200	300, 600, 900, 1200
Danger of formation on spontaneous combustion of coal ($O_в$)	From dangerous to non-hazardous	Dangerous (Cy = 1); Not dangerous (Cy = 2)
Water formation of the formation (W_n)	0–15	0; 15
Propensity of the soil rocks of the formation to pouches (P)	From non-fluffing to strongly bulging	Non-budding (ky = 1) Weakly speaking (ky = 2) Strongly bulging (ky = 3)

The range of qualitative and quantitative parameters of the technological scheme for mining is presented in Table 3.

Table 3. Range of changes in qualitative and quantitative parameters of the technological scheme for mining.

Mining and technological conditions	Value Range
Speed of holding, m/day (V)	10…20 (up to 50–60)
Location relative to the front of the clearing works (Po)	In the zone and outside the zone
Scheme of carrying out (C)	Single-Paired
The method of conducting (spruv)	Combine
Method of attachment (Skr)	Arched, Combined and Anchor
The method for transporting the rock mass (Cr)	Conveyors, Trolleys
Method of performing special measures (See)	No, There are

The vector of mining technological conditions has the form:

$$T2 = (\boldsymbol{L}_{л}, \boldsymbol{L}_{ст}, \boldsymbol{X}, \boldsymbol{T}_p, \boldsymbol{M}_n, \boldsymbol{R}, \boldsymbol{B}) \quad (3)$$

The operating conditions of the anchor support are characterized by the following mining parameters (Table 4), [3].

A methodical approach for the creation of advanced technological schemes for mining workings is presented in Figure 2.

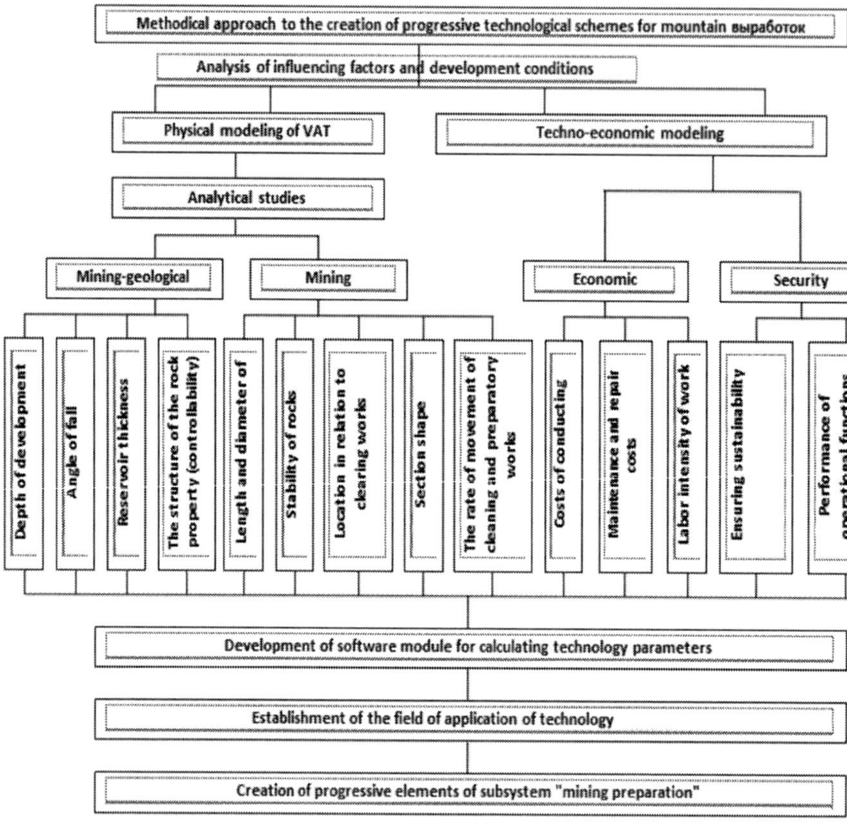

Fig. 2. Solving of the problem of creating progressive technological schemes.

Table 4. Parameters of the use of anchor support in mining technological developments.

Anchor parameters	Countries				
	Germany		England	Australia	USA
Diameter, mm	25–30.4		22	19–21	16–19.1
Length in the rock, m	2.1–2.4		2.1–2.3	1.5–2.3	1.4–2.3
Estimated load capacity (kN/anchor)	360–530		310	220–320	150–220
Section shape of the output	Arched	Rectangular	Rectangular	Rectangular	Rectangular
The density of the installation (anchor / m^2) roof/side	1.0-2.0	1.4-2.2	1.1-3.0	0.5-0.7	0.4-0.7
	0.6-1.9	0.5-1.2	0.3-0.9	0.11-0.23	0.09-0.15

3. Conclusion

1) To ensure an effective increase in the volume of mine workings, a methodical approach has been developed to solve the problem of creating a progressive technology for conducting and maintaining preparatory mine workings.
2) To ensure an effective increase in the volume of mine workings, a methodical approach has been developed to solve the problem of creating a progressive technology for conducting and maintaining preparatory mine workings.
3) To ensure the protection and increase the efficiency of underground mining, the main provisions of the underground coal mining technology are formed, based on the improvement of the subsystem of "mining and preparatory work", which consist in finding the patterns of development of geomechanical processes for the development of optimal technological schemes for carrying out mine workings with an anchor fastening.
4) The procedure for the development of technological schemes for conducting mine workings with the use of anchorage was taken into account, taking into account the stressed state of the rock massif:

- Study of regularities of interaction between anchor support and rock massif, with geomechanical processes taking place in host rocks and under the influence of technological factors.
- Determination of the parameters for fixing the mine workings taking into account the impact of mining and time factors.
- Creation of a cadaster model for establishing the area of rational application of anchorage of mine workings and zoning.
- Development of progressive methods and means for attaching control to the state of the array to improve the quality of the strengthened rocks prone to changing mining mining conditions, taking into account the technogenic stress-strain state of the massif.

References

[1] V. Demin, Proceedings of the Conference Actual problems of the mining and metallurgical complex of Kazakhstan-na-Karaganda, 128 (2017)
[2] *Handbook on the securing of capital and preparatory mine workings* (Nedra, Moscow, 1982)
[3] Zh Vareha, A Saranchukov, V. Demin, Proceedings of the International Scientific and Practical Conference Actual Problems of the Mining and Metallurgical Complex of Kazakhstan, 102 (2007)

MATEC Web of Conferences **155**, 01032 (2018) https://doi.org/10.1051/matecconf/201815501032
IME&T 2017

SIMULATION MODELING OF INTELLIGENT CONTROL ALGORITHMS FOR CONSTRUCTING AUTONOMOUS POWER SUPPLY SYSTEMS WITH IMPROVED ENERGY EFFICIENCY

Ruslan Gimazov[1], *Stanislav* Shidlovskiy[2,*]

[1] Tomsk Polytechnic University, School of Energy & Power Engineering, 634050, Tomsk, Russia
[2] Tomsk State University, Faculty of innovative technologies, 634050, Tomsk, Russia

Abstract. The paper considers the issue of supplying autonomous robots by solar batteries. Low efficiency of modern solar batteries is a critical issue for the whole industry of renewable energy. The urgency of solving the problem of improved energy efficiency of solar batteries for supplying the robotic system is linked with the task of maximizing autonomous operation time. Several methods to improve the energy efficiency of solar batteries exist. The use of MPPT charge controller is one these methods. MPPT technology allows increasing the power generated by the solar battery by 15 – 30%. The most common MPPT algorithm is the perturbation and observation algorithm. This algorithm has several disadvantages, such as power fluctuation and the fixed time of the maximum power point tracking. These problems can be solved by using a sufficiently accurate predictive and adaptive algorithm. In order to improve the efficiency of solar batteries, autonomous power supply system was developed, which included an intelligent MPPT charge controller with the fuzzy logic-based perturbation and observation algorithm. To study the implementation of the fuzzy logic apparatus in the MPPT algorithm, in Matlab/Simulink environment, we developed a simulation model of the system, including solar battery, MPPT controller, accumulator and load. Results of the simulation modeling established that the use of MPPT technology had increased energy production by 23%; introduction of the fuzzy logic algorithm to MPPT controller had greatly increased the speed of the maximum power point tracking and neutralized the voltage fluctuations, which in turn reduced the power underproduction by 2%.

1 Introduction

One of the main problems for autonomous robotics industry is power supply of robots. One of the possible solution is supplying robots with solar batteries. This method has a number of disadvantages connected with low efficiency of modern solar batteries. However, there are methods to improve the energy efficiency of solar batteries. The use of MPPT charge controller is among these methods.

* Corresponding author: shidlovskiy@tpu.ru

Maximum Power Point Tracking (MPPT) is one of the ways to improve the energy efficiency of photovoltaic modules and wind-energetic turbines by obtaining the maximally possible power on output of these systems. MPPT is realized by use of digital devices that analyze the current vs. voltage curve (CvV) in order to determine the optimal voltage-current pair, providing the maximum output power.

The most common MPPT algorithm is the perturbation and observation algorithm. In this method, a MPPT device changes the input resistance by a small amount, thus changing the voltage of the solar installation, and measures the power; if power increases - the controller continues to change the voltage in the same direction until the power stops to increase. Figure 1 shows the principle of perturbation and observation algorithm.

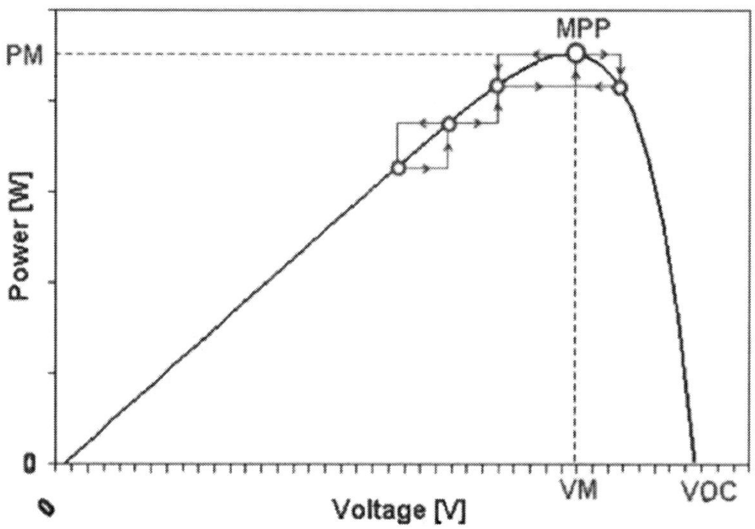

Fig. 1. The principle of the maximum power point tracking according to the perturbation and observation algorithm.

The disadvantages of the perturbation and observation algorithm are the power fluctuations and fixed time of "climbing" (the maximum power point tracking) [1]. Decrease of the "climbing" time leads to the increase in fluctuation amplitude, which leads to the power underproduction. Decrease of the power of fluctuation amplitude leads to the increase of time of the maximum power point tracking. These problems can be solved by using a sufficiently accurate predictive and adaptive algorithm [2, 3].

The urgency of solving the problem of improved energy efficiency of solar batteries for supplying the robotic system is linked with the task of maximizing autonomous operation time.

The aim of this work is to develop an autonomous solar-powered supply system with improved energy efficiency.

The development tasks include system design and simulation.

2 Development of the autonomous power supply system

An autonomous robot with the solar-based power system should consist of lightweight and high-strength materials and electronic circuitry that consume little energy. Lower level of energy consumption and lower weight of the structure make the use of solar batteries more promising.

There are several options for building the solar-based power systems of robots [4]. The paper considers the development of the combined power system based on solar batteries.

As we have noted above, use of MPPT charge controller can improve energy efficiency of the solar-based power supply system. It is possible to realize such device for an autonomous robot by means of a single-board microcontroller (Arduino, Mbed, etc.). The idea of constructing the charger based on single-board microcontroller for the photovoltaic plant is as follows: Buck-converter, which is controlled by a microcontroller, is used as a charger. When the voltage/current of the photovoltaic module increases, the PWM generator increases pulse frequency, which in turn leads to increase of the output current.

3 Simulation modeling of the system

An autonomous robot with the solar-based power system should consist of lightweight and high-strength materials and electronic circuitry that consume little energy. Lower level of energy consumption and lower weight of the structure make the use of solar batteries more promising.

We used MATLAB / Simulink environment to create a model of an autonomous power system.

In order to evaluate effectiveness of the MPPT algorithm and the fuzzy logic-based MPPT algorithm, in addition to the controller itself we designed photovoltaic system including a solar battery, charge controller, accumulator and load. Figure 2 shows the developed model.

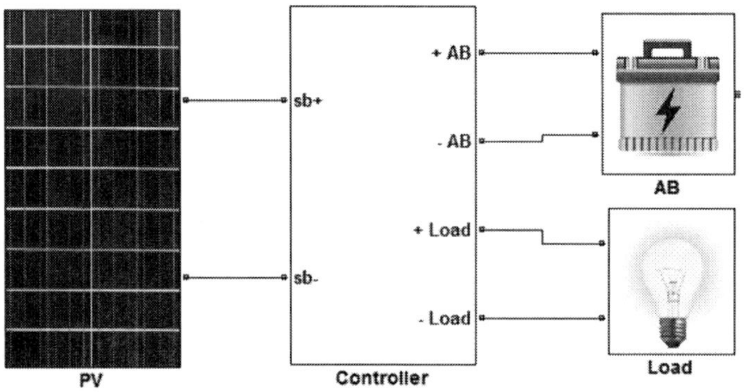

Fig. 2. Model of photovoltaic system.

Perturbation and observation MPPT algorithm is realized in the controller unit. Figure 3 represents this model.

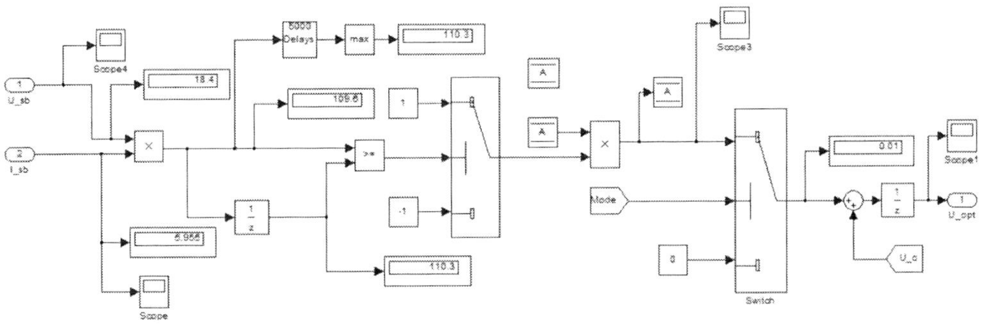

Fig. 3. Model of a common MPPT controller.

Use of the MPPT controller increases the energy production of solar batteries and increases the rate of battery charge. Figure 4 shows comparative graphs of the rate of battery charge (no load).

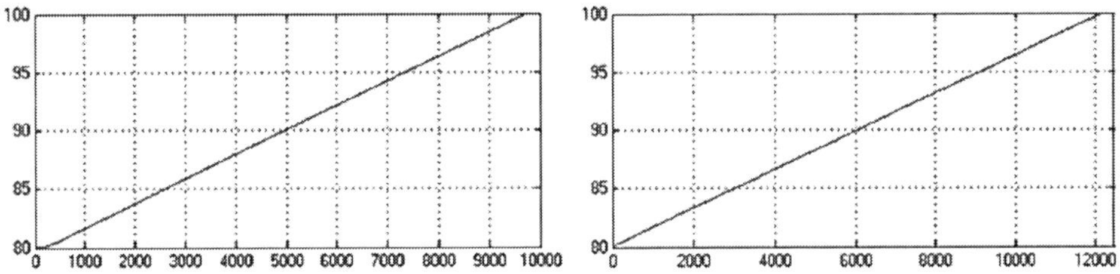

Fig.4. Comparative graphs of the rate of battery charge: left – with use of MPPT; right –with the set voltage.

Figure 4 shows that the rate of battery charge has increased significantly with the MPPT controller. Figure 5 represents the mode of the maximum power point tracking of the developed model.

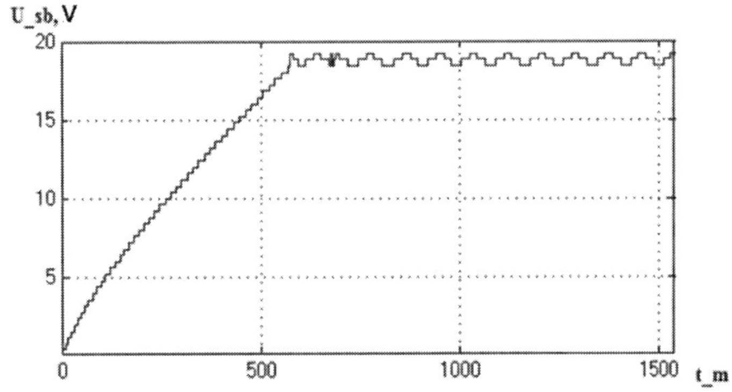

Fig. 5. Mode of maximum power point tracking.

Figure 5 shows that during the algorithm realization, the voltage and thus the power begin to fluctuate; it leads to underutilization of power generated by the solar battery. As we have noted above, this problem can be solved by using the fuzzy logic-based adaptive algorithm. Figure 6 shows a model of the MPPT controller with a fuzzy regulator.

As input linguistic variables of the fuzzy controller, we use the ratio of solar power change to solar voltage change (E), as well as the behavior of E (dE):

$$E(k) = (P_{sb}(k)-P_{sb}(k-1))/(V_{sb}(k - V_{sb}(k-1)) \quad (1)$$

$$dE(k) = E(k)-E(k-1) \quad (2)$$

The output variable of the fuzzy logic unit is a scaling factor of the step–size MPPT algorithm. Dependence of the output variable value on the input linguistic variables is defined by the set of rules given in table 1.

Table 1. Rules of the fuzzy controller.

E↓ \ dE→	**low**	**middle**	**high**
low	low	low	low
middleL	midL	midL	low
middleH	midH	midL	low
high	high	midH	midL

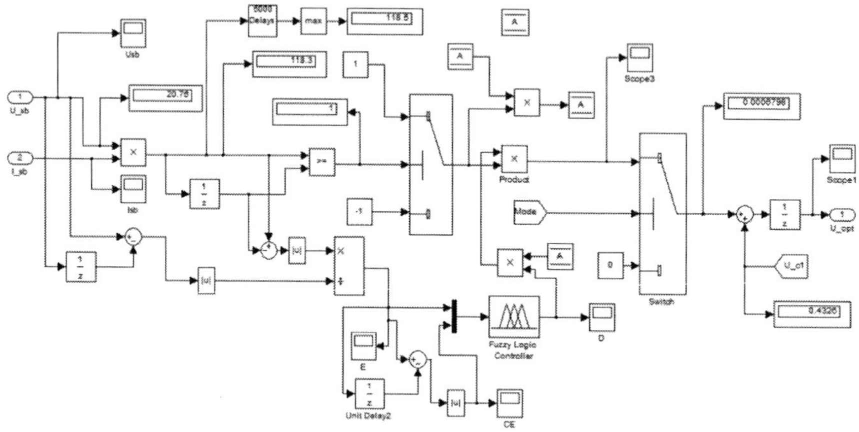

Fig. 6. Model of a fuzzy logic-based MPPT controller.

4 Study of the influence of the fuzzy logic unit on the MPPT controller

To study the influence of the fuzzy logic unit on the solar photovoltaic plant, we have set the following initial parameters of the model:

- Normal conditions: temperature 25^0 C, solar insolation level W = 1000 kWh/m^2, the angle of incidence $\alpha = 90^0$.
- Consumer load is turned off.

Figures 7, 8 represent results of modeling.

Figure 7 shows that the use of the fuzzy logic unit in the MPPT controller has greatly increased the speed of the maximum power point tracking and reduced voltage fluctuations, which in turn neutralized the power fluctuations (Figure 8).

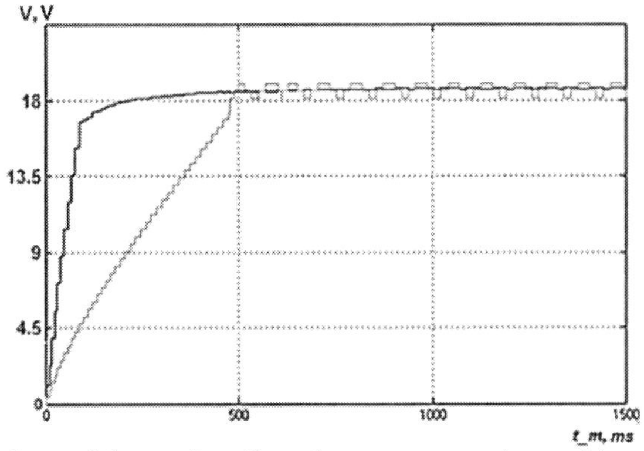

Fig. 7. Voltage comparison of the modes of maximum power point tracking: green – normal MPPT algorithm; blue – fuzzy logic-based algorithm.

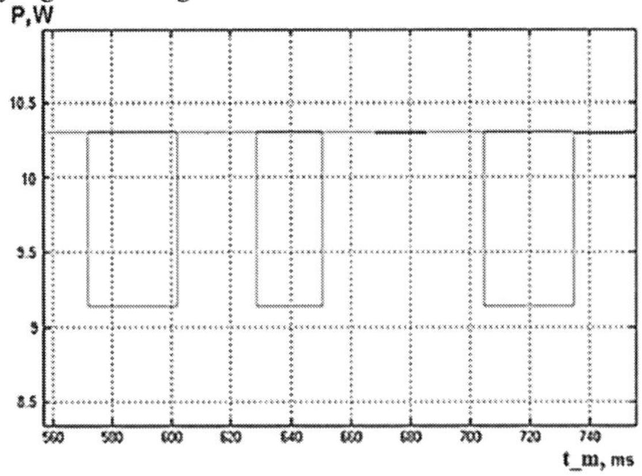

Fig. 8. Power comparison of the modes of maximum power point tracking: green - normal MPPT algorithm; blue - fuzzy logic-based algorithm.

5 Conclusion

The described work resulted in development of a solar-based power supply system of autonomous robots. Effectiveness of the developed system is improved by the use of the MPPT charge controller with a fuzzy logic-based adaptive algorithm. Effectiveness of the perturbation and observation MPPT algorithm with adaptation was studied by means of simulation modelling, which proved the model adequacy. The comparative characteristics of the controller with the fuzzy logic and without it were obtained. The developed fuzzy control system increased the speed of the maximum power point tracking and neutralized the power fluctuations.

Simulation modelling established facts that the use of MPPT technology increased energy production by 23%, introduction of the fuzzy logic-based algorithm into the MPPT controller had greatly increased the speed of the maximum power point tracking, neutralized the voltage fluctuations, which in turn reduced the power underproduction by 2%.

Acknowledgment

The following research has been financially supported by Russian Foundation for Basic Research (RFBR) 16-07-01138 A "Intelligent Reconfigurable Control Systems, Navigation and Image Processing for Autonomous Mobile Robots."

Reference

[1] V.A. Russkin, S.M.Semenov, R.K. Dixon, Bulletin of TPU. **4**, 78 (2016)
[2] S.V. Shidlovskiy, *Automatic control. Reconfigurable structures.* (Tomsk State University, Tomsk, 2006)
[3] D.V. Shashev, MATEC Web Conf. **79,** (2016)
[4] D. Lovin, *How to create android robot by own hands* (DMK-Press, Moscow, 2007)

SHORT-TERM FORECASTING OF ELECTRIC ENERGY GENERATION FOR A PHOTOVOLTAIC SYSTEM

V.T. Dinh, *Yuhao* Yan[a]

Tomsk Polytechnic University, 634028, Tomsk, Russia

Abstract. This article presents a short-term forecast of electric energy output of a photovoltaic (PV) system towards Tomsk city, Russia climate variations (module temperature and solar irradiance). The system is located at Institute of Non-destructive Testing, Tomsk Polytechnic University. The obtained results show good agreement between actual data and prediction values.

1 Introduction

The renewable technologies provide a feasible alternative of generating electricity which inexhaustible, clean and can be used in a decentralized way. Considering the geographical position and climatic characteristic of Tomsk Oblast, Russia, there is a good potential of renewable energy development in PV systems [1]. The prediction and optimization of electrical performance of PV system provides essential references values on maintaining the actual load stability as well as the system design considerations of GCPV system.
The analysis of difference models showed that: the time series data of PV power generation output have chaotic characteristics, so they can be predicted. PV power generation prediction generally includes solar irradiance forecast, temperature forecast and model output statistics [1], and the key is solar irradiance forecast. According to the range of forecast time, solar power generation prediction can be divided into short-term forecast (1 hour–10 days), medium-term forecast (11–30 days), long-term prediction (1–5 years), and the most useful prediction is short-term prediction.

All the PV power generation prediction models include two stages: data collection; and data processing. The role of data collection is to provide input information for the prediction model. As for PV power prediction, the most important information are meteorological parameters and historical PV power output data. The meteorological parameters come from measuring device (such as weather sensors, etc.) and weather forecast information.
Recently, there are few methods that involve in PV system outputs prediction. For example by employing Artificial Neural Network (ANN) which utilizes the solar radiation, ambient temperature and module temperature as it inputs whereas voltage and current are predicted as outputs. Besides, the evolutionary programming (EP) algorithm schemes is implemented

[a] Corresponding author: sxsmymm@gmail.com

for adjusting the optimal value of the fill factor in order to achieve more accurate outputs of PV systems.

In this paper we demonstrate the necessity of short-term time series prediction of generated power using mathematical approach based on fundamental equations.

2 Theoretical aspects of forecasting of PV electric energy

2.1 About photovoltaic system

The tested PV system is located at Institute of Non-destructive Testing, Tomsk Polytechnic University, Tomsk, Russia. The station includes 3 kW solar battery and 2 kW wind-driven electric plant [2]. The station is shown in Figure 1.

The station consists of:

- solar generator ARPS-250 (10 elements);
- wind generator 1000 W (2 elements);
- voltage transducer =48/~220V, 50Hz.

Specification:

- generator capacity, no more than 5 kW;
- output voltage ~220 V, 50 Hz;
- surface area of the PV modules, no more than 230 m^2;
- effective area of the station, no more than 300 m^2;
- height of wind generator tower 6 m.

Operating conditions:

- temperature, from -50 to 50° C.

Fig. 1. Wind - solar station at Tomsk, Russia.

2.2 Statistical model of PV power output

The PV array power generates the DC component output, so the PV inverter performs the conversion of the variable DC output of the PV modules into a utility frequency AC current that can be fed into the commercial electrical grid or used by a local, off-grid electrical network. Therefore, the conversion efficiency of the PV inverter will affect the AC component of output power [3]. The maximum power output of the PV system can be expressed by

$$P_S = \eta_{IVN} \times P_A = \eta_{IVN} \times FF_A \times V_A \times I_A = \eta_{IVN} \times M_P \times M_S \times P_M$$

where,

η_{IVN} is PV inverter efficiency;

P_A and P_M are the PV array and PV module power output, respectively;

I_A and V_A are the PV array current and voltage, respectively;

The structure of EP based on the output power of photovoltaic system is shown in the Figure 2.

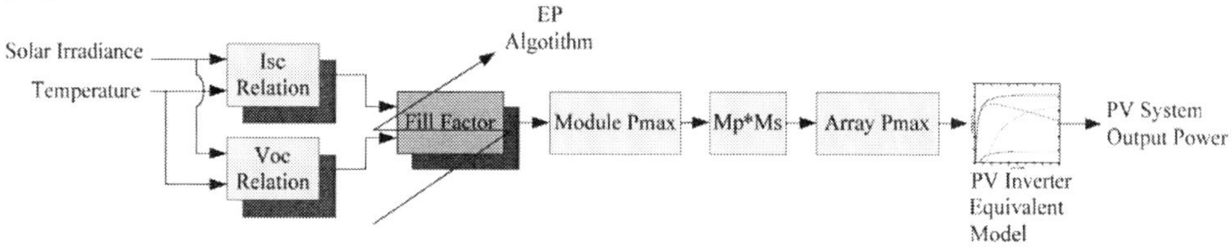

Fig. 2. Structure of EP based on the output power of photovoltaic system model.

2.3 The solar irradiation model

To calculate the sun's position in the sky at any location and time, two angles should be known: Azimuth angle and the Altitude angle. Also the sunrise and sunset times will help determining the start and end points of azimuth angle. In this paper we follow the convention of azimuth angle starting from zero degree at North Pole. Accordingly east, south and west poles are at 90°, 180° and 270°, respectively.

Azimuth angle (Figure 3) depends on latitude and time of year. The following equation is used to find the angle at any location [4].

$$\gamma = \frac{1}{2}\cos^{-1}\left[\frac{\sin\delta.\cos\phi - \sin\phi.\cos\delta.\cos H}{\cos\alpha}\right]$$

where,

H = $H = 15 \times (LST - 12)$ hour angle;
LST – local solar time;
δ – declination angle;
ϕ – latitude of the place;
α – Elevation angle.

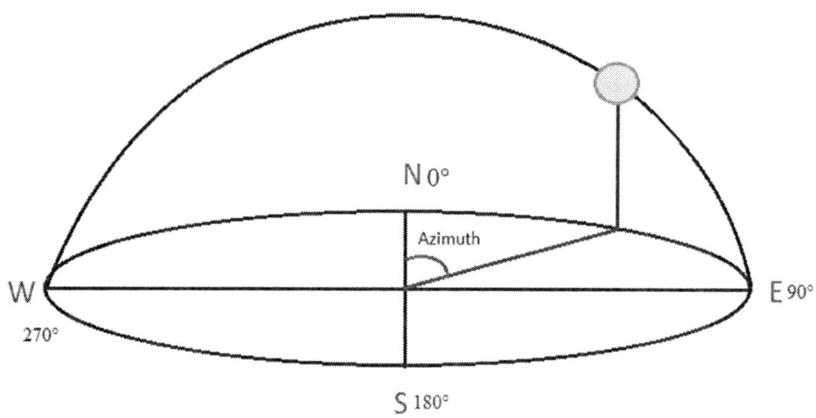

Fig. 3. Azimuth Angle.

In order to point the dependence of extraterrestrial solar radiation by the time of the year, the following equation were used [5]:

$$G_{on} = G_{sc} \cdot \left(1 + 0.033 \cdot \cos \frac{360 \cdot n}{365} \right)$$

where,

Gon is the extraterrestrial radiation measured on the plane normal to the radiation on the n-th day of the year;

Gsc is the solar constant, *Gsc* = 1367 W/m2;

n is the day of the year.

2.4 Temperature forecasting value

The detailed values of temperature forecasts over a 10 day period are provided by varies websites, in this paper we get the values from website wunderground.com by using python and packages Beautiful Soup 4 and Requests (Figure 4). The received values are organized with an array.

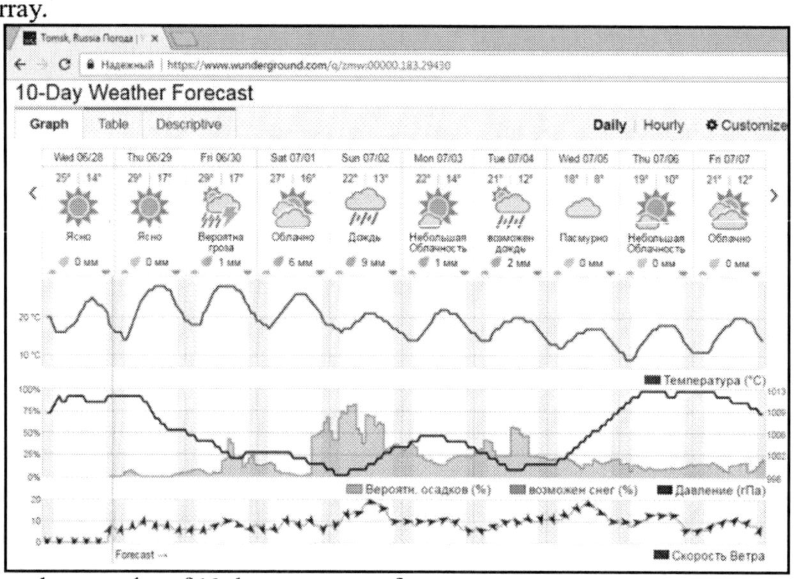

Fig. 4. Example screenshot of 10-day temperature forecast.

3 Results and discussion

Software solarEnergy was created on the basis of the MATLAB 2012b. The main window screenshot is shown in Figure 5. The size of PV panel and daily average consumption are set in the field "Real size of panel" and "Daily average consumption" of the software. Plots of prediction value of output and balance are shown in the plot area (Figure 6).

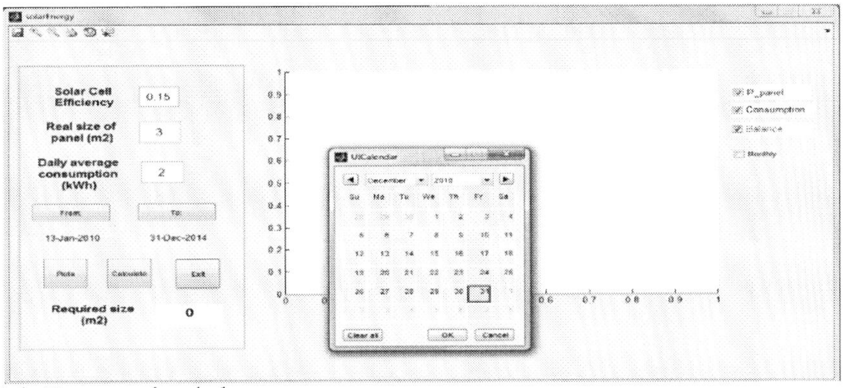

Fig. 5. SolarEnergy main window.

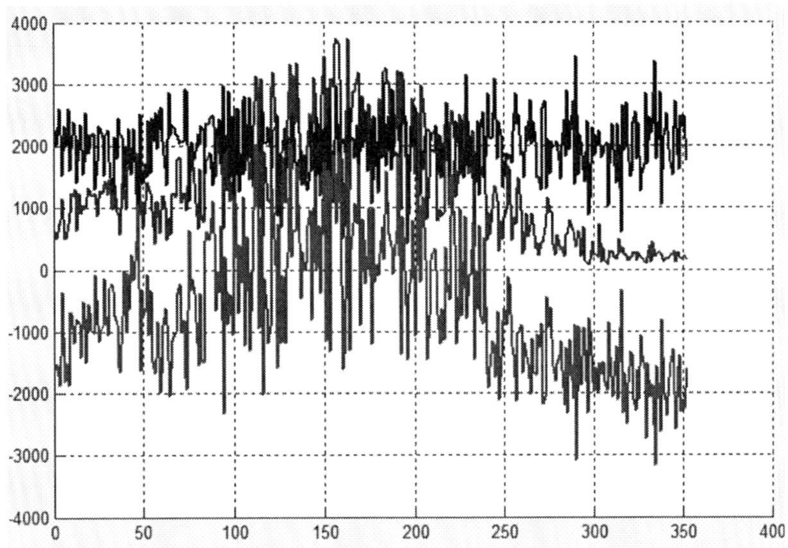

Fig. 6. Example of plot of power output and balance.

4 Conclusion

The performance of grid connected PV system is primarily influenced by the solar irradiance and module temperature. This paper demonstrated the prediction on PV system outputs based on mathematical approach for the PV system, which is located at Tomsk Oblast, Russia. The proposed approach provides an effective method for prediction and feasible in time under the same accuracy. Based on the results, it shows a good correlation between actual and predicted of electrical performance on grid connected PV system. In the future we plan to continue the research, including prediction using neutral network and machine learning methods to improve accuracy and range of forecasting, as well as design applications to wind farms and other renewable power sources.

References

[1] Z.W. Zheng, Y.Y. Chen, M.M. Huo, B. Zhao, ICSGC **12**, 601 (2011) doi: 10.1016/j.egypro.2011.10.081
[2] A.V. Yurchenko, A Ohorzina, IFOST **7**, 1 (2012) doi: 10.1109/IFOST.2012.6357698
[3] M P Thekaekara, A J Drummond, Natl. Phys. Sci **229**, 6 (1971)

doi: 10.1038/physci229006a0
[4] S.A. Khalil, A.M. Shaffie. RSER **27**, 853 (2014) doi: 10.1016/j.rser.2013.06.038
[5] L.T. Wong, W.K. Chow, Applied Energy **69**, 191 (2011) doi: 10.1016/S0306-2619(01)00012-5

ASIA SCIENCE AND TECHNOLOGY PORTALS

Phonepaseuth Solinthone[1,2*], *Tatiana* Rumyantseva[1], *Anachack* Phongtraychack[1]

[1]National Research Tomsk State University. 634050, Tomsk, Russia
[2] Ministry of Science and Technology, Vientiane, Lao PDR

Abstract. The article presents the information about science and technology portals in Asian countries and their role in dissemination of science and technology-related information. The article describes a new Lao Science and Technology Web Portal launched by the Ministry of Science and Technology of Lao PDR in 2017. The authors analyze goals, content, functions, and benefits of the Lao Science and Technology Web portal as well as problems suppressing its further development.

1 Science and Technology Portals in Asian Counties

The level of science and technology development in Asian countries varies. Several years ago, India, China and the West Asian civilizations were most notable for their contribution to science and technology development. At present, Japan is likely to be the most notable country in Asia in terms of its technological and scientific achievements; it is a world known country for electronics and automobile products. However, in recent years, China and India have also once again become major contributors to science and technology [1].

The Japanese Government launched the Asia Science and Technology Portal (ASTP) through Japan Science and Technology Agency (JSTA). The ASTP is designed to promote international research collaboration and researcher mobility. The portal contains information on researcher profiles, funding, job vacancies and international meetings. The Asia Science and Technology Portal was developed to support searching of S&T related information in ASEAN+6 countries (Australia, China, India, Japan, New Zealand and ROK), to promote international joint research and research personnel exchange, to enhance human networking related to S&T, and to provide fresh and practical information in an effective manner [2].

The ASTP project is carried out as an initiative in science and technology diplomacy of Japan; it is planned and proposed by the Cabinet Office and is operated and managed by JSTA. Sixteen participating countries contribute to the ASTP project.

Initially, the government of Japan proposed creation of an "Open Access Database" for promoting personnel exchanges and joint research among the countries of the region at the "Informal ASEAN+3 (China, Japan and Korea) Ministerial Meeting on Science and Technology" held in Malaysia in August 2006.

* Corresponding author: phonepaseuth.slt@gmail.com

In addition, the Report on Science and Technology Diplomacy submitted by the Government of Japan in February 2010, elaborates on the creation of a mutual database for the Asia region as the basis for creation of a global human resources network [3].

As a first step for achieving these goals, JSTA created the Asia Science and Technology Portal based on the initiative of the Cabinet Office.

To consolidate an international framework of the 16 countries in East Asia, the Cabinet Office has contacted, through diplomatic channels, each government for their cooperation and support. Further, Japan introduced the concept and mission of ASTP at the second ASEAN-Japan Cooperation Committee on Science and Technology in Vientiane, Laos (May 20, 2010). Ten ASEAN members include Brunei, Cambodia, Indonesia, Laos, Malaysia, Myanmar, the Philippines, Singapore, Thailand, and Vietnam [4].

The target group of ASTP includes:

1. Young and promising researchers who wish to conduct joint international research.
2. Researchers who wants to initiate a collaborative research in ASEAN+ six countries;
3. Researchers and policy makers wishing to conduct research on science and technology policies in ASEAN+ six countries.

ASTP provides the following information:

- Research Opportunities (Research Funding, R&D Job Postings).
- Meetings and Events in Asia.
- Articles and Commentaries related to Joint International Research implemented in the East Asian Region.

Links (Research Results, Human Resource Development & Research Cooperation, User Facilities)

- R&D Personnel Directories.
- Universities, Public Research Institutes, Academic Societies.
- Life in Foreign Countries (Housing, Education, Visas, etc.).
- Science and Technology Policies and Indicators in Participating Countries [3].

Information can be searched by country, subject area and keyword. RSS feeds are available for daily updates. Users may communicate through posting comments.

Currently, some Asian countries still have a limited access to data in the field of science and technology. Different organizations attempt to develop specialized web-portals and disseminate science and technology related information [5].

2 Development of the Lao Science and Technology Web Portal

In 2017, Ministry of Science and Technology of Lao PDR launched the Lao Science and Technology Web Portal. The portal has the following goals:

- open possibilities for international and domestic collaborations in research for young and promising scientists.
- promotion of S&T cooperation and stability.
- fostering S&T growth and high level of employment.
- providing resources not only for scientists but also for policy makers (E-book portal).

The portal represents a website that provides a single point of access to applications and information. From a user perspective, a portal is a web site with pages that are organized by tabs or some other form of navigation. Each page contains a nesting of sub-pages, or one or more portlets with individual windows that display different information from static HTML content to complex web services. A page may contain multiple portlets, giving users access to different information and tools in a single place. Users can also add their own pages, adding portlets of their choosing, and changing the Look and Feel of the interface.

Lao Science and Technology Web Portal is intended to facilitate the transition process from a common to digital education. The portal has three main areas. The first area deals with education. It includes information about science and technology, various tips and recommendations for students and scientists, and e-books. The second domain is designed to introduce to the users' e- markets and supplier directories. This way we hope to integrate e-commerce to ever province in Laos via business-to-business (B2B) or business-to-government (B2G) cooperation. Finally, it gives an opportunity for a province to introduce its products online. The official domain is "stportal.most.gov.la".

The Ministry of Science and Technology is looking for recognition, for ease of operations, for productivity gains and cost savings, or a combination of all of these advantages; the portal will succeed or fail depending on the perceived benefits to the Ministry. Theoretically, every member of the science and technology community in Laos should benefit from the portal. It should make it easier and more efficient for people to carry out their role in the institution.

The ST web portal provides managers with information and support for effective decision-making, and provides feedback on daily operations, output, or reports (generated through accumulation of transaction data). The program will help our organization to reform and move towards the next steps necessary for developing Lao S&T system.

3 The Goals of the Lao Science and Technology Web portal

Currently, there is a number of limitations pertaining to how S&T information is being managed, reported and promoted in Lao PDR. The authors have identified the following key limitations:

1. Limited capacities to analyze S&T information.
2. S&T information is not detailed enough to meet the users' needs.
3. Lack of business integration with science, technology and innovation.
4. Lack of skilled staff and human resource capacities engaged in research.
5. Lack of quality data dissemination on a regular basis.
6. Poor financial support for research in science and technology.
7. Limited access to international resources and data.
8. Poor infrastructure, which leads to slow internet connectivity and its malfunctioning in remote places.
9. Absence of Institute for Science and Technology.

Based on the limitations noted above, the following targets have been identified for the Lao Science and Technology Web Portal:

1. Open up possibilities for international and domestic collaboration in research for young and promising scientists.
2. Promotion of S&T cooperation and stability.
3. Fostering S&T growth and high levels of employment.
4. Providing resources not only for scientists but also for policy makers (E-book portal).
5. Integration of business (E-Market).

The portal represents a website that provides a single point of access to applications and information. From a user perspective, the portal is a web site with pages that are organized by tabs or some other form of navigation. Each page contains a nesting of sub-pages, or one or more portlets with individual windows that display different information from static HTML content to complex web services. A page can contain multiple portlets, thus giving users access to different types of information and tools in a single place. Users can also customize their view of a portal by adding their own pages, adding portlets of their choice, and changing the Look and Feel of the interface.

Lao Science and Technology Web Portal is intended to facilitate the transition process from a common to digital education. The portal has three areas of action. The first area deals with education. It includes information about science and technology, various tips and recommendations for students and scientists, and e-books. The second domain is designed to introduce e- markets and supplier directories to the users. This will help to integrate e-commerce to every province in Laos via business-to-business (B2B) or business-to-government (B2G) cooperation. Finally, it gives an opportunity for a province to introduce its products online.

4 Content of the Lao Science and Technology Web Portal

Lao Science and Technology Web Portal contains the following sections:

1. News and Information on science and technology.
   - Selecting the journal articles. The selected articles must be related to environmental research published in peer-reviewed open-access high-impact journals. The articles should be recent (published within the past year) and cover a variety of topics such as science and technology, pollution, endangered species conservation, climate change, and others.
   - Text adaptation. The articles are then rewritten in the Lao language using common vocabulary, real-world examples and comparisons where appropriate.
   - The draft text is then edited and approved by the scientists who published the original research paper. The finalized text is then edited for aesthetics and readability by creating an appealing layout and adding illustrations.
2. E-book portal overview. Science and technology portal contains E-book Supplemental Materials for Lao people. The main idea is to collaborate with schools, universities and institutes form Laos and other countries as well as to collect and create e-books for Lao people, especially in the field of agriculture because about 70% of Lao people are farmers using low-level technology.
3. E-Market overview. Integration of business and ecommerce with each Lao province. It means that every province can upload their products to the web portal by means of central level-super admin.

5 Functions of the Lao Science and Technology Web Portal

The main functions of the Lao Science and Technology Web Portal include the following:

1. Search. Giving the people who visit the web portal the ability to easily find the information that they are looking for is a key component in creating a user-friendly website. Website navigation that is easy to use and understand is essential to user-friendliness, but sometimes website visitors need more than intuitive navigation to find the content that they are seeking. This is where a website search feature can be essential. Information can be searched by a subject area and a keyword.
2. RSS feeds. RSS (Rich Site Summary; originally RDF Site Summary; often called Really Simple Syndication) uses a family of standard web feed formats to publish frequently updated information: blog entries, news headlines, audio, and video. RSS feeds enable publishers to syndicate data automatically. RSS feeds are available for daily updates. RSS Feeds function is shown in Figure 1.

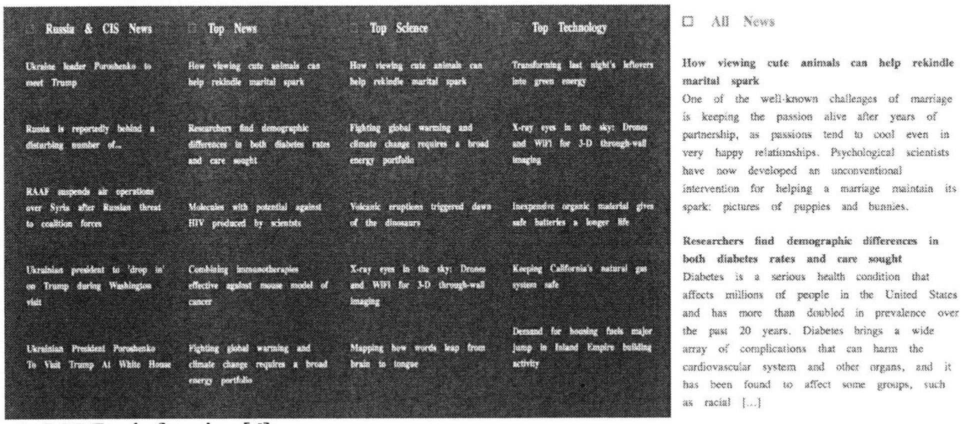

Fig. 1. RSS Feeds function [6].

3. Social media. The Social Media Icons Widget lets you add icons for the most popular social networks to sidebar or other widget area. The icons are linked to social media profiles, giving readers access to your latest status updates of the following social networks: Facebook, Twitter, and Gmail. Social media function is shown in Figure 2.

Fig. 2. Social media function [6].

4. E-books. An electronic book (or e-book) is a book publication available in a digital form, consisting of text, images (or both), readable on the flat-panel display of computers or other electronic devices. Although sometimes defined as "an electronic version of a printed book", some e-books exist without a printed equivalent. Education (Universities, Public Research Institutes, and Academic Societies) support books (Research Results, Human Resource Development & Research Cooperation, User Facilities) for Students and researchers. This E-book project is supported by the Ministry of Science and Technology, Ministry of Education and Sport of Laos, University, Institutes, and international partners. Interface of E-book collection supported by National Research Tomsk State University is shown in Figure 3.

Fig. 3. Interface of E-book collection supported by National Research Tomsk State University [6].

5. E-market. This section provides information on how to use electronic marketplaces, or e-markets, and supplier directories. We focus on e-markets and supplier directories, which have a focus on business-to-business (B2B) or business-to-government (B2G) cooperation.
   - It is open to several buyers and several sellers.
   - It is a trading platform, the e-market itself does not sell nor buy goods or services on the platform.
   - Supplier directories support companies in establishing new business relationships but no actual trade takes place at these platforms.
   - E-markets and supplier directories are also called B2B Internet platforms. Such platforms include all Internet-based technical solutions that aim at facilitating the establishment of new trading relationships between companies or at supporting existing relationships. The focus is put on trade to or between consumers.

6 Benefits of the Science and Technology Web Portal to Lao PDR

Benefits of the Science and Technology Web Portal to Lao PDR include the following:

1. Promotion of S&T policies and indicators in Laos.
2. Improved levels of customer interaction with an increase in customer services.
3. Information storage and viewing.
4. Promoting research opportunities.
5. Digital education (universities, public research institutes, academic societies) and networks (Research Results, Human Resource Development & Research Cooperation, User Facilities) for students and promising scientists.
6. Promoting S&T in Laos PDR.
7. Integration of business intelligence (E-Market).

7 Problems suppressing further development of the Lao Science and Technology Web Portal

We have identified some problems, which suppress further development of the Lao Science and Technology Web Portal. They include the following:

1. Proprietary data, access to which is controlled for commercial, national security, or competitive reasons.
2. Lack of comprehensive directory of resources, their content, or access methods.
3. Lack of IT skills on the district, village and provincial level.
4. Lack of requisite software and content in local language.
5. Poor financial support for research in the field of science and technology.
6. Limited capacities to analyze information in S&T.
7. Poor infrastructure, which leads to slow internet connectivity and its malfunctioning in remote places.

8 Conclusion

The Lao Science and technology web Portal intends to foment the transition process from a common education into a digital education. The second main area is e- market aimed to acquaint users with electronic marketplaces and supplier directories, and to introduce e-commerce in each Lao province via business-to-business (B2B) or business-to-government (B2G) cooperation. The third main area of the Lao Science and Technology web portal is making scientific research more accessible to the general audience and particularly to children. In addition, the web portal provides managers with information and support their effective decision-making, provides feedback on daily operations, output, or reports (generated through accumulation of transaction data). The portal will foster reforming and moving forward in developing Lao S&T system.

Acknowledgements

The research presented in this paper was supported by the Russian Foundation for Basic Research (grant No 16-29-04388/17).

References

[1] Science and technology in Asia. URL: https://en.wikipedia.org/wiki/Science_and_technology_in_Asia#South_Asia
[2] Introduction of Asia Science and Technology Portal (ASTP). URL: http://asti.dost.gov.ph/press-room/575-japan-science-and-technology-agency-jst-had-launched-the-asia-science-and-technology-portal-astp
[3] Background of ASTP. URL: https://www.jst.go.jp/pr/info/info737/index_e.html.
[4] ASEAN Membership URL: http://asean.org/asean/asean-member-states/.
[5] Japan Science and Technology Agency. URL: http://www.jst.go.jp/EN/
[6] Lao Science and Technology Web Portal URL: http://stportal.most.gov.la/

SOLAR SIMULATOR ON THE BASIS OF POWERFUL LIGHT-EMITTING DIODES

Oxana Aldoshina[1*], *Vyacheslav* Yugay[2], *Nurbol* Kaliaskarov[2], *Ulan* Esenjolov[2], *Saltanat* Nesipova[2]

[1] National Research Tomsk Polytechnic University, Lenin Avenue, 2, 634050, Tomsk, Russia
[2] Karaganda State Technical University, 56, Mira, Karaganda, 100000, Kazakhstan

Abstract. The Article describes the construction design and experimental results of the Solar Simulator, based on powerful LEDs with different radiation spectrum. There was established that the block from 25 light-emitting diodes provides radiation intensity sufficient for carrying-out of AM1.5G standard requirements. Fine tuning of the resultant spectrum and the required illumination for all radiation spectral components were obtained through light-emitting diodes current control.

1 The object of study

Solar Simulators provide illumination that is similar to natural sunlight in accordance with the requirements of the standard. They are intended for solar battery parameters estimation. Conditions for solar spectrum AM1.5 are more often used during the simulation operation.

There are several types of sources of light that are used for implementation of Solar Simulators. Xenon arc lamps are the most commonly used source of flat spectrum for today. There is a need of complicated optical system for this source that should provide a steady light during the testing and leads to low Energy Performance of this test system. Recently the medium power light-emitting diodes (around 1.5W) have been already applied for sunlight simulation, which intensity has 90% of the required standard intensity [1].

In this Article, we study and evaluate the features of Solar Simulators based on modern high power LEDs. In addition, there was conducted research of hybrid LEDs and sources in the form of halogen lamps with a B class spectrum usage [2].

Solar Simulator was designed with the use of light-emitting diodes, which are set to the printed wire board in the form of hexagon-shaped star. Most of the radiant power, about 10W, is formed by Bridgelux white light-emitting diodes. These light-emitting diodes provide necessary emission in 500…600 Nm spectrum and produce main part of the required optical power for 400…500 Nm spectral band [3,4].

The rest of power 3W is produced by Luxeon blue light-emitting diodes. LedenginLZ4 10W light-emitting diodes were used as a light source in 600-700 Nm and 700-800 Nm wavelength band, while Osram IR LEDs were used in 800-900 Nm and 900-1100 Nm band. Picture 1 shows the radiation spectrum of the used LEDs.

* Corresponding author: oxi80-07@mail.ru

2 Development

Printed wire boards with light-emitting diodes were placed in "comb" form, in order to provide optimal flow homogeneity, and set on the aluminum square plate, which was installed on the blown aluminum radiator (Figure 1). Each color of the radiation spectrum was regulated by an independent current, which was provided by the switched power supply TTI QL564-TP and Mastech HY-5003.

The spectral distributions of the power of the LEDs were measured by multichannel analyzer Hamamatsu PMA-12 and a spectrometer AvantesAvaSpec 2048 (Figure 2).

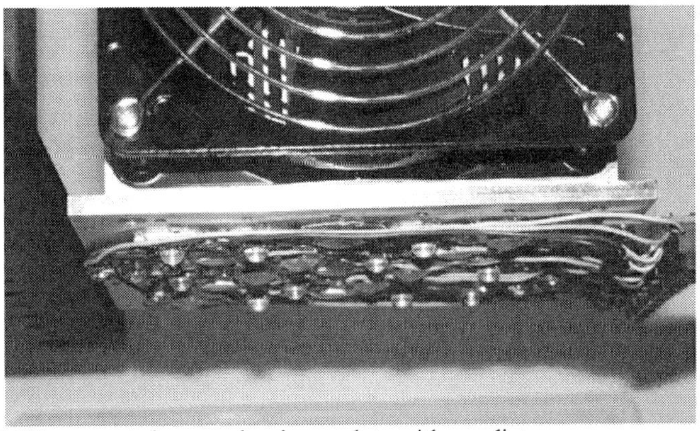

Fig. 1. Matrix of LEDs mounted on an aluminum plate with a radiator.

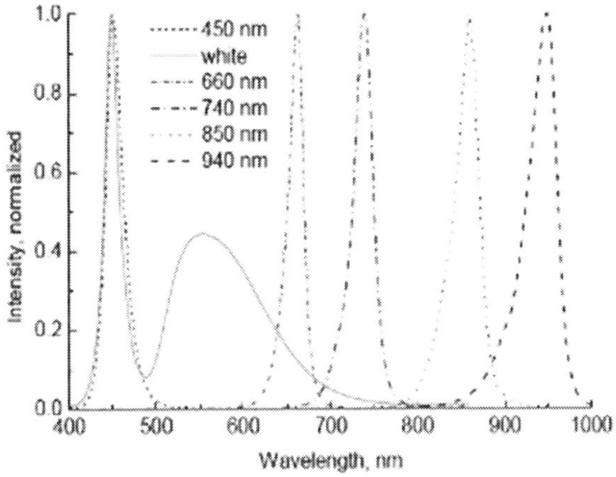

Fig.2. Radiation spectra of used LEDs.

The body of the Simulator was focused on the carrying-out of the AM1.5 requirements. Spectral and power characteristics of the Simulator with the selected light-emitting diodes were represented in Table 1. Six spectral bands and required illumination band were shown in the first two table columns. Prospective powers of the light-emitting diodes in the six spectral bands were shown in the right columns. The required illumination of each diodes group was determined on this value basis (the third line from below). After multiplying of this result to the projected sensitivity of the testing silicon photodiode (second line from below), the required photocurrent density came out, which was shown in the last line of the Table 1.

Table 1. Spectral and power characteristics of the Simulator.

Range, nm	mW/ cm²	The power of LEDs in the spectral ranges (%)					
		450 nm	white	660nm	740nm	850nm	940 nm
400 - 500	18.4	93	33	-	-	-	-
500 - 600	19.9	3	44	-	-	-	-
600 - 700	18.4	-	23	99	6	-	-
700 - 800	14.9	-		-	98	7	-
800 - 900	12.5	-		-	-	6	5
900 - 1100	15.9	-		-	-	-	97
Required illumination, mW/ cm²		3.7	46.0	9.1	152	13.4	15.4
Estimated* sensitivity, A / W		0.5	0.5	0.6	0.7	0.68	0.57
Photocurrent Density, mA / cm²		0.8	19.0	4.9	9.1	8.3	9.0

* Approximate values based on tabular data.

3 Measurement results

It was further investigated the distribution of radiation flux at various distances using motorized 2D platform (stands 8MT195 and 8MT295) and photodiode (OsramBPW34B) a Thorlabs power meter(PM100D). In Figure 3 and Figure 4 presents the results of experimental determination of the intensity of light radiation at a distance of 12 cm from the plane of installation of the simulator.

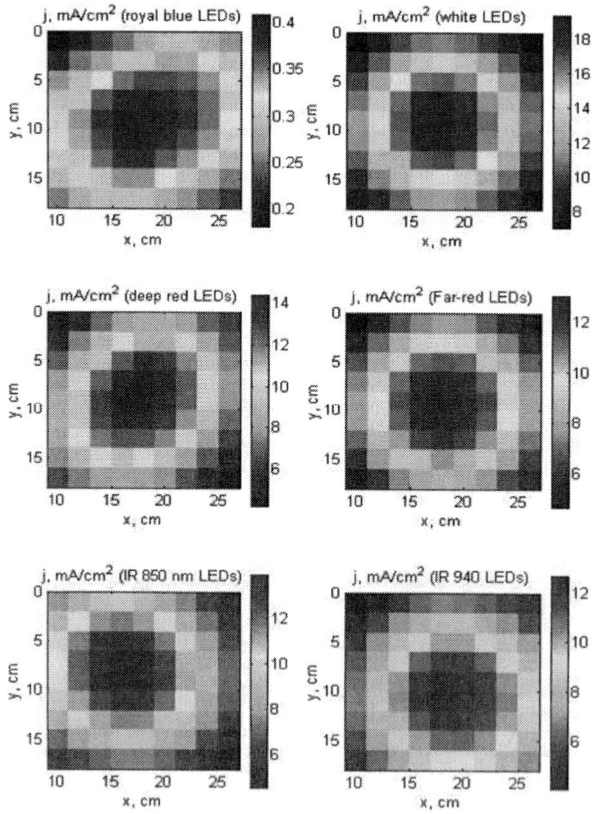

Fig. 3. Dependence of the photocurrent induced by a silicon photodiode OsraBPW34B at the measurement point.

The photocurrent of the photodiode Osram BPW34B Si was measured in two positions at a distance of 2 cm from each other. Arrays of various colors are switched one after the other in order to assess whether sufficient photocurrent for the illumination.At a distance of 12 cm from the plane of installation of the led array white led generates a current density of more than 18 mA / cm2.In accordance with the requirements AM1.5G it is a necessary level of illumination in the appropriate spectral range.The excess radiation power present in all of the above light-emitting diodes with a wavelength of about 600 nm. The contribution of blue LEDs is not sufficient for perfect compliance AM1.5G.However, they are not allowed by the standard, required for the spectrum in the range of 0.75 to 1.25.

As can be seen from Figure 4, from all of the LEDs operating at maximum rated current, photocurrent of the silicon photodiode exceeds 70 mA / cm2.From this result follows that in this case, the photon flux is much higher than required in the condition of AM1.5G.

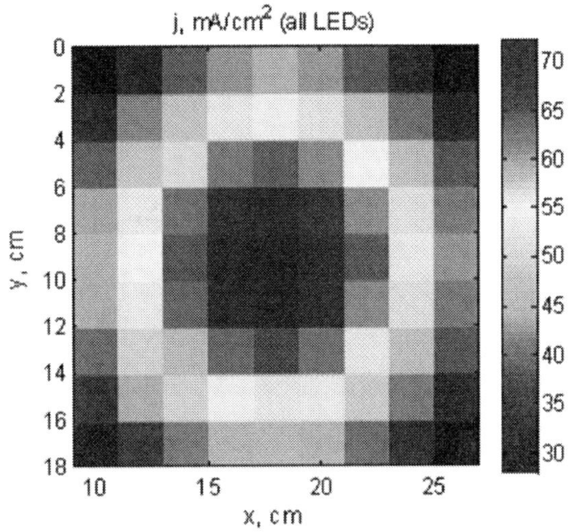

Fig. 4. Dependence of the photocurrent induced by the silicon photodiode OsramBPW34B in the case when all the LEDs operate at a rated current.

The peak of illumination of the developed solar`s simulator was measured for all six LED groups. The summarized results of the measurements are presented in Figure 5.

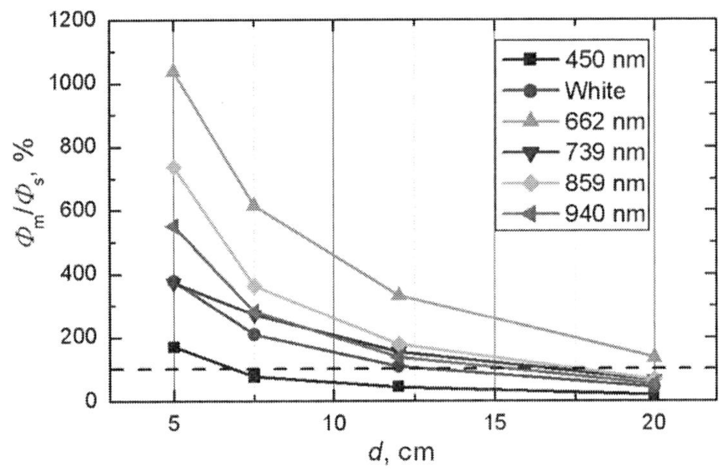

Fig. 5. Peak of illumination depends on distance to simulator.

For the most powerful diode group (white color) the required peak light exceeded respectively 4 and 2 times. Large excess capacity also has for the diode groups with wavelengths above 600 nm, 450 nm. These LEDs exhibit lower peak illumination, however, the main contribution in the wavelength range 400 - 500 nm is provided by white LEDs.

During the study of the illumination parameters and the spectral homogeneity of the simulator, the optimal distance was determined experimentally, which was approximately 8.2 cm from the lens of the LEDs. For the best approximation to the requirements of AM1.5G, optimal LED currents were also found. The spectral inhomogeneity of illumination at a higher spatial resolution was determined using a crystalline silicon photodiode. The result is shown in Figure 6.

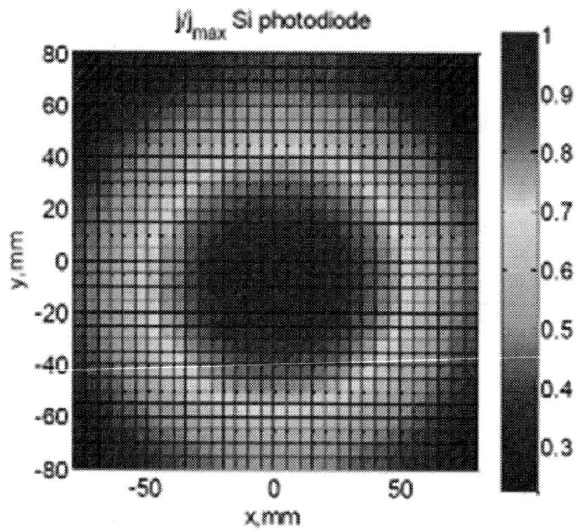

Fig. 6. Photocurrent distribution at a distance of 8.2 cm for a crystalline silicon photodiode detector.

For a circular area with a diameter 2 cm deviation of the photocurrent is less than 2 %, which corresponds to the A-class heterogeneity. For B and C classes, this area grows up to 3 cm and 4 cm, respectively.

4 Conclusion

To build a solar simulator of A class that meets the requirements AM1.5G, it is sufficient to use the 6 kinds of high power LEDs (10 W or more than 5 groups, 3 watts each).

In the developed simulator to produce the desired illumination of the usable area in the form of a circle with a diameter of several centimeters is used 25 LEDs and 39 watts of electric power.

References

[1] S.H. Jang, M.W. Shin, Current Applied Physics **10**, 537 (2010).
[2] M. Bliss et al., Solar Energy Materials & Solar Cells **93**, 825 (2009).
[3] A. Okhorzina, A. Bikbulatov, A.Yurchenko, N. Bernhard, O. Aldoshina, MATEC Web of Conf. **79,** 01010 (2016)
[4] E V Yurchenko, A D Mehtiev, V V Yugai, F N Bulatbayev. IOP Conf. Series: Materials Science and Engineering 81 (2015) 012114 doi:10.1088/1757-899X/81/1/012114

COMPUTER SIMULATION AND MODELLING SYSTEM OF NON-STATIONARY HEAT EXCHANGE PROCESSES

Aleksandr Pilipenko*, *Sergei* Petrov

Orel State University named after I.S.Turgenev, 302020, Oryol, Russia

Abstract. This article suggests the solution to the up-to-date task of improving and increasing the energy efficiency of buildings and quality of functioning of systems, controlling boilers and heating devices, by using computer simulation and modelling of heat exchange processes of the building and operation of the heating systems. The authors present proven (checked) mathematical models, applicable for external protections, rooms, heaters, boilers, insulated and non-insulated piping. The authors offer an automated system of computer simulation and modelling, which not only allows us to solve mathematical models, but also to complement them with information from a real object and improve them by introducing corrective factors and coefficients.

1 Introduction

The "Smart House" systems, aiming at improving energy efficiency, are being developed for more than 10 years in Europe, and new materials are being created to achieve heat and energy conservation [1]. However, it is worth considering that many of years of research on the territory with gentle climatic conditions may not be applicable in the Russian Federation, where climatic conditions differ and vary in the range from -50 to +50 °C. In this regard, the task of mathematical and computer modelling of non-stationary processes of heat exchange, which will help us to investigate the processes of mixing, convection, radiation and thermal conductivity in pipe railings and piping enclosures, protection enclosures (protective cover), walls, halls, pipelines, heating systems and boilers, is relevant and of vital importance [2]. In this paper, the authors make an attempt to describe the system of computer modelling of heat-exchange processes in various media and, under certain conditions, which will help to analyze the heating systems, rooms, buildings and other objects in details.

2 Mathematical models

Researches in the field of processes of conduction, convection, and radiation of heat in outdoor enclosures (possessing heat capacity or not) within the system are possible due to differential equations (1–3) and the boundary conditions (4–7):

* Corresponding author: a@pilipenko.info

$$\frac{\partial t_1(x,\tau)}{\partial \tau} = a_1 \frac{\partial^2 t_1(x,\tau)}{\partial x^2} \tag{1}$$

$$\frac{\partial t_2(x,\tau)}{\partial \tau} = a_2 \frac{\partial^2 t_2(x,\tau)}{\partial x^2} \tag{2}$$

$$\frac{\partial t_3(x,\tau)}{\partial \tau} = a_3 \frac{\partial^2 t_3(x,\tau)}{\partial x^2} \tag{3}$$

$$\frac{\partial t_1}{\partial x|_{x=0}} = -\frac{\alpha_B}{\lambda 1}(t_B - t_{CT.B}) \tag{4}$$

$$\frac{\lambda_1 \partial t 1}{\lambda_2 \partial x\, |_{x=\delta_1}} = \frac{\partial t_2}{\partial x\, |_{x=\delta_1}} \tag{5}$$

$$\frac{\lambda_2 \partial t_2}{\lambda_3 \partial x\, |_{x=\delta_2}} = \frac{\partial t_3}{\partial x\, |_{x=\delta_2}} \tag{6}$$

$$\frac{\partial t_3}{\partial x|_{x=\delta_3}} = -\frac{\alpha_H}{\lambda_3}(t_{CT.H} - t_B) + \frac{q_H}{\lambda_3} \tag{7}$$

where t_1, t_2 denote temperature of primary and secondary coolants in the heating element,°*C;* $t_{CT.B}$ denotes temperature of the inner (internal) surface of the protection enclosures (protective cover), °C; $t_{CT.H}$ denotes temperature of the outer (external) surface of the protection enclosures (protective cover), °C; $t_{ok.b}$ is temperature of the inner (internal) surface of the window, °C; λ_1, λ_2, λ_3 denote coefficients of thermal conductivity of protection enclosures (protective cover, guard, guarding rail) layers, Watt/m·°C; $\alpha_н$ denotes the coefficient of heat transfer from the outer surface of the protection enclosures, Watt/m·°C; $\alpha_в$ denotes the coefficient of heat reception (heat absorption) of the inner surface of the protection enclosures, Watt/m·°C; τ is time, h; $q_н$ denotes specific heat flow rates (heat flux), Watt/m^2; δ_1, δ_2, δ_3 denote thickness ratio of the protection enclosures (protective cover) layers, m; a_1, a_2, a_3 denote thermal diffusivity (conductivity) coefficients of the protection enclosures (protective cover) layers, m^2/h; $t_в$ is internal air temperature, °C.

Investigation of the processes of heat convection and radiation in heaters (heating apparatus) and non-insulated (non-isolated) pipe-lines is usually conducted on the basis of finding the solutions of differential equations (8–10) with boundary conditions (11–12).

$$c_1 m_1 \frac{\partial t_1}{\partial t} + c_1 Gl \frac{\partial t_1}{\partial x} = \alpha_1 F_1 (t_C - t_1); \tag{8}$$

$$c_c m_c \frac{\partial t_c}{\partial \tau} = \alpha_1 F_1 (t_1 - t_c) - \alpha_2 F_2 (t_C - t_B); \tag{9}$$

$$c_B m_B (t''_B - t'_B) + \alpha_2 F_2 \int_0^2 (t_c t_S) dx = \frac{\alpha Q}{d\tau}; \tag{10}$$

$$x = 0, t_1(x,\tau) = t'_1 \tag{11}$$

$$x = l_1 t_1(x,\tau) = t_1'' \tag{12}$$

where c_1, c_2 denote specific heat of primary and secondary coolant (heat-transfer agent), joules/kg·°C; c_c denotes specific heat of material of the heater wall, joules/kg·°C; t is temperature of the coolant (heat-transfer agent) after mixing, °C; t' denotes temperature of the coolant (heat transfer agent) at the inlet of the heat exchanger (converter), °C; t'' denotes temperature of the coolant (heat transfer agent) at the outlet of the heat exchanger (converter), °C; c_B specific heat of air in the room, joules/m·°C; F_1, F_2 are the heating surfaces, m^2; l is the pipe-line length, m; Q is the heat flow, Watt; α_1, α_2 denote coefficients of heat transfer from the coolant towards the pipe wall and from pipe wall towards the environment, Watt/m·°C; G denotes primary and secondary coolant consumption, kg/h; m_1, m_2 denote weight of the primary and secondary coolant in the heating device, kg; m_c, $m_в$ denote mass of the walls and outer medium (environment), kg; t_C denotes temperature of the pipe-line wall, °C.

Investigation of the processes of convection and heat conduction in isolated heat conductors becomes possible with an application of equations (13–15) with boundary conditions (16–18).

$$c_1 m_1 \frac{\partial t_1}{\partial t} + c_1 Gl \frac{\partial t_1}{\partial x} = \alpha_1 F_1 (t_C - t_1); \tag{13}$$

$$\frac{\partial t_c}{\partial \tau} = a_c \frac{\partial^2 t_c}{\partial y^2} \; 0 \le y \le \delta \tag{14}$$

$$\frac{\partial t_u}{\partial \tau} = a_u \frac{\partial^2 t_u}{\partial y^2}; \quad \delta_1 < y\delta_2 \tag{15}$$

$$\frac{\partial t_c}{\partial y|_{y=0}} = \frac{\alpha_1}{\lambda_c}(t_1 - t_c); \tag{16}$$

$$\frac{\lambda_c}{\lambda_u} \frac{\partial t_c}{\partial y|_{y=\partial_1}} = \frac{\partial t_u}{\partial y|_{y=\partial_1}}; \tag{17}$$

$$\frac{\partial t_u}{\partial y|_{y=\partial_2}} = \frac{\alpha_2}{\lambda_u}(t_u - t_н \tau); \tag{18}$$

where λ_c, λ_u – denote the pipe wall and thermal insulation thermal (heat) conductivity coefficients, Watt/m×C; a_c, a_u – denote thermal diffusivity (conductivity) coefficients of the material of the wall and and insulation, m^2/h; t_u – denotes external air temperature, C.

Investigation of the behaviour of heated surface of boilers and the convection and radiation processes is possible thanks to formulas (19–20), if we take into consideration boundary conditions (21–22).

$$c_1 m_1 \frac{\partial t_1}{\partial t} + c_1 Gl \frac{\partial t_1}{\partial x} = \alpha_1 F_1 (t_C - t_1); \tag{19}$$

$$c_C m_C \frac{\partial t_C}{\partial \tau} = ql - \alpha_1 F_1 (t_1 - t_C); \tag{20}$$

$$x = 0; (x,t) = t_1'; \tag{21}$$

$$x = l; (x,\tau) = t_1'' \tag{22}$$

where q – is a specific heat flow.

The analysis of such processes of heat transfer and exchange as radiation, convection and mixing, depending on changes in the volume of air in the room, is usually implemented with a help of using equation (23).

$$c_B V \frac{\partial t_B(\tau)}{\partial \tau} = Q_{OT}(\tau) + Q_u(\tau) + Q_b(\tau) - \alpha_B F_B (t_B - t_{CT}) -$$

$$- \alpha_B F_H (t_B - t_{CT.B}) - \alpha_B F_{OK} \times (t_B - t_{OK.B}) - c_B L (t_B - t_H) \tag{23}$$

where V – is the volume of air in the room, m^3; L – denotes ventilation indoors, kg/h; F_B - denotes the area of internal protections, m^2; F_H – denotes the area of external protections (walls), m^2; F_{ok} – is the area of the window, m^2; Q_{OT} is heat flow from the heaters of the system of heating, Watt; Q_u – denotes heat flow due to solar radiation, Watt; Q_b – denotes production of domestic heat, Watt.

Moreover, all the designed and applied mathematical models have been successfully tested and compared with real objects on the basis of boilers in Oryol. A plan of the experiment, including 9 tests in each of ten experiments, has been developed, and the sample size has been calculated and justified.

3 Computer modelling

The computer modelling system, based on LabVIEW graphical programming media, has been implemented, which originally has been developed as the system for mathematical calculations and computer modelling. The advantage of using LabVIEW in the simulation and modelling process is in availability of functions (virtual appliances and gadgets), allowing to solve differential equations in the period of time, similar to real, with a help of Runge-Kutta method (the second order) [3, 4]. The system, that performs computer simulations and modelling, sensor data collection and analysis of results, as well as defining the correction coefficient and including it into a mathematical model, has been created during the work progress.

Collection of information from the sensors is the function, which can be switched off, and can be replaced by manual typing of information, performed by the user. However, the authors believe that the product will be useful for engineers, working in the field of automation, as well as builders, technologists and energy transmission and transfer engineers, who can work directly on the analyzed object and, after having been connected to the existing sensors through data collection device NI USB- 6008 [5, 6], can enter and input the actual data from the object into the computer simulation and modelling system.

The front panel of the program has been designed in the form of five paged (riffled) tabs, thus, helping us to control work in a proper way at every separate stage (level). On the first tab (Figure 1), the key parameters, such as mathematical models, wall and pipelines materials, heating systems, etc. are being selected with a help of the database. Moreover, on the first tab there is a panel of graphs, that represent the generalized solution of mathematical models. On the second tab there are graphs for each individual parameter. On the third tab the solution of mathematical models in the form of an array representation (Figure 2), as well as differential equations with substituted known values and corrective coefficients, are being displayed. On the fourth paged through window the graphs of values, obtained from sensors, are being displayed. The fifth tab displays the correction coefficients and the graphs, which have already undergone the procedure of correction.

After having been launched, the program tests the ports and, if the sensors and ADCS (analog digital converters) are connected, you can choose the necessary options and parameters from the data collection devices. After selecting the options and parameters you can begin viewing the results of computer modelling, and information from sensors by switching tabs at the bottom of the window.

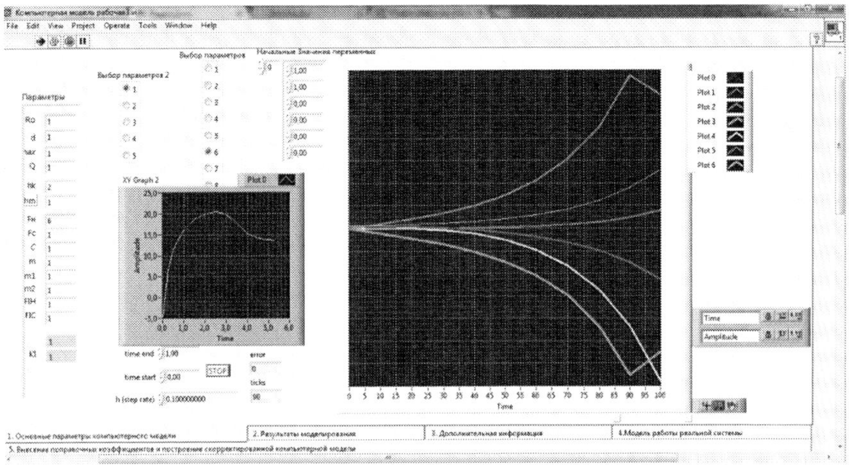

Fig. 1. The sreenshot of the main tab in computer modelling system.

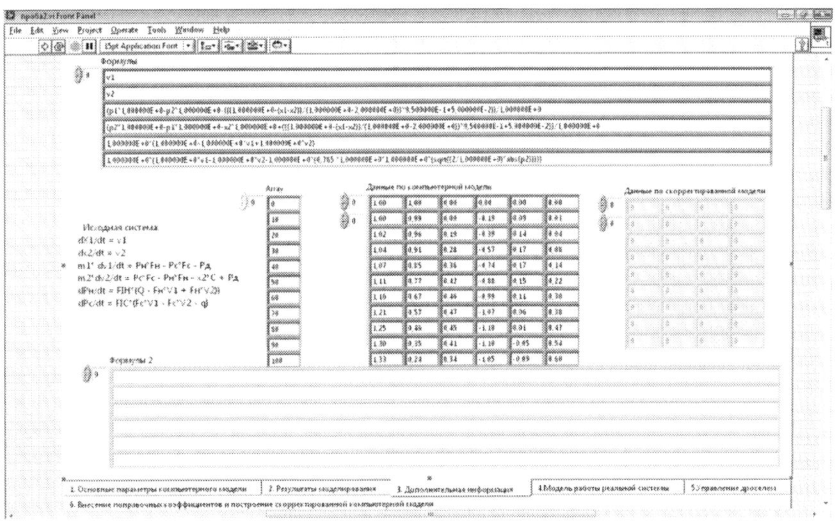

Fig. 2. The sreenshot of the tab with solution of mathematical models.

4 Conclusion

The use of the developed system of computer simulation and modelling allows us not only to simulate and model heat transfer processes in heating equipment and rooms, but after conducting measurements and imposing corrective coefficients also allows us to predict temperature fluctuations and measurements. Studies and scientific research in the field of heat exchange processes of the building is carried out before connecting any type of utilities, conveniences and facilities, as well as while conducting the processes of automation and modernization of old, obsolete rooms and buildings. The thorough and detailed analysis of heat exchange processes and selection of relevant parameters of the equipment operation allows us to increase the energy efficiency of buildings and to improve the quality of functioning of regulation and control systems. The work was carried out within the framework of the basic part of the State task for 2017 - 2019, project code 1.5265.2017

References

[1] S.Z. Gu, J.W. Yang, J.R. Liu, China Soft Science **1**, 1 (2013)
[2] S.P. Petrov *Analysis and synthesis of control systems of thermal energy transfer process in the cogeneration system of central heating* (Publishing House "Orlik & K", Orel, 2009)
[3] U.G. Pirumov, *Numerical methods* (Drofa, Moscow, 2003)
[4] A.V. Pilipenko, Fundamental and applied problems of technique and technology **6-2**, 58 (2012)
[5] G. B. Wiersma, *Environmental Monitoring* (CRC Press, New York, 2004).
[6] Liu Gaoqiang, Wu Weilin , MATEC Web of Conferences **61**, (2016) doi: 10.1051/matecconf/20166106006

HIGH-PERFORMANCE ADAPTIVE NEURO-FUZZY CLASSIFIER WITH A PARAMETRIC TUNING

Sergey Gorbachev [a], *Vladimir* Syryamkin

National Research Tomsk State University, 634050, Tomsk, Russia

Abstract. The article is devoted to research and development of adaptive algorithms for neuro-fuzzy inference when solving multicriteria problems connected with analysis of expert (foresight) data to identify technological breakthroughs and strategic perspectives of scientific, technological and innovative development. The article describes the optimized structural-functional scheme of the high-performance adaptive neuro-fuzzy classifier with a logical output, which has such specific features as a block of decision tree-based fuzzy rules and a hybrid algorithm for neural network adaptation of parameters based on the error back-propagation to the root of the decision tree.

1 Introduction

Foresight research is becoming a widespread tool for solving multicriteria problems of expert data analysis that allow revealing technological breakthroughs and strategic perspectives of scientific, technological and innovative development [1]. Foresight as a system of methods is being constantly developed and improved, and for the last twenty years, considerable experience of their practical application has been accumulated [2]. The analysis of publications shows that a combination of qualitative and quantitative methods is effective [3]. Moreover, constant complication of a system of methods requires development of intellectual approaches that ensure scientifically based transparent conclusions.

Thus, when evaluating the effectiveness of technologies, traditional statistical methods have their drawbacks, especially when two or more classes are connected by a chain of internally connected sampling objects, have a non-spherical form, are linearly inseparable sets, or when the density or volume of classes is different [4–6]. When classes intersect, contact or chain, the classification model is blurred, so the choice of mathematical methods for solving the problem of automatic classification is preceded by fuzzy procedures [7]. The theory of fuzzy sets operates with such concepts as a weighted membership of the object under investigation to the set and it offers a rather flexible apparatus for a formal description of similar situations. Therefore, the goal set transfers from question "Does the object belong to the class?" to identifying the degree of membership of an object to a

[a] Corresponding author: spp03@sibmail.com

particular class. Summarizing, we can say that a fuzzy approach to the solution of classification problems in certain cases allows us to separate clusters of complex shape, thereby opening up new possibilities for interpreting the results of classification.

Comparing the reliability of the results of classification obtained by different authors allows us to single out the most effective classifiers, which include neural networks [4, 8] and methods of cluster analysis [9, 10]. The best results were obtained using a three-layer perceptron and a probabilistic neural network, as well as Kohonen self-organizing maps. At the same time, in the studies known to us, classification of the effectiveness of scientific and technological solutions and technologies has not been considered, and a neural-fuzzy network has not been used to classify scientific projects.

The purpose of this study is to develop high-performance adaptive algorithms for neural-fuzzy inference when solving multicriteria problems of expert analysis (foresight), with high classification accuracy and logical inference based on a compact set of the classification rules.

2 A method for constructing an adaptive neural-fuzzy classifier

The construction of classical adaptive neuro fuzzy inference systems (ANFIS) is associated with formalization of input parameters and target indicators of the effectiveness of scientific (innovative) projects and technologies in the form of a vector of interval values (fuzzy interval); the hit in each interval of this vector is characterized by some degree of uncertainty [11]. This procedure is called "fuzzification". Based on primary data, experience, and intuition experts usually do not have difficulties in making quantitative evaluation of the boundaries (intervals) of possible (permissible) values of parameters and the range of their most possible (preferred) values. For example, experts can characterize the primary data using triangular fuzzy numbers with the following membership function $\mu A\ (x)$ of the fuzzy set A:

$$\mu_A(x)=\begin{cases}0,\ if\ x\le a_{\min}\\ \dfrac{x-a_{\min}}{c-a_{\min}},if\ a_{\min}<x<c\\ \dfrac{a_{\max}-x}{a_{\max}-c},if\ c\le x<a_{\max}\\ 0,\ if\ x\ge a_{\max}\end{cases} \quad (1)$$

This expression simulates the following expert position: "parameter A is approximately equal to c and uniquely enters the segment $[a_{min},\ a_{max}]$".

In practice, an analytical representation of the membership function is often used [7], for example, the Gaussian membership function has the form:

$$\mu_A(x)=e^{-\frac{(x-c)^2}{2\sigma^2}}, \quad (2)$$

where c is a center and σ is a standard deviation (width).

To form the base of fuzzy rules, the authors proposes to use the fuzzy decision tree [12, 13], which can be considered as a hierarchical network with direct signal propagation. Therefore, we propose to use a hybrid algorithm for back propagation of the error to the root node of the tree in order to adapt its parameters (centers and width of membership functions and parameters of rules consequents on tree leaves).

3 Results and Discussion

Let us consider a set of data D describing multidimensional feature objects (innovative technologies), where in each line there are m numerical values for the attributes $x_1,...,x_n$ and one class C_l among the set $C = \{C1, C2\}$.

Let x_k j be the value of the k-th input variable of the j-th observation;

C_j is a value of output variable of j-th observation;

N is a number of observations;

n is a number of input variables.

3.1 Initialization of fuzzy model (stage *1*)

Based on data fuzzification, we set the initial parameters of the term membership functions and construct a fuzzy solution tree whose leaves have output parameters r_{i1} and r_{i2} for each i-th branch. The structure of such a tree sets the basis for fuzzy rules such as a singleton with a complexly constructed consequent (conclusion) of the following form:

$$Rule\ i:\ IF\ x_1 = A_{1i}\ and\ x_2 = A_{2i}\ and\ ...\ and\ x_n = A_{ni}\ THEN\ y_1 = r_{i1}\ and\ y_2 = r_{i2}, \quad (3)$$

where A_{ki} is the linguistic term, by which the k-th input variable is estimated.

Figure 1 shows the structural-functional diagram of the constructed adaptive neural-fuzzy classifier with a logical conclusion that implements the mapping $F : \Re^5 \to \Re^2$. The module solves the problem of classifying multidimensional feature objects based on the construction of a fuzzy decision tree (stage *1*) and fine-tuning the parameters of membership functions and rules consequents with hybrid learning algorithms (stage *2*). It consists of *7* layers with the following basic functions of the nodes.

Layer 1 (input).

As inputs, indicators are considered, according to which experts evaluated innovative technologies, for example, the degree of technology influence on the solution of global problems that arose in the framework of the most acute global crises (ecological, food, geopolitical, and energy).

Nodes in the *1*-st layer accurately convey the input values of the attributes in the *2*nd layer.

Layer 2 (fuzzification).

Conversion of numerical values of attributes into linguistic terms in order to reduce information and present it in an understandable form for an expert. Each k-th input variable is estimated by the linguistic term A_{ki} .

This layer simulates the above-described membership functions.

Layer 3 (aggregation) determines the degree of truth conditions for each of the fuzzy rules, it is a non-adaptive layer of *AND*-neurons that simulate the logical connection *AND* by the multiplication of the membership functions:

$$\omega_{ij} = \prod_k \mu_{ki}(x_{kj}) = \prod_k \exp(-\frac{(x_{kj} - c_{ki})^2}{2\sigma_{ki}^2}),\ i=1,...,m, \quad (4)$$

where i is a rule number corresponding to the branch of the decision tree, in this case $i=1, ..., 10,$

m is a number of fuzzy rules.

The multiplication is computed over all the variables x_k on the i-th branch. ω_{ij} can be interpreted as a weight, or a degree of truth, of the premises of the i-th rule (the branch of the tree) in the j-th example.

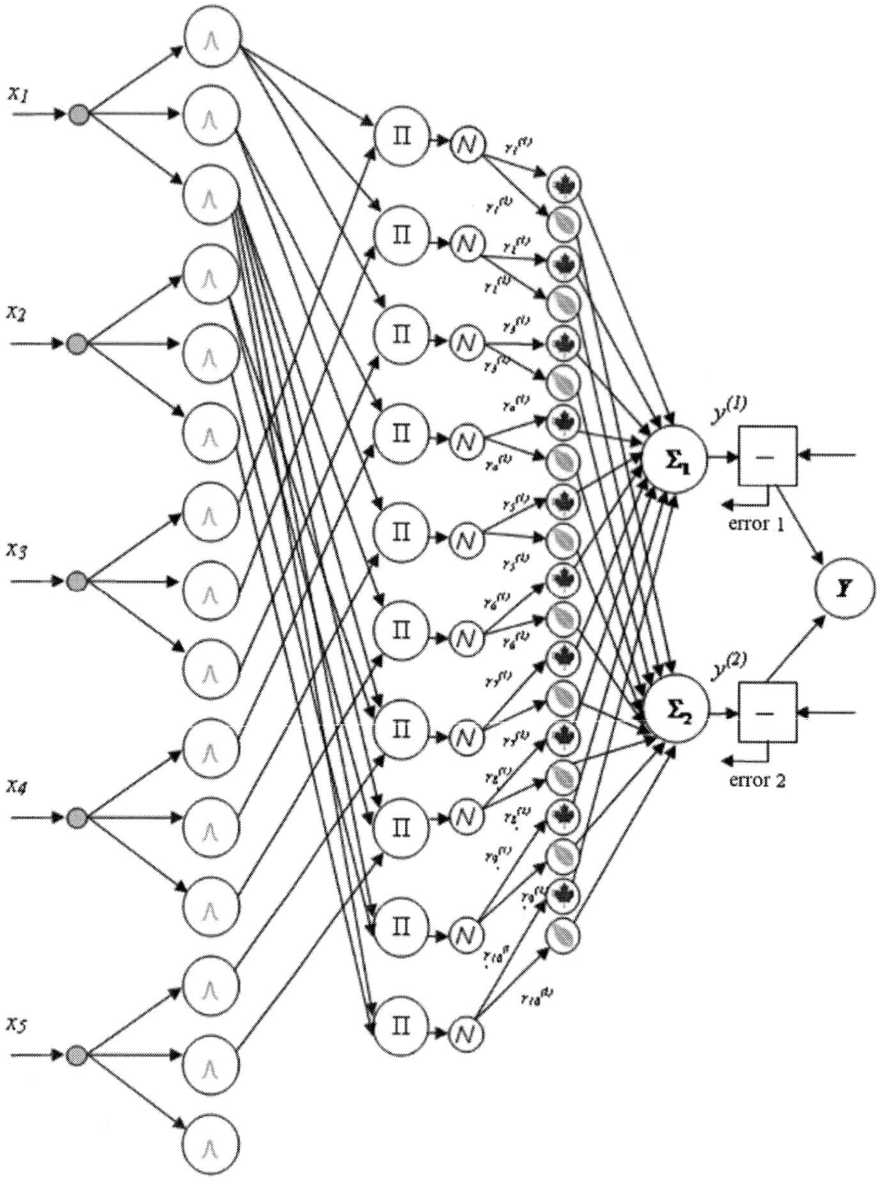

Fig. 1. Structural-functional scheme of the adaptive neural-fuzzy classifier based on a fuzzy decision tree.

Layer 4. Normalization of the degree of rules execution.

The nodes of this layer are nonadaptive; they calculate the relative degree (weight) of the fuzzy rule execution among all the rules in the *j*-th example using the formula:

$$\overline{\omega_{ij}} = \frac{\omega_{ij}}{\sum_{k=1}^{m} \omega_{kj}} \quad (5)$$

Layer 5 (activation). On the leaves of the tree, it determines the degree of truth of the consequents (conclusions) of the rules for each class using weight parameters r_i^1 and r_i^2 which are initialized, for example, by the method of finding the nearest neighbor.

Note that the procedure for initializing the values of the membership parameters (centers and widths), rules antecedents and rules consequents is preliminary, and they are subject to fine-tuning at the next stage.

For each class, the adaptive nodes of layer 5 calculate the degree of truth of the rule conclusions when passing the j-th signal:

$$y_{ij}^1 = \overline{\omega_{ij}} r_i^1, \; y_{ij}^2 = \overline{\omega_{ij}} r_i^2, \; i=1,\dots,m. \tag{6}$$

Layer 6 (accumulation) combines all the degrees of truth of the conclusions of each rule to obtain the membership function of the output variable. The nonadaptive nodes of this layer for each class form the output value of the variables y_j^1 and y_j^2, representing the degree of truth of assigning the j-th signal to the class as a sum of the degrees of truth of all rules:

$$y_j^1 = \sum_{i=1}^{m} y_{ij}^1, \; y_j^2 = \sum_{i=1}^{m} y_{ij}^2. \tag{7}$$

Layer 7 (defuzzification). After fine-tuning the parameters (stage 2), this layer calculates the total output Y_j and the forecast class l_j of the j-th signal by the method of the right modal value (based on the maximum degree of truth):

network output $Y_j = \max(y_j^1, y_j^2)$,

forecast class $l_j = \arg\max_{l=1,2} \{y_j^l\}$.

The accuracy of fuzzy inference is achieved by the joint work of rules pertaining to one class.

3.2 Adaptive parametric tuning of the neuro-fuzzy classifier (stage 2)

Let us consider the constructed network of fuzzy inference (Figure 1). It has 5 input variables, each of which is fuzzy in 3 linguistic terms. The rule base is formed by the weighted branches of the decision tree that connects layers 2 and 3. It is obvious that exactly one rule corresponds to each node of the layer 3. Thus, the rule base includes 10 rules.

We describe a gradient algorithm for learning the parameters of the centers and the widths of the Gaussian membership functions and the rule consequents. The error of the fuzzy decision tree function is defined as a differentiable function in terms of the root-mean-square deviation of E [14]:

$$E = \frac{1}{2N} \sum_{j=1}^{N} (Y_j - C_j)^2 \tag{8}$$

where Y_j is the output value obtained as a result of fuzzy inference,
C_j is the output value from the training sample,
N is a number of observations.

A necessary condition for minimizing the error is that its derivatives are zero in respect to Gaussian parameters (center c_{ki}, width σ_{ki} of the Gaussian function of membership of the k-th input attribute on the i-th branch) and to the parameters of the rule consequents. As a result, we obtain the following general rule for updating parameters by the back propagation method:

$$r(t+1) = r(t) - \eta \frac{\partial E}{\partial r}. \tag{9}$$

By the rule for differentiating a composite function, we write the partial derivative:

$$\frac{\partial E}{\partial r}=\frac{\partial E}{\partial Y}\frac{\partial Y}{\partial \mu}\frac{\partial \mu}{\partial r}=\sum_{j=1}^{N}(Y_j-C_j)\frac{\partial Y}{\partial \mu}\frac{\partial \mu}{\partial r}. \quad (10)$$

We get the following rules for updating the parameters:

$$r_i^1(t+1)=r_i^1(t)+\frac{\eta}{N}\sum_{j=1}^{N}(Y_j-C_j)(\prod_k \mu_{ki}(x_{kj}))^2. \quad (11)$$

The expression $\prod_k \mu_{ki}(x_{kj})$ is the degree of truth ω_{ij} of the *i*-th rule premises (branch of the tree) in the *j*-th example, which is computed by the layer 3. Finally, we have:

$$r_i^1(t+1)=r_i^1(t)+\frac{\eta}{N}\sum_{j=1}^{N}(Y_j-C_j)\omega_{ij}^2 \quad (12)$$

The condition for termination is reaching the counter of iterations of a given value or getting the error value less than the set value.

4 Conclusion

1. Instead of a classical *4*-layer network, we can use a neural-fuzzy system in order to solve the problems of classification of complex objects in a multidimensional weakly formalizable space of features using expert data. The structure of this neural-fuzzy system automatically determines the base of fuzzy rules, reverse cycle-iteration of neural network by the back propagation method, and may be applied for setting parameters (center and width of the membership function and weight parameters rules consequents on tree leaves).

2. The synthesized knowledge base can be interpreted as a partition of the space of influencing factors into regions with blurred boundaries, within which the response function assumes an indistinct value. The number of such fuzzy regions is equal to the number of rules.

3. The proposed neural network method of adapting the parameters of the decision tree (as a fuzzy hierarchical system) based on the back propagation of the error from the tree leaves to the root node, improves the accuracy of the fuzzy decision tree classification, without changing the tree structure while preserving the interpretation.

4. The method proposed increases the speed of operation of the classical algorithm because of the optimized network scheme and the reduced number of calculations. In comparison with another known method of synthesis of neural-fuzzy networks using the training sample by introducing rules into its structure, our method is advantageous because it does not require a parametrically complex construction and cumbersome model with a large number of neurons.

5. The described structural-functional scheme of an adaptive neuro-fuzzy classifier with a logical conclusion, built based on a combination of hierarchical methods (decision trees) with fuzzy logic and neural networks, has a significant cognitive potential for modeling sensations, perceptions, pattern recognition, learning, and memorizing patterns with the purpose of identifying knowledge from the data. It allows the complex use of strong qualities of fuzzy systems (interpretability of accumulated knowledge) and neural networks (the ability to learn from data).

6. From the point of view of expert methods, a new intellectual logically transparent foresight tool for foresight research has been obtained.

7. Such hybrid neural-fuzzy networks have great potential for practical use in intellectual systems for analyzing, managing and supporting multi-criteria decision-making process in the case of processing multidimensional weakly formalized expert data

describing advanced technologies, scientific and technical (innovative) objects, and territories.

Acknowledgements

The article is prepared within the framework of Competitiveness Enhancement Programme of National Research Tomsk State University and grant of the Russian Foundation for Basic Research No 16-29-12858.

The authors are grateful to Tatiana B. Rumyantseva from the National Research Tomsk State University for her assistance in preparing the paper.

References

[1] A. Sokolov, Fores. **1**, 8 (2007)
[2] M. Keenan, I. Miles *The Handbook of Technology Foresight* (Edward Elgar, 2008)
[3] N.V. Morozova, Bul. of the Chuv. Univ. **3**, 178 (2014)
[4] S.V. Gorbachev, V.I. Syryamkin, *Neuro-fuzzy methods in intelligent systems for processing and analyzing multidimensional information* (Tomsk State University, 2014)
[5] D.V. Shashev, S.V. Shidlovskiy, *J. Phy: Conf. Ser.* **881** (2017)
[6] D.V. Shashev, *MATEC Web Conf.* **79** (2016)
[7] D.A. Vyatchekin, *Fuzzy methods of automatic classification* (UE «Technoprint», 2004)
[8] V.I. Smirnov, *Forecasting and classification of economic systems under conditions of uncertainty by methods of artificial neural networks* (Orenburg State University, 2004)
[9] D.S. Khaydukov, Phil. of mathem: act. probl. **2**, 4 (2009)
[10] V.V. Boyarshinova, Eltyshev D.K. *Automated control systems and information technologies* (Publishing House of the Central Scientific and Technical University, 2015)
[11] D. Rutkovskaya, M. Pilinsky, L. Rutkovsky, *Neural networks, genetic algorithms and fuzzy systems* (Moscow, 2006)
[12] C.Z. Janikow, *Fuzzy Processing in Decision Trees* (University of Missouri, 1993)
[13] S.V. Gorbachev, *Designing a Hybrid Neural-Fuzzy System Based on Soft Computing* (Tomsk State University, 2014)
[14] V.V. Kruglov, M.I. Long, R.Y. Golunov, *Fuzzy logic and artificial neural networks* (FIZMATLIT, Moscow, 2001)

THE NUMERICAL SIMULATION OF THE SHIELDED HELMHOLTZ COILS SYSTEM MAGNETIC FIELD

Ivan Zatonov*, *Pavel* Baranov

National Research Tomsk Polytechnic University, 634050, Tomsk, Russia

Abstract. This article describes the calculation of the magnetic field in the axial direction for the Helmholtz coil. The inhomogeneity of the magnetic field for this coil system is calculated. Authors defined the functional connection of the magnetic flux density in the system center from the antimagnetic shield thickness in two ways: iron and Finemet alloy shield. The dependency graph of the magnetic field inhomogeneity in the center of shielded Helmholtz coils system from the axial coordinate value is estimated

1 Introduction

The measurement of the weak magnetic field is crucial for the science and technology application[1–3]. For the weak magnetic field measures the magnetic-test coils, flux-gate meters, non-isotropic magnetoresistive transducers, squids are used.

For the sensor of the weak magnetic field adjustment and calibration measurement it is required to have the magnetic field with high homogeneity. For the calibration devices with the inaccuracy less than 1% field inhomogeneity should not exceed 0.1–0.3% in the magnetic field measurement device space.

In the function of homogeneity magnetic field sources solenoids the Helmholtz coils [4] (Figure 1), square coils system, Spheroidal coils [5], multicoil system [6] and others coil systems are used.

The presence of the Earth magnetic field and others magnetic disturbance makes the process of weak magnetic field calibration more difficult and sometimes impossible and it also could lead to inaccuracy appearance. The article presents the analysis of the magnetic field homogeneity with the magnetic flux density equals to 50 μT with the magnetic shield from the iron and Finemet alloy material.

2 The calculation of the magnetic field axial component

The calculation of the magnetic field inhomogeneity requires the evaluation of the magnetic flux density in the axial direction.

* Corresponding author: iaz5@tpu.ru

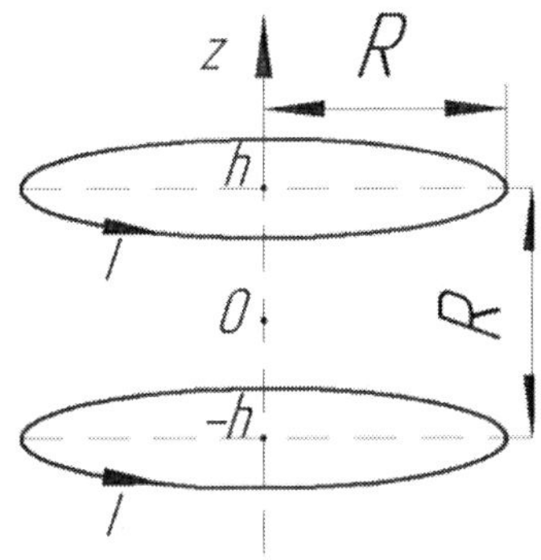

Fig. 1. Helmholtz coils.

According to the Biot-Savart law the magnetic flux density of the Helmholtz coils on the distance 'z' from the coordinates center can be calculated by the following formula:

$$B(z) = 0.5\mu_0 INR^2 \left\{ \left[(z-h)^2 + R^2\right]^{-\frac{3}{2}} + \left[(z+h)^2 + R^2\right]^{-\frac{3}{2}} \right\} \quad (1),$$

where μ_0 – is vacuum permeapility equals to $4\pi\times10^{-7}$ H/m; I – current in the wire; N–number of winding; R – coils radius; h – distance between coils

For the determination of distance 'h' in is necessary to find the solution the conditions (2). The second member of the Taylor expansion in the point z=0 must be equals to zero:

$$B^{(2)}(0) = 0. \quad (2)$$

The solution of the formula (2) shows that the creation of the inhomogeneity magnetic field requires coils to be at the distance of R/2 from the center. Taking into account formulas (1) and (2) we have the finite expression for determination of the Helmholtz coils magnetic flux density in the axial direction:

$$B(z) = 0.5\mu_0 INR^2 \left\{ \left[\left(z - \frac{R}{2}\right)^2 + R^2\right]^{-\frac{3}{2}} + \left[\left(z + \frac{R}{2}\right)^2 + R^2\right]^{-\frac{3}{2}} \right\} \quad (3)$$

Knowing the magnetic flux density spread nature in the direction in the line of 'z' and the value of flux density in the system center it becomes possible to calculate the field inhomogeneity using the following formula:

$$\delta_z = \frac{|B(0) - B(z)|}{B(0)} * 100\% \quad (4),$$

where $B(0)$ – the magnetic flux density in the system center.

This calculation considers the ideal situation when the cross section of the coils tends to zero. In the reality coils have some cross section that influence the spread of the magnetic field. The finite element analysis was carried out in order to take into account coils cross section influence as well as external magnetic field and the magnetic shielding.

3 Finite-element analysis

For the finite-element analysis 2D axisymmetric space dimension model was created in Comsol software. The principle geometry of this model is presented on the following Figure 2.

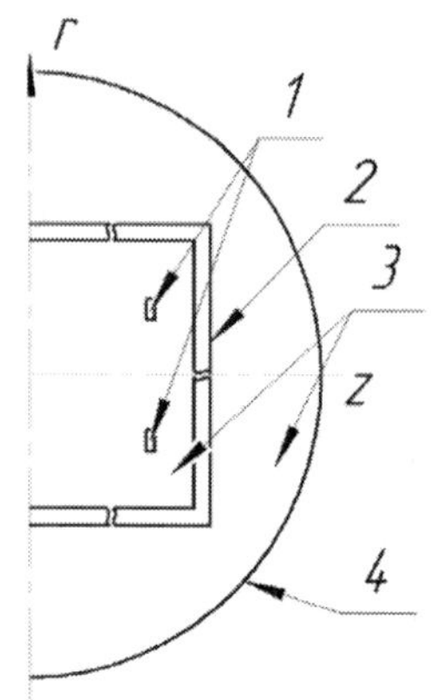

Fig. 2. The principle geometry of the model, where 1 – Helmholtz coils; 2 – magnetic shield; 3 – air domains; 4 – boundaries of the calculation domain.

The magnetic field with stationary computation is used in this model. Parameters of the calculation are presented in the table 1.

Table 1. Parameters of the model.

Parameters	Parameter value
Coils radius R, mm	50
distance h, mm	25
Wire section, mm^2	Square cross section, 0.04
Number of turns	10
Winding current, mA	1

Primarily the value of magnetic flux density in the axial direction without shield and external magnetic field was calculated. According to the simulation results until the value of current in the wind of the coil is equal to 1 mA, the magnetic flux density in the system center is 1795×10^{-7} T (Figure 3).

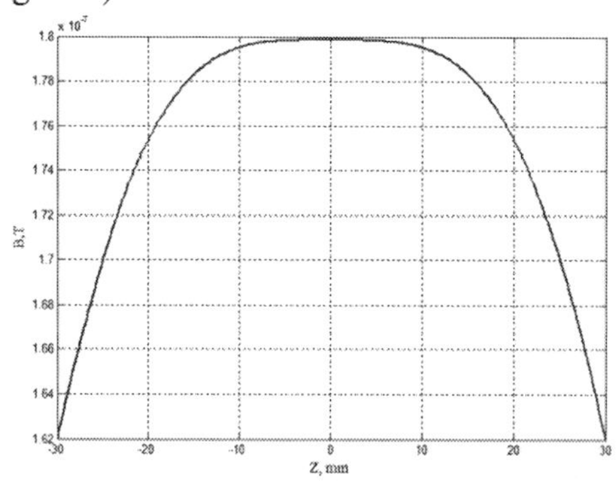

Fig. 3. The magnetic flux density in the axial direction without shield and external magnetic field.

The magnetic field inhomogeneity is calculated by the formula (4). According to the calculation, Helmholtz coils with the parameters from the table 1 provide the needed homogeneity in the interval [-8.5; 8.5]mm (Figure 4).

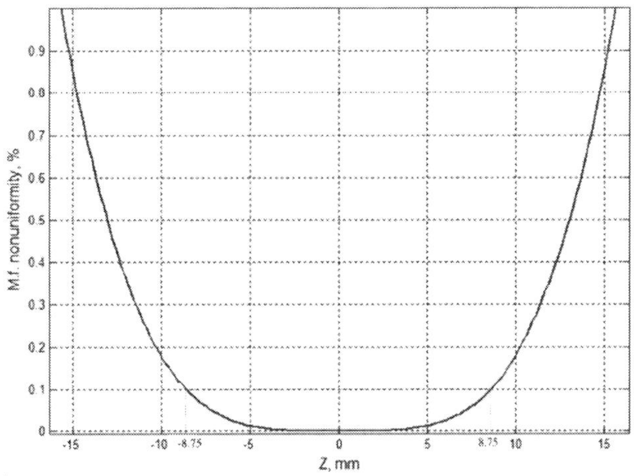

Fig. 4. Magnetic field inhomogeneity for the Helmholtz oils system.

4 The testing of the shield material and thickness

4.1 Iron shield

For the shielding of the permanent magnetic field the iron shield is used [7]. To define the magnetic shield thickness that allows to provide magnetic shielding of the external magnetic field under the flux density value lesser then 10 nT authors carried out the parametrical analysis to define the dependence of the flux density in the system center with the coil current equals to zero and the iron shield thickness.

Input parameters were:the value of the external magnetic field equals to 50 μT; the shield is cylindrically shaped with the r=100 mm and the height equals to 200 mm.

According to the calculation the raise of the iron thickness leads to the non-linear reducing of the induction (Figure 5).

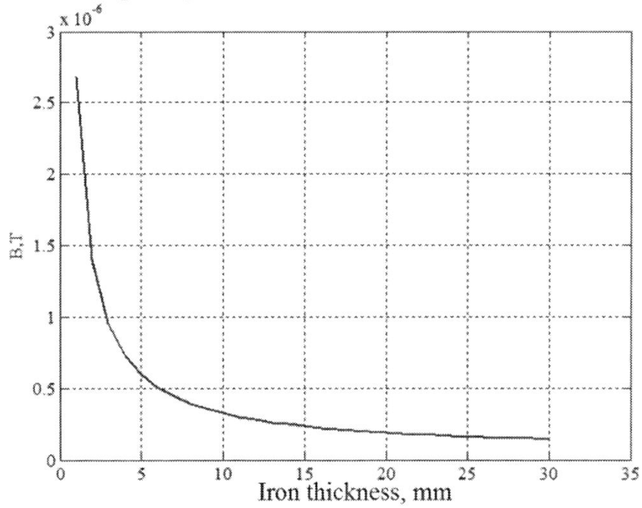

Fig. 5. The value of the magnetic flux density in the shielding system center in the dependence from the shield thickness.

The Figure 5 shows that if the shield thickness equals to 30 mm then the external magnetic field falls to the value of 1,5 μT, that exceeds the value of the field created by the Helmholtz coils with the same parameters.

The resulting magnetic field of the Helmholtz coils system and the external magnetic field is presented on the Figure 6.

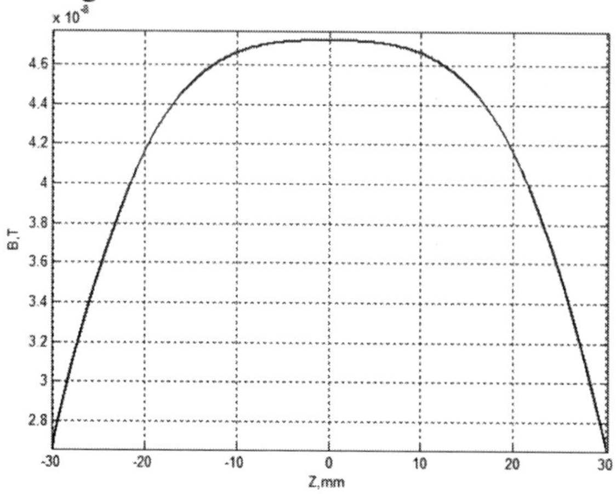

Fig. 6. the magnetic field inhomogeneity with the presence of the external magnetic field and 30 mm thickness iron shield.

The calculation showed that iron magnetic shield with the thickness 30 mm allows to provide the needed homogeneity higher than 0.1 % only in the interval [-2.5; 2.5]mm. But the iron shield of that thickbess do not allow to provide the sheilding of the external magnetic field under the value of 10 μT. Thus the using of the iron shield in that case is inappropriate.

4.2 Finemet shield

The alloy Finemet is usually used as the magnetic shielding material [8]. The distinctive characteristic of the alloy is its high magnetic permeabilty that defines it as great ferromagnetic shield.

The Figure below shows the dependence of the magnetic flux density in the system center from the Finemet thickness.

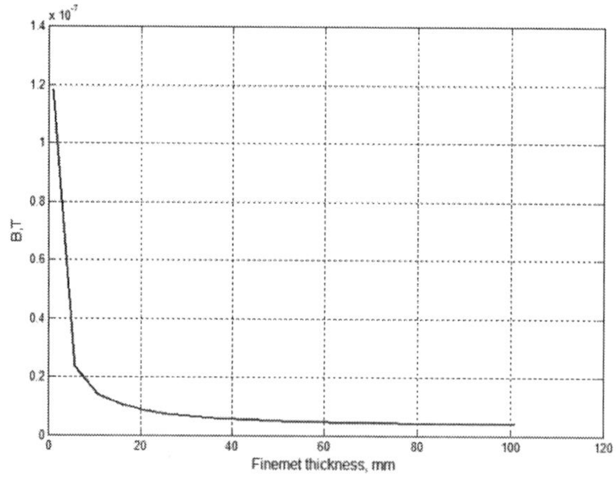

Fig. 7. The dependence of the magnetic flux density in the system center from the Finemet thickness.

The comparison of the Figures 5 and 7 shows that in the conditions of using the parameters from the table 1 Finemet is more preferable for the magnetic shielding for this model. The parametrical analysis of the shield thickness and the current in the coils windings influence on the magnetic field inhomogeneity. The simulation results are presented on the Figure 8.

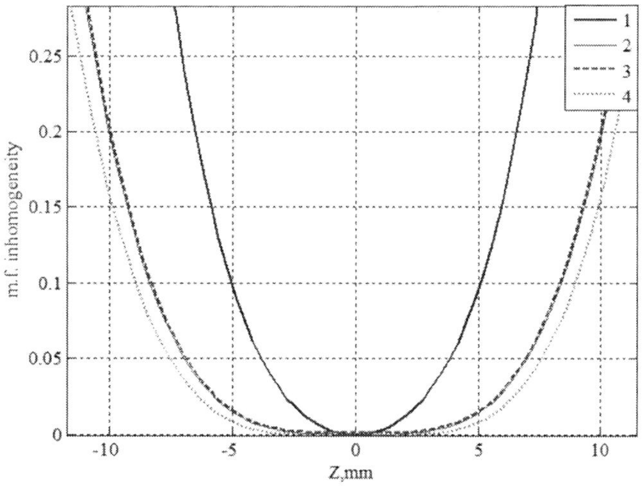

Fig. 8. Helmholtz coil magnetic field inhomogeneity in the situation of availability of external magnetic field and shielding material, where 1 – I=1mA, shield thickness=1 mm; 2 – 5mA,1mm; 3 – 1mA,5mm; 4 – 5 mA, 5 mm.

According to the Figure 8, it is apparent that the plot 2 and 3 are coincided and it shows that step-up of the coil in the winding influence the augmentation of the magnetic field homogeneity interval. The system with the current in the coils winding equals to 5 mA and the shield thickness – 5 mm has the optimal magnetic field homogeneity. But the magnetic shield with the thickness 5 mm do not allow to provide magnetic field shielding in the center of the system to the value of 10nT.

According to the Figure 7, to provide the magnetic shielding to the value 10nT it is necessary to set the 18 mm thickness of the magnetic shield.

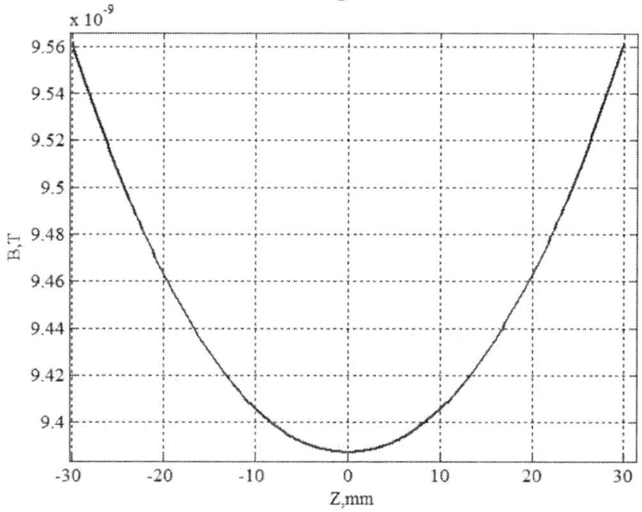

Fig. 9. The magnetic field shielding.

The Figure 9 shows that the value of the magnetic flux density in the direction of 'z' in the interval [-30; 30] mm is lesser than the 10nT.

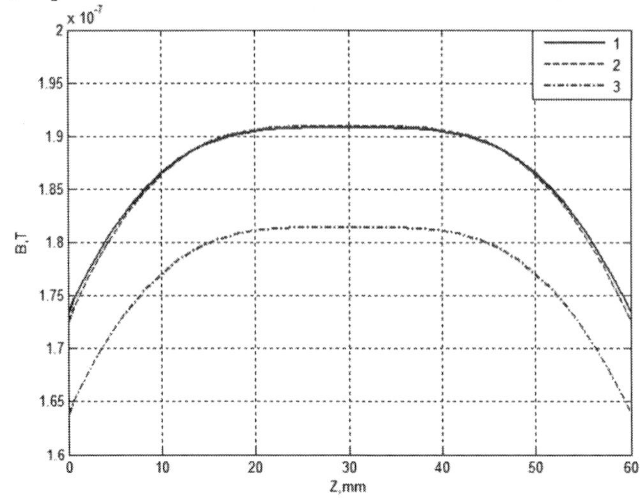

Fig. 10. The magnetic flux density in the axial direction, where 1 – the external magnetic field is absent, the current in the coils winding – 1 mA; 2 – the external magnetic field equals to 50μT, the current in the coils winding1.05mA; 3 – the external magnetic field equals to 50μT, the current in the coils winding 1 mA.

The Figure 10 demonstrates that in the presence of the external magnetic field the magnetic flux density in the axial direction averagely falls on more than 5%. The increasing of the current value in the coil from the 1mA to the 1.05 mA allows to receive the original characteristic.

The Figure 11 below shows the distribution lines of the magnetic flux density in the model.

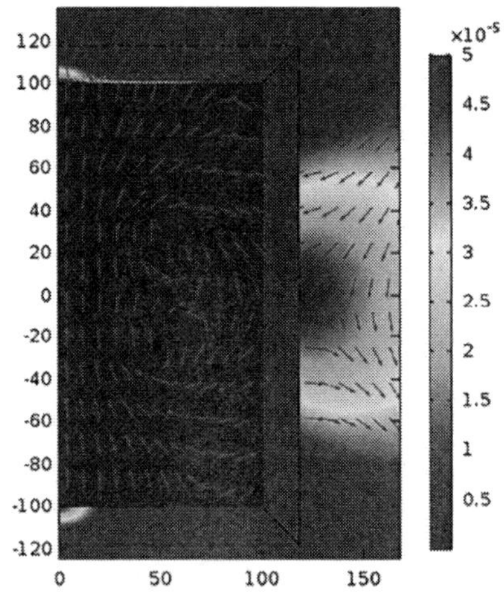

Fig. 11. The distribution of the magnetic field. Shield thickness 18 mm; current in the coils wining – 1.05 mA.

The red arrow line shows that magnetic flux density is closed by the magnetic circle in the shield material. It is explain by the high value of Finemet material magnetic permeability.

5 Conclusion

1. The comparison of the Figures 3 and 10(plot #1) shows that in the absents of external magnetic field the magnetic shield leads to the magnetic flux density increasing in the conditions of the task.
2. The use of the ferromagnetic shield in the case of this simulation allowed to provide external magnetic field shielding to the value lesser than 10nT.
3. The use of the calibrating coils allows to improve the homogeneity on the magnetic field and to create the required magnetic flux density in the shielding system.

Acknowledgments

The research is funded from Russian Science Foundation (RSF), Grant Number 17-79-10083.

References

[1] P. Ripka, *Magnetic science and technologies* (Artech house, Boston,2001)
[2] Can, H. & Topal, U. *J.* Supercon.& Nov. Magnet. **28**(3),4 (2015)
[3] M.W. Johnson, M.H.S. Amin, S. Gildert, et.al., Nature **473**,194 (2011)
[4] V.E. Baranova, P.F. Baranov, Mech.&Mach., 4 (2014)
[5] Y. Öztürk, B. Aktaş, Rev. Scien. **667**,12009 (2016)
[6] V. Ogay, P.Baranov, A. Stepankova, Mat.Scienc. &Eng. **66**, 6 (2014)
[7] A. Keshtkar, A. Maghoul and A. Kalantarnia, In.J. Com.Sc.&El.Eng. **3** (4), 3(2011)
[8] Hitachi metal [Online] URL: http://www.hitachimetals.com

SIMULATION OF THE BETATRON MAGNETIC FIELD AT THE ELECTRON BEAM DISPLACEMENT IN COMSOL MULTIPHYSICS

Andrey Kolomeytsev, *Vladimir* Smolyanskiy* and *Igor* Stepanov

National Research Tomsk Polytechnic University, 634028, Tomsk, Russia

Abstract. This article describes the betatron magnetic field simulation in Comsol Multiphysics software package. The model makes it possible to evaluate the betatron focusing properties at the moment of electron displacement onto the target. The simulation results coincide with the experiment described earlier. The study shows the vertical size of the target affects the focal spot size. However, a target size decrease leads to low output dose, since not all electrons reach it. Therefore, the vertical size of the target should be greater than or equal to the vertical dimension of the accelerated beam.

1 Introduction

Betatrons have found wide application in terms of defectoscopy, medicine and X-ray industrial tomography. The main difference between betatrons and other radiation sources is a narrow radiation pattern of the beam and, consequently, a lower radiation hazard in non-stationary applications, portability and high energy of the accelerated beam. In particular, this makes it possible to introduce betatrons into modern tomographic complexes for the testing of large-sized objects, including containers and vehicles. Betatrons are also used in electronic equipment radiation testing with the aim of predicting the service life in space conditions [1–10]. However, such systems have high requirements to the magnetic field characteristics to ensure radiation stability. Improving the betatron characteristics is a continuous process. This is due to the ever-growing demands of the radiation source market.

Close to 10^{12} electrons are injected into betatron in one acceleration cycle. Only less than 1% reach the target and generate bremsstrahlung. There are many experimentally verified reasons for such an assertion. Among them may be the following:

- Insufficient energy of beam displacement.
- Low density of electron space charge.
- Low synchronization accuracy between the beam displacement and the accelerating magnetic flux.
- Insufficient stability of voltage pulse parameters.
- Non-optimal geometry and mutual arrangement of electrodes.

* Corresponding author: aak65@tpu.ru

Each factor influence evaluation entails a large number of complex experiments. However, it can be avoided by using computer-aided engineering systems. Computer simulation makes possible to more deeply study the internal processes in the betatron magnetic field during operation and save considerable time and resources.

This paper is devoted to the study of the betatron magnetic field during the electron beam displacement Comsol Multiphysics software package is chosen as a simulation environment.

2 Materials and methods

The betatron field has axial symmetry. The electron moves in the magnetic field along the potential minimum line. Its orbit has the form of a circle. This trajectory is called the equilibrium orbit. In this orbit, the axial forces hold the electron in the median plane and the radial component of the Lorentz force is compensated by the centrifugal forces. Such an orbit exists if the induction of the magnetic field B_0 at the points of this orbit at any time instant is 2 times less than the average induction of the magnetic field B_{av} inside the orbit.

When accelerated electrons reach a predetermined energy, they should either be directed at the target to form bremsstrahlung or be withdrawn from the vacuum chamber in the form of a direct electron beam. In either case, the electrons must be displaced from the equilibrium orbit. For this purpose, special displacement schemes are used in inductive cyclic accelerators. The most widely used method of electron displacement from an equilibrium orbit to a target is to change the magnetic field distribution in the accelerator interpolar space. It is achieved by passing current through special displacement coils.

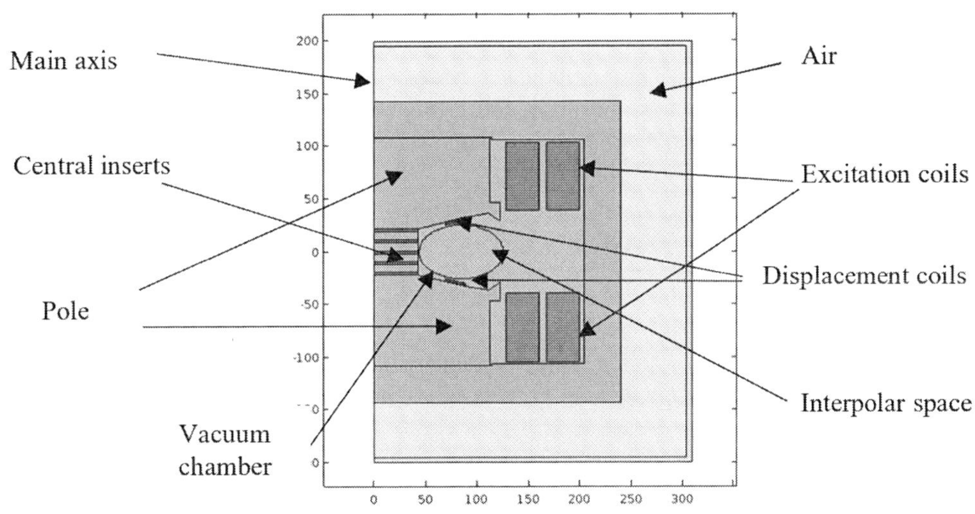

Fig. 1. Magnetic circuit elements.

The paper considers a magnetic field during the electron beam displacement in a small-size betatron MIB-10 with an output electron energy of 10 MeV. Figure 1 shows the magnetic system main elements. The operating frequency is 150 Hz. The injector is installed inside the vacuum chamber. Because of thermionic emission, an electron flux equivalent to the current strength of 0.6–0.8 A is pulled out from the cathode surface. Under the influence of the potential difference between the anode and the cathode, the electrons are accelerated to energy of 40 keV and fall into the interpolar space. In the interpolar space, electrons tend to an equilibrium orbit and are uniformly distributed along it under the action of the Coulomb force. Further, the electrons are accelerated in the growing magnetic field. At the end of the acceleration cycle, a current is applied to the displacement coils, which distorts the magnetic field in such a way that the equilibrium orbit radius begins to increase until the electron beam reaches the target established at the anode of the injector. The injector with the target is located at a radius of 110 mm relative to the betatron main axis.

To increase the equilibrium orbit radius and the electron beam displacement to the external target, it is necessary to reduce the field in the orbit region. The change magnitude in the dB_0 field required to change the radius of the orbit by a value of $dR = R_m\text{-}R_0$ is given by:

$$dB_0 = B_0\left(\frac{R_0}{R_m} - \left(\frac{R_0}{R_m}\right)^n\right) \tag{1}$$

Loop coil that create a displacing magnetic field are located above and below the equilibrium orbit between the vacuum chamber walls and the surfaces of the betatron poles. The most common scheme for beam displacement on a target is an oscillating (discharge) circuit with a storage capacity of C. Neglecting the active energy losses in the discharge circuit, the amplitude current value can be determined as:

$$Im = \omega_0 \cdot C \cdot U_c \tag{2}$$

where $\omega_0 = \frac{\pi}{t_c}$ – is the natural oscillation frequency of the discharge current, t_c – is the displacement pulse duration , U_c – is the capacitor voltage, V [11].

The discharge efficiency drops sharply with an insufficient displacement current. This is easily explained by the fact that the electron beam does not reach the target and bremsstrahlung does not arise. In addition, the current change rate affects on the focal spot dimensions. A more detailed study of this effect will allow creating sources with a variable focal spot size that will not depend on the accelerated electron energy.

The electron kinetic energy is always consistent with the magnetic field induction. In other words, there is a relationship between magnetic induction on an equilibrium orbit and the electron energy. This relationship can be described by an empirical formula:

$$(W(W + 2E_0)^{\frac{1}{2}} = 300B_0R_0 \tag{3}$$

where R_0 – is the equilibrium orbit radius, m; B_0 – is the magnetic induction in the equilibrium orbit, T; E_0 – is the electron rest energy.

Knowing the equilibrium radius, we can calculate the induction value in the equilibrium orbit at the time when the electron energy reaches 10 MeV.

$$B_0 = \frac{(W(W+2E_0)^{\frac{1}{2}}}{300R_0} \tag{4}$$

All the values obtained give us enough information to create and calculate the betatron computer model.

3 Results and discussion

The simulation results show how the induction of the magnetic field is distributed in the magnetic circuit. As shown in Figure 2, the largest induction in the central block. A further increase in the current in the inductive coils will lead to saturation of the magnetic inserts in the central block. As a result, the average induction at $R < R_0$ will grow slower than the induction at $R > R_0$, which will lead to decrease radius.

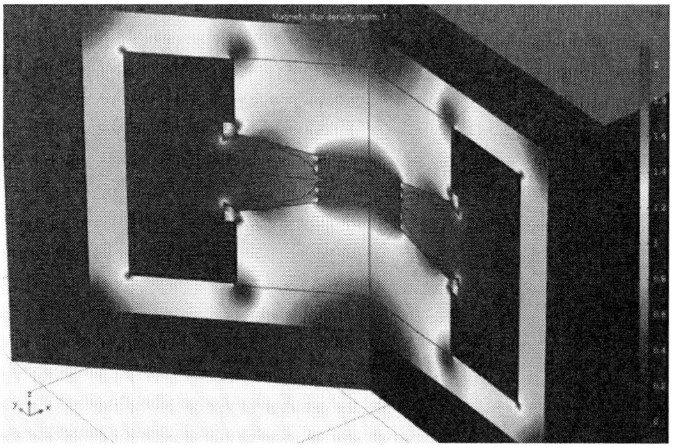

Fig. 2. Magnetic flux density before displacement start.

The further current increase in the excitation coils is undesirable, since the target is at a larger radius than the equilibrium orbit. This means that the electrons have acquired the maximum possible energy from the accelerating field and it is time to displace them onto the target. Let us consider the potential function at this time point.

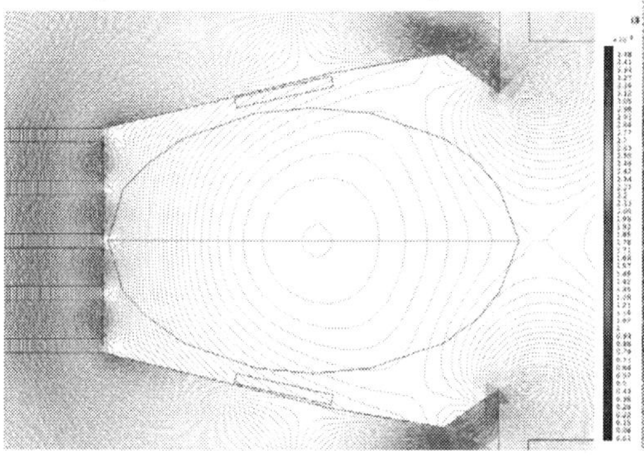

(a) I=0

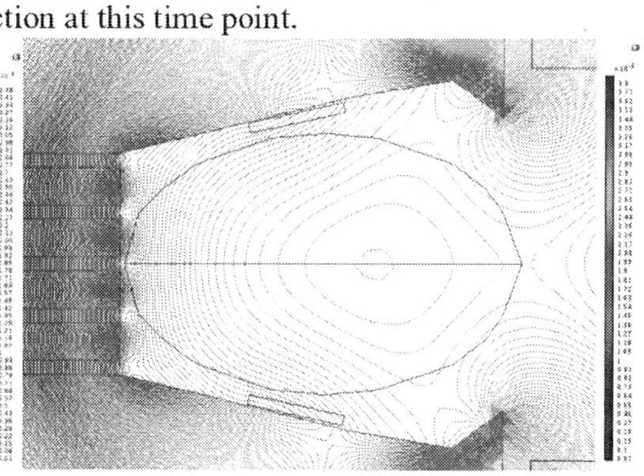

(b) I=0.5Im

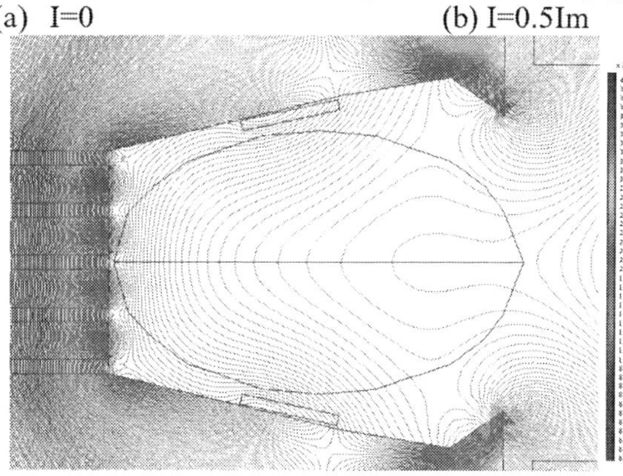

(c) I=Im

Fig. 3. Potential function while displacement.

As can be seen from Figure 3*a*, the isolines of the potential function inside the accelerator chamber are concentrated around the equilibrium orbit. Experiments show that the diameter of the electron beam is 2–3 mm. This diameter is obtained when the balance of centrifugal and magnetic forces and the forces of the Coulomb interaction. A current increase in the displacement coils (Figure 3*b,c*) leads to an expansion of the equilibrium orbit. In the wake of this, the electron beam orbit expands. This is because the magnetic flux at $R < R_0$ increases, and at $R > R_0$ it remains unchanged (Figure 4). The change in the ratio of the average induction and induction on the radius leads to an expansion of the orbit.

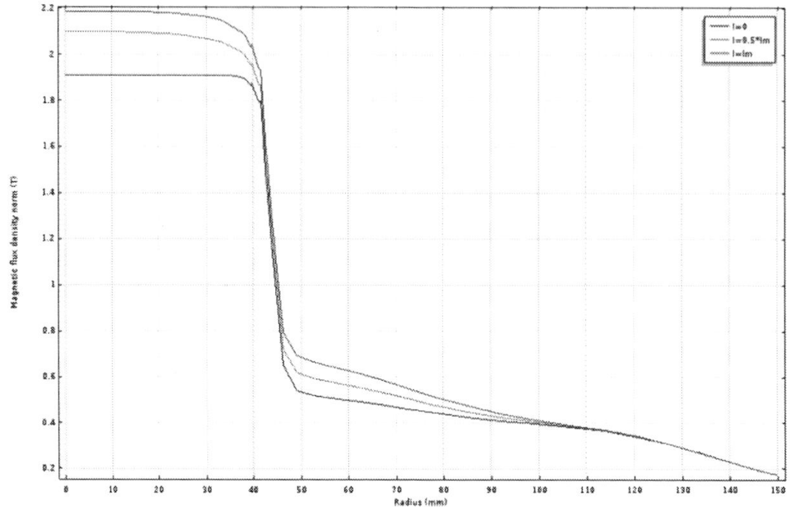

Fig. 4. Magnetic flux density against radius at median plane.

Ability to focus and hold the electron beam at all stages of the work is crucial for betatron operation. The potential function or rather the potential well depth also determines the betatron focusing properties. The simulation results show that the vertical focusing is preserved (Figure 5.) with the equilibrium orbit displacement. This means that there is no precondition to increase the vertical beam size during the electron beam displacement to the target.

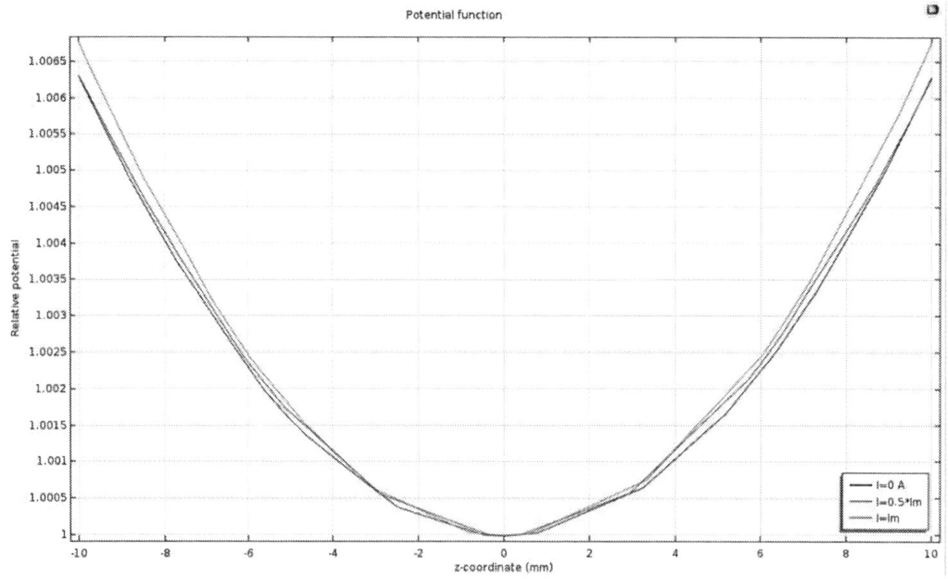

Fig. 5. The potential function against axial coordinate at the potential minimum.

Research [12] shows that the target vertical size directly affects the size of the focal spot. However, a decrease in the target leads to a decrease in the output dose, since not all electrons reach the target. Therefore, the vertical size of the target must be greater than or equal to the vertical dimension of the accelerated beam.

The graph of the radial focusing is shown in Figure 6. As the displacement current increases, the potential barrier at $R > R_0$ decreases. The radial oscillations of the electrons increase. Too rapid increase in the displacement current will cause electrons fly past the target and fall on the anode box. The output dose will be significantly reduced.

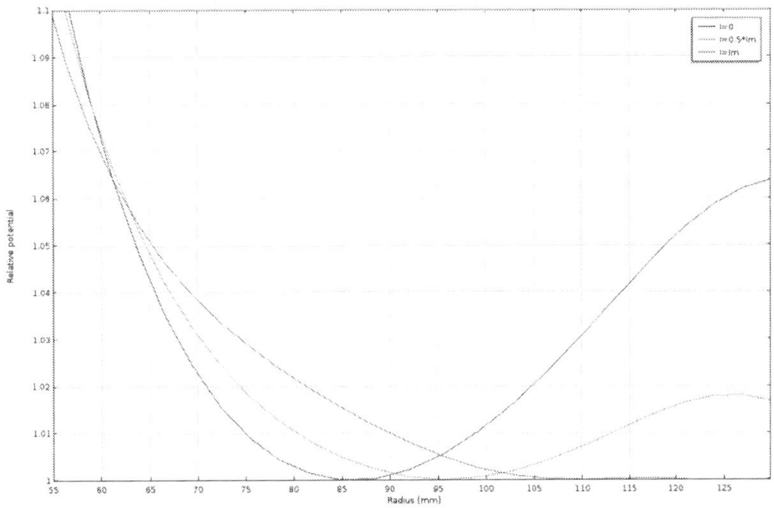

Fig. 6. The radial component of the potential function at median plane.

The simulation results allow us to estimate the qualitative and quantitative characteristics of the betatron magnetic field in the electron beam displacement process. These data will be used to optimize and find new solutions in improving the betatron efficiency.

4 Conclusion

In this paper, a betatron computer model is presented. This model describes the betatron magnetic field at the moment of the displacement of electrons from the equilibrium orbit. Increasing the beam displacement efficiency will allow us to achieve the characteristics of sources based on linear accelerators.

The experiments validate the simulation results. Betatron operation simulate helps to better understand the ongoing physical processes. This simulation allows us to quickly find the optimal design solution when developing new accelerator models.

It should be noted that despite the low prevalence of betatrons with respect to linear accelerators, they have a much greater potential for development. Achievement of the set goals will allow expanding the device application area of this device.

Acknowledgments

The research was supported by the Russian Science Foundation (Grant #17-19-01217).

References

[1] L. De. Chiffre, S. Carmignato, J-P. Kruth, R. Schmitt and A. Weckenmann, CIRP An - MT **63**, 655 (2014)

[2] N. Schuhmann and H. Okruch, DGZfP Proc, **BB67** (1999)
[3] V. Smolyanskiy, M. Rychkov and V. Borikov, MATEC WoC **102,** 01033 (2017)
[4] L. Bonaccorsi, F. Garescì, F. Giacobbe, F. Freni, F. Mantineo, R. Montanini and A. Sili **105,** 33 (2013)
[5] A. du Plessis, S. G le Roux and A. Guelpa, CS in NDT&Ev **6**, 17 (2016)
[6] N. Reims, T. Schoen, M. Boehnel, F. Sukowski and M. Firsching, Proc of SPIE **921209**, (2014)
[7] W. Holub and U. Hassler, NDT in Canada, (2013)
[8] E .Vaynberg et al., 16th World Conf. NDT. **1,** (2004)
[9] G. Herman, *Fundamentals of Computerized Tomography.* (The Graduate Center, CUNY, New York, 2009)
[10] M. Stein, V. Kasyanov, V. Chaklov, J. Macleod, P. Marjoribanks and S. Hubbard, WCNDT (2004)
[11] V. Moskalev and V. Chaklov, *Betatron: monography* (TPU, Tomsk, 2009)
[12] V. Kasyanov et al., Instr. & Exp. Tec's. **41,** 41 (1998)

DEVELOPMENT OF SOCIAL SYSTEMS IN THE CONTEXT OF PROGNOSIS

Svetlana Kvesko[1*], *Mikhail* Kornienko[1], *Nadezhda* Petrunina[1], *Nataliya* Kabanova [2]

[1]National Research Tomsk State University, 634050, Tomsk, Russia
[2]National Research Tomsk Polytechnic University, 634050, Tomsk, Russia

Abstract. The paper dwells on the prognosis of the social systems development. The prognostic analysis is based on the systemic approach to the issue; it shows that the conditions of the transference to the knowledge society determine the formation of the brand new features in social prognoses. The paper states the specific features and conditions of social prognosis, which ensure the validity of control actions. Besides, the current analysis provides recommendations on how to devise social prognostic strategies (within the frame of a) transference to the knowledge society and b) environmental instability). Finally, the functional load of social prognosis in the contemporary conditions is outlined.

1 Introduction

The process of interacting with such systems is characterized by non-linearity, which lies in the realm of system response, i.e., the system can react to the impact factor disproportionately to the intensity of the latter in accordance with the non-linear law. Besides, social systems, as the open-end ones, involve the incipience of continuous alterations, which, in turn, leads to the high complexity of prognostic schemes.

Multiple ambiguities of the social environment aggravate the management complexity caused by non-linearity of social processes within the frame of transference to the knowledge society, while the potential social systems' development determines the spectral breadth of a potential trajectory, as well as the complexity of problems arising the reform. Within the frame of current circumstances, the dynamic- and tendency-oriented structuring of relevant prognoses becomes one of the hot and the most complicated tasks to perform. The specificity of social systems development in the context of transference to the knowledge society seems significant enough to become the subject of current research.

The possibility of structuring prognostic schemes for the social dynamic pattern has been causing a lot of debate throughout decades. Thus, Karl Popper – being in denial of objective social development regulations – argued that history does not make sense at all whereas any kind of social prognosis can lead to nothing but a mass collapse [1]. Alvin Toffler, American philosopher and social-theorist, in his attempts at systematizing the society's future, anticipates the gigantic changes lying ahead of us. Toffler's futuristic extrapolations concern the exponential sophistication of social systems, the increase in the

* Corresponding author: svetla_kvesko@mail.ru

multiplicity of developmental alternatives, and the increase in the decision-making complexity [2, 3].

One of the first theoretically grounded and significant attempts at social prognosis was made by Daniel Bell in his influential book *The Coming of Post-Industrial Society* [4]. According to Bell, his work had the purpose of outlining the trends aimed at the delineation of humankind community in XXI century. The theory of post-industrial society made Bell one of the most renowned prognostic theorists in the West; it defines the framework of social research, yet, sets no strict constraints.

The importance of structuring the social system prognostic trajectories remained a subject of interest for a whole bunch of significant thinkers. The modern phase of social development is characterized by the presence of a big number of global threats that evoke the unheard-of risks and dangers for the whole humankind and thus transform the prognostic solutions into improbable matters [5–11].

The socially conscious approach to global threats, as well as to the controversies of the modern world, has been accompanied by the buildup of a specific type of society – that is, the knowledge society predetermined by the development [12, 13].

Furthermore, the role of knowledge in the economic phenomena was intensely explored by Joseph Schumpeter [14]. In 1970s Daniel Bell suggested a notion of postindustrial society that centers around scientific knowledge and employs the latter as a base for social transformations [4]. In the contemporary world, the tools and methods used for application of scientific knowledge are changing rapidly [15, 16]. According to Peter Drucker, knowledge alters the whole function of society as a complex inter-related system [12, 16].

It is important to state that – under the conditions of transference to the knowledge society – the task of structuring the long-term social prognoses becomes increasingly complicated. In the XXI century, the current state of global social context is being subject to radical transformations that initiate the rising tide of diverse global risks, global dangers, and global threats, namely, nuclear, chemical, biogenetic, environmental, economic, and existential ones. In the process of such transformations a new social context is emerging – where the risk-production is looming large in all the social undertakings. The risk-production industry undermines the traditional practice aimed at social prognosis and social emergency prevention [17]. The complexity of social prognostic process is aggravated by the fact that prognoses as phenomena are capable of placing significant influence on the real world. Besides, the existent methods appear to be incapable of taking into account the external (i.e., environmental) ambiguity, as well as the behavioral ambiguity of systems per se [18, 19].

2 Development of social systems in the context of prognosis

Still, the existent models of complex systems are not consistent in providing clear-cut prognosis of the probable outcome due to various ambiguities that reduce the accuracy of the model, namely:

- ambiguity evoked by absence or lack of knowledge for the influential factors / processes;
- ambiguity evoked by incommensurability of numerical estimations for the descriptive values of a system's dynamic pattern;
- ambiguity caused by non-linearity, as well as the presence of numerous equilibrium states and systemic attractors;
- ambiguity evoked by lack of accuracy in the employed vocabulary and consequent inability to render the facts.

Due to the listed factors, the decision-making that concerns the issue of trajectorial selection is accompanied by a number of problems. Firstly, it is difficult to figure out the sufficient parameters that would allow a predictor to determine the systemic condition and ensure its manageability. The mentioned process bears the name of formalization; the latter, being applied within the realm of social systems – characterized by heterogeneous components (e.g., technological, biological, political, etc.), requires a steady set of parameters in order that the situational context may be properly described. Secondly, information overload is transfiguring into as problem caused by the rising tide of open access information. The tendency to push the limits in the process of considering all the existent parameters and factors leads to information overdose. Any decision that is made loses its importance even before the step due to the low speed of incoming data processing – compared to the speed of environmental and systemic transfigurations. Thirdly, the phenomenon known as emergence, or 'synergetic effect' unfolds another significant problem. The Interacting systems tend to create a higher-level system possessing its own peculiar properties not characteristic of the sub-systems involved. The precise and accurate prognosis based on systemic properties and traits becomes impossible because of the freshly emerged unique supra-system.

The ambiguities described above aggravate the complexity of management process being stipulated by non-linearity of social processes – whereas the potential variance of social systems development determines the wide range of probable spectra, as well as the complexity of problems that tend to occur. This peculiarity was pointed out and described by a number of scholars; thus, Arnold Toynbee highlighted the numerous developmental alternatives typical of human communion at the moment known as bifurcation point, i.e., the turning point transferring the system to the realm of instability [20,21]. Under these circumstances, the exploration of social prognoses and the ways of their outlining becomes one of the hottest and the most complicated tasks that need to be solved in order that social systems may become adaptable to the non-linear environment [22].

L. Zadeh – one of the synergetic paradigm pioneers – formulated the so-called incompatibility principle according to which the system's complexity is directly proportional to the probability of its precise description [23]. However, the postulate of ultimate unpredictability for developmental trajectories under non-linear conditions (suggested by I. Prigogine – yet another prominent pioneer of contemporary synergetics [24]) is lately being widely criticized. We believe that the Kurdiumov's weird attractors scholarship presents a more reliable basis for modeling in social terms [25]. Since the spectrum of a probable outcome is limited, the systemic behavioral pattern cannot be totally unpredictable and is consequently subject to adequate description and subjective anticipation – not in terms of unmistakably predetermined outcome but in terms of belonging to a limited spectrum of multiple conditions.

Transference to the knowledge society currently becomes particularly noteworthy. Knowledge brings changes into behavioral patterns and the whole identity of social environment. High responsiveness and peculiar ambiguity of social environment has long ago become one of the characteristic companions of the global information space. The simultaneous use of information and communication technology (ICT) provided facilities for certain qualitative alterations in the manner of social interaction and adaptation.

Information revolution displays the correlation between ICT and social development process – as it has triggered the formation of macro trends based on the expansion of interactive capacity within social systems. To a considerable extent, the mentioned systems appeared due to the rising integrative processes. Further, the latter catalyzed globalization. Nevertheless, the absence of framed determinism in a social behavioral pattern expands the developmental tracks spectra in the context of forming brand-new social structures. Such structures tend to be characterized as flexible and thus subject to efficient decision-making, which, from the viewpoint of the authors, breaks the boundaries of social prognosis and social management.

Within the framework of changeable environment and management troubles, social systems are in desperate need of boosting their flexibility and adaptability, which notably depend on prognostic strategies and sheer probability of anticipating social evolutionary tracks and environmental dynamic pattern.

Growing environmental instability has lately become a crucial factor that predetermines the specificity of the way social systems work and develop. At the same time, it does not simply foster the brand new opportunities – but evokes a bunch of additional problems faced by social systems. One of the chief and central problems is promoting and furthering the adaptability of social systems within the complex and changeable environment. The system can be described as sufficiently adaptable when it is permanently exposed in the light of vital alterations that allow the system to survive and develop whereas the adaptation per se pertains to the process of a system's merge with the new conditions. Traditionally, adaptive systems can be divided into self-adjusting and self-organizing ones. While responding to external alterations, the former change their functional parameters without in-system rearrangements while the latter are able to rearrange their internal structure. For social systems, structural changes can be expressed via modifications in the management frameworks, systemic interrelations, etc.

The adaptable system is able to struggle with the crises caused by its internal dynamic pattern, as well as by the environmental changes. Basically, the social system adaptability predetermines the probable spectrum of its behavioral patterns within unstable environment. There exists a bunch of approaches to the analysis of social adaptability phenomenon. From the standpoint of the first approach, the adaptable system is characterized as single entity and noted for its consistency – for which read, the system's gist it traced throughout all its levels. It seems appropriate to compare an adaptable system to a fractal one where one and the same geometrical pattern is generally reproduced on a reduced scale. Another approach suggests that system be regarded as a living body. In the same way that cells, however different in terms of their functional load, still store all the vital information, the divergent social subsystems contain information about the system and remain under control of the ultimate goal to survive and improve.

Although evident and significant differences between the two mentioned approaches can be easily traced, there is a trait they hold in common: both of them claim unity to be the essence of social system structuring. Unifying factors can be various: the functional load / goal, the accepted type of interconnections and interrelations, the method of information exchange, etc.

Obviously, the social system adaptability is currently provided by numerous factors. The key factor can be traced in the cumulative nature of information processing, in the knowledge transfer, and in the informational navigation within a social system. The most beneficial seem to be the systems that employ the group dynamics method based on the self-organization paradigm. Such systems tend to be highly adaptable thanks to their capability to generate, select, and store novelties and significant alterations – thus providing permanent self-upgrade.

As noted by Ikujito Nonaka [26], the essence of upgrade processing can be presented as a sequence of four steps:

- emergence of chaos / instability within a system;
- chaotic aggravation and controversy focusing;
- emergence of self-organized subsystems and interaction adjustments;
- knowledge transference.

We believe that ability to coin novelties is a primary sign of a system's diversity – as was formulated by William Ashby in *An Introduction to Cybernetics* [27]: "In order to create the system capable of solving a diverse task, we need to make the system's diversity wider than that of the task – or to ensure that the system has the potential to generate it." In case of social system that implies that under the influence of multiple external factors a system is able to succeed in the adaption process only providing the expansion of control modifiers. Still, the close analysis of various systems demonstrates that their sustainability, adaptability, and improvement potential are directly proportional to the degree of systemic diversity. The sequential changes in the social self-renovation process are nothing but a mechanism of diversity boost oriented towards the expansion of a system's capacity to resolve the adaptability issues.

3 Conclusion

Vast number of adaption issues in the realm of social systems is stipulated by the complexity of the dynamic pattern characterized by the following traits:

- evolutionary development;
- instability;
- self-management and self-reproduction;
- procedural asymmetry.

Nevertheless, it is not the external factors that determine the social adaptation outcome – but their internal dynamic pattern, which is primarily predetermined by the system ergodicity. The presence of immanent aims and criteria in the functional and developmental scopes of social systems allows us to argue that the internal interactions play the key role in sketching the direction of evolutionary process.

The multiplicity of choices at bifurcation points (being a crucial feature of non-linear complex systems) increases due to social systems, which is predetermined by the fact that informational, cultural, and psychological correlations prevail over physical interactions. What is more, the multiplicity of systemic, changeable, and subjective aims / criteria elicits additional changeability and thus leads to the predominance of the unpredictable over the anticipated. The listed peculiarities ground the necessity to maintain social systems on the level of sufficient flexibility and adaptability – to ensure the functioning of the latter within unstable conditions.

Globalization brought on changes in social structure and functional load; social systems nowadays are getting increasingly flexible and sensitive to any kind of external alteration. Systemic evolution is permanently accompanied by the increase in complexity and necessity to adapt to the open-end, ever-transforming environment. For social system that stipulates the rising significance of innovative enterprises. We argue that sustainability of social systems is directly linked to innovations based on co-adaptation and co-evolution.

In this particular case co-adaptation can be expressed as equilibrium in the assimilation and accommodation within social systems, as well as systemic mutual adaptation. What is more, the identity of social adaptation processes is undergoing a number of changes; social systems tend to replace genesis (i.e., adaptation to the given condition) with environmental enhancement. This tendency displays itself via diversity of:

a) flexible and changeable interconnections;
b) ways of mutual coordination;
c) high heterogeneity of internal structure;
d) self-regeneration and self-reproduction.

Thus, co-adaptation provides the context for co-evolution that involves the sequence of coordinate, indissolubly adjusted alterations capable of taking place at any evolutionary level of a social system.

Consequently, the increasing management complexity in the realm of modern social systems is determined by their specificity – on the one hand. On the other hand, there goes another factor, namely, the high speed of social dynamic pattern changeability predetermined by the transference to the knowledge society placing restrictions on forecasting horizon and on allowed time of control actions. In contrast, the mentioned processes stipulate the necessity to figure out effective prognostic-based methods of managing social systems that would speed up the adaption to the unstable environment. Self-organization procedure provides the optimal directions in the process of managing social systems.

Interaction with such systems implies non-linearity displayed by the level of system response to contributing factor, which is not proportionate to the intensity of the latter but complies with the non-linear law. The base for self-fulfillment or self-destruction of a prophecy lies in the frame of positive vs. negative public reaction.

In conclusion, whatever social prognostic trajectory may be, it is always hypothetical, i.e., it concerns mainly the developmental trends, not the actual future states of a social system. Within the frame of transference to the knowledge society, due to multiplicity of a probable outcome, because of the prognosis instability with regard to both external and internal impact an essential part of social prognoses is ruled out invalid. Nevertheless, even invalid prognoses carry a significant functional load: any social prognosis, be it valid or not, reveals the probable multiplicity of social developmental alternatives, allows to evaluate the consequences of decision-making process, and serves as a warning of high risks and disastrous developmental tracks.

References

[1] K. Popper, *The Open Society and Its Enemies* (Feniks, Moscow, 2015)
[2] A. Toffler, *The Third Wave* (William Morrow, NewYork, 2014)
[3] S. Kvesko, T. Shinn, MATEC Web of Conf. **79,** 01048 (2016)
[4] D. Bell, *The Coming of Post-Industrial Society. A Venture in Social Forecasting* (Basic Books, New York, 2013)
[5] E. Vasendina, I. Plotnikova, A. Levitskaya, S. Kvesko, IOP Conf. Series: Mater. Sc. and Engin. **110,** 012070 (2016) doi:10.1088/1757-899X/110/1/012070
[6] A. Toffler, *Future Shock* (Bantam Books, New York, 2002)
[7] U. Beck, *Risk Society* (Toward a New Modernity, London: SAGE. 1992)
[8] M. Castells. *The Rise of the Network Society* (Blakwell Publishers, London, 2010)
[9] A. A. Kornienko. Proc. Soc. and Behav. Sc. **166** (2015)
[10] F. Janszen. *The Age of Innovation* (Prentice Hall, London, 2000)
[11] V. I. Syryamkin, D. S Zhdanov. Telecomm. and Radio Engin. (2012)
[12] P. Drucker, *Post-capitalist Society* (HarperCollins, New York, 1993)
[13] F. Machlup, *The Production and Distribution of Knowledge in the United States.* (Princeton University Press, Princeton, 1962)
[14] J. Schumpeter, *Theory of Economic Development* (Progress, Moscow, 2012)
[15] V. Bestuzhev-LadaI, *Prognostic Grounding of Social Innovations* (Academia, Moscow, 2011)
[16] P. Drucker, *Innovation and Entrepreneurship* (Collins, New York, 2012)
[17] A. Tolstoukhov, Issues of Philosophy **8**, 49 (2003)
[18] A. V. Yurchenko, MATEC Web of Conf. **79**, (2016)
[19] N. V. Chicherina, V. Kazakov, O. Yu. Korneva, The European Proceedings of Social & Behavioural Sciences (EpSBS) **19**, 349 (2017) doi: 10.15405/epsbs.2017.01.47
[20] N. V. Chicherina, V. A. Tolkacheva, The European Proceedings of Social & Behavioural Sciences (EpSBS) **19**, 683 (2017) doi:10.15405/epsbs.2017.01.90

[21] A. J. Toynbee, *A Study of History* (Progress, Moscow 2011)
[22] S. B. Kvesko, S. E. Kvesko, N. Kabanova, D. P. Shamrova, MATEC Web of Conf. **79**, 01062 (2016)
[23] L. A. Zadeh, IEEE Trans. Syst. Man. Cybern. **1**, 28. (1973)
[24] I. Prigogine, I. Stengers, *Order out of Chaos. Man's New Dialogue* (Progress, Moscow, 2014)
[25] S. P. Kurdiumov, E. N. Knyazeva, *Laws of Evolution and Self-Organization in Complex Systems* (Nauka, Moscow, 2015)
[26] I. Nonaka, *Creating Organizational Order out of Chaos: Self-renewal in Japanese Firms* (University Press, New York, 2008)
[27] W. R. Ashby, *Introduction to Cybernetics* (Nauka, Moscow, 2014)

AIR TEMPERATURE CONTROL SYSTEM IN A BUILDING ON THE BASIS OF MATHEMATICAL MODELLING

Aleksandr Pilipenko*, *Sergei* Petrov

Orel State University named after I.S.Turgenev, 302020, Oryol, Russia

Abstract. In the article the authors solve a crucial objective of reducing the cost of heating services through the introduction of a system, controlling boilers and heating devices, based on modelling of the building and predicting heating systems functioning and operation. The authors describe the method of mathematical modelling of heating systems of the building and represent the developed mathematical models, pledged and installed into the developed automated system. The authors have offered an automated system of dispatch control of the boilers with the elements of computer simulation and modelling.

1 Introduction

Development of intelligent (intellectual) systems of management and control of the buildings and nonstationary technological processes is an urgent and relevant objective now. One of the most common nonstationary processes in buildings is the process of air temperature regulation with an account of heat losses. In this regard, the development of the automated control systems of air temperature in the building with elements of forecasting, based on mathematical modelling is a promising and prospective research area.

The existing dependences for transfer functions of the links of the systems under consideration and research are not always convenient for complex integrated control models [1–3]. It is important to have a model, that would be able to manage and control temperature of the building by changing the coolant flow (consumption of a heat-transfer agent) and its temperature. Assuming that the heating system is established and stabilized, and values of air temperature in different rooms are similar, it is useful to represent it in the form of a long pipe in order to develop a mathematical model of heating system (Figure 1), through which the heat transfer with $T(x, t)$ temperature, depending on (x) distance and (t) time flows.

* Corresponding author: a@pilipenko.info

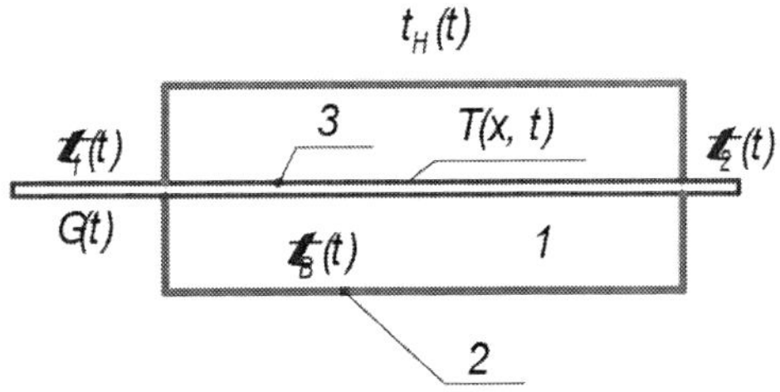

Fig. 1. The diagram, showing the principle scheme of a heated building and heating systems:
1 – denotes the heated building; 2 – denotes the external enclosures (rails) of the heated building;
3 – denotes the tube, which is equivalent to a heating system of the building.

The heating system is often treated as a set and range of parallel and consistently included pipelines and heating appliances; the latter can be represented by the length of the pipe, having larger section and the perimeter. It is anticipated that the area and perimeter of the living section, as well as pipeline length, which is equivalent to the heating system are defined with a help of experimental data on derived from the work of the system, i.e. the suggested model can be configured in relation to a real running system of heating.

2 Mathematical model

We get the differential equation (1) in partial equations with boundary and initial conditions (2) from the condition of thermal balance of the pipe after transformations.

$$cG(t)\frac{\partial T}{\partial x} + cS\frac{\partial T}{\partial t} + kfT = kfT_B, \tag{1}$$

$$T(0,t) = \tau_1(t); \;\; T(x,0) = T_0(x), \tag{2}$$

where c is the volumetric heat capacity of water, joules /(m^3×K); G (t) denotes the heat carrier (coolant channel) flow (consumption) rate in the heating system, m^3/h; S, f denote the area (m^2) and the perimeter (m) of the living section of the equivalent pipe respectively; k is the coefficient of heat transfer through the walls of the pipe, watt/(m^2×°C); $T_B(t)$ is the average temperature of internal air,°C; $\tau_1(t)$ denotes water temperature at the inlet of the heating system,°C; $\tau_2(t)$ is temperature of water at the outlet of the heating system (in the process of heading out of the system),°C.

Applying direct and inverse Laplace transform [4], we can write down the solution to the task for semi-infinite pipe in the following form:

$$\begin{aligned} T(x,t) &= \exp(-kft/(cS))T_0(x - S^{-1} \times \\ &\times \int_0^t G(\varphi)d\varphi + (kf/(cS)) \cdot \int_0^t \exp(-kf(t-\varphi)/(cS)) \cdot T_B(\varphi) \cdot 1(x - S^{-1}) \times \\ &\times \int_\varphi^t G(\theta)d\theta)d\varphi + S^{-1} \cdot \int_0^t \exp(-kf(t-\varphi)/(cS)) \cdot G(\varphi)\tau_1(\varphi)\delta \cdot (x - S^{-1} \times \\ &\times \int_\varphi^t G(\theta)d\theta)d\varphi, \end{aligned} \tag{3}$$

where $1(x) = \begin{cases} 0, & x \le 0; \\ 1, & x > 0. \end{cases}$

$$z = x - S^{-1}\int_{\varphi}^{t} G(\theta)d\theta$$

By means of replacing the variable we transform expression (3); let's take into consideration the fact, that when **t** are large enough, influence of initial conditions is missing, besides, we are interested in a limited range of values $x \in (0,l)$ (l denotes length of the equivalent pipe, m)

$$T(x,t) = (kf/c) \cdot \int_{0}^{x} \exp(-(kf/(cS))\cdot(t-\varphi(z,x,t)))$$

$$T_B(\varphi(z,x,t))/G(\varphi(z,x,t))dz + \exp(-(kf/(cS))(t-\varphi(0,x,t))) \cdot \tau_1(\varphi(0,x,t)), \quad (4)$$

where $\varphi(z,x,t)$ and $\varphi(0,x,t)$ are the roots of equations respectively $x - S^{-1} \cdot \int_{\varphi}^{t} G(\theta)d\theta = z$ и $x = S^{-1} \cdot \int_{\varphi}^{t} G(\theta)d\theta$

In practice, measurement of the heating systems parameters is done discretely. Let's set in the settings that all values are measured in the moments of time ($t = nh$), where h is the sampling interval. Controlling values are also measured in these moments, and we consider that the inlet temperature of straight flowing water varies between the neighboring time points nh, $(n+1)h$, linearly, and consumption (flow) in this interval remains constant and unchangeable and equal to G(nh). The sampling interval (h) is usually assumed to be equal to several hours (from 1 to 6 h) because of the large heat storage capacity (heat accumulation) of buildings. Time of the passage of water through the heating system ranges from 10 to 30 min; whence, we may come to the conclusion that Sl/G < h. Taking into consideration the suggestions and formulae, represented above, we obtain an expression for the $\varphi = nh - S(l-z)/G((n-1)h)$ value. Thus, we write the formula (4) for return water, i.e. when x = (l)

$$\tau_2(nh) = T(l,nh) = (kf/c) \cdot \int_{0}^{l} \exp(-kf(l-z)/cG((n-1)h))) \times \quad (5)$$

$$\times T_B(nh - S(l-z)/G((n-1)h))dz + \exp(-kfl/(cG((n-1)h)))\tau_1(nh - Sl/G((n-1)h))$$

As the $T_B(nh - S(l-z)/G((n-1)h))$ and $\tau_1(nh - Sl/G((n-1)h))$ values are not measured, so, we get them by means of linear interpolation in the interval ((n-1) h, nh), of the known values $T_B((n-1)h), T_B(nh)$ $\tau_1((n-1)h), \tau_1(nh)$. Taking into account the last of the ratios (5), after evaluating the integrals, we get the following expressions:

$$\begin{aligned} &\tau_2(nh) = T_B(nh)(1-\exp(-kfl/(cG((n-1)h)))) + (T_B((n-1)h) - \\ &- T_B(nh)(1-\exp(-kfl/(cG((n-1)h))))cS/(hkf) - (T_B((n-1)h) - T_B(nh)) \times \\ &\times \exp(-kfl/(cG((n-1)h)))Sl/(hG((n-1)h)) + \exp(-kfl/(cG((n-1)h))) \times \\ &\times (\tau_1(nh) + (\tau_1((n-1)h) - \tau_1(nh))Sl/(hG((n-1)h))) \end{aligned} \quad (6)$$

In order to predict the air temperature inside the building we have used (applied) the results, obtained in [5], where the building has been seen as a single container link and impacts on the building have been approximated by the zero degree splines. For a particular point (moment) in time ((n +1)/h) we have the following equation (7):

$$\begin{aligned} &T_B((n+1)h) - T_H((n+1)h) - Q((n+1)h)/(q_0V) = \\ &= \exp(-h/\beta/\beta)_B(nh) - T_H(nh) - Q((n+1)h)/(q_0V)), \end{aligned} \quad (7)$$

where T_H is temperature of external air, °C; Q denotes the heat supply into the building joules/hour; q_0 denotes the specific heat losses of the building, joules/(m^3 h° C); V is the volume of the building on the outer measurement, м3; β denotes the coefficient of an accumulation. Let's write down the formula for the amount of heat, given (transmitted) to the building by the system of heating,

$$Q(nh)=cG((n-1)h)(\tau_1(nh-Sl/G((n-1)h))-\tau_2(nh)). \qquad (8)$$

The argument of the τ_1 value takes into account the transport delay of the coolant in the process of passing through the system of heating. By means of interpolating τ_1 linearly let's substitute the expression for $Q((n+1)h)$ from formula (8) into formula (7), using beforehand the formula (6) for defining $\tau_2((n+1)h)$. In order to simplify the result we have introduced the following nomenclature (notations)

$$A_1=(1-exp(-h/\beta))(1-cS/(h\,k\,f))\,c/(q_0V);$$

$$A_2=(1-exp(-h/\beta))\,cSl/(q_0\,V\,h);$$

$$A_3=exp(-h/\beta);$$

$$A_4=c(1-exp(-h/\beta))\,/\,(q_0V);$$

$$A_5=cS\,/\,(h\,k\,f);$$

$$A_6(nh)=Sl/(h\,G\,(nh)).$$

If we take into consideration these notations, the formulae for defining $T_B((n+1)h)$ and $\tau_2((n+1)h)$ will get the following form:

$$\begin{aligned}\tau_B((n+1)h) &= \{1 + A_1G(nh)(1-\exp(-kfl/(cG(nh)))) + A_2\exp(-kfl/(cG(nh)))\}^{-1}\times\\ &\times\{T_B(nh)(A_3-(A_4-A_1)G(nh)(1-\exp(-kfl/(cG(nh)))) + A_2\exp(-kfl/(cG(nh)))) +\\ &+ T_H((n+1)h) - A_3T_H(nh) + A_4G(nh)\tau_1((n+1)h)\times\\ &\times(1-\exp(-kfl/(cG(nh)))) - A_2(1-\exp(-kfl/(cG(nh))))(\tau_1((n+1)h)-\tau_1(nh))\};\end{aligned} \qquad (9)$$

$$\begin{aligned}\tau_2((n+1)h) &= T_B((n+1)h)(1-A_5+(A_6(nh)+A_5-1)\exp(-kfl/(cG(nh)))) + T_B(nh)\times\\ &\times(A_5-(A_5+A_6(nh))\exp(-kfl/(cG(nh)))) + \exp(-kfl/(cG(nh)))(\tau_1((n+1)h) +\\ &+ (\tau_1(nh)-\tau_2((n+1)h))A_6(nh).\end{aligned} \qquad (10)$$

Formulae (9), (10) predetermine the connection between τ_2, τ_1, G, T_B, T_H in the dynamic regime (mode). As T_H we should use and take the given above temperature of external air, taking into account both fast and slow heat loss, as well as the influence of wind and solar radiation/2/.

In the stationary mode we can get the known dependences [1, 2] out of expressions (8), (9).

$$(\tau_2-T_B)/(\tau_1-T_B)=exp(-kf\,l/(cG)), \qquad (11)$$

$$q_0V(T_B-T_H)=cG(\tau_1-\tau_2). \qquad (12)$$

Dependences (11), (12) can be used in order to define the initial approximation of values $k\,f\,l/c$ and q_0V/c in accordance with experimental data in stationary conditions. One must use the step-by-step approach (method). At each step magnitudes k, β, q_0 can be adjusted and adopted to the changing operating conditions.

3 System of control

For the realization of the automatic control system with computer modelling LabVIEW graphical programming medium and cRIO controller from National Instruments have been chosen [6]. The advantage of using the LabVIEW graphical programming language in modelling is in the presence of functions (virtual appliances and gadgets), allowing and helping to solve differential equations in the period of time close to real time, by applying Runge Kutta method: Second order, as well as Laplace method.

The developed automated system consists of the hardware-software complex on the basis of cRIO (Figure 2), installed in the residential building and on the lower level of the boilers, as well as of the dispatching control panel (Figure 3), which is installed on distribution stations and boilers.

The work of lower-level automated control system comes to fulfilling two basic functions, such as: PID-regulation in accordance with the manually set values of the operator from the dispatching console; and regulation in a close to real time mode according to the specified database signs, based on the results of comparison of the data of simulations and modelling, as well as specific information from the sensors.

Fig. 2. Low-level system installation, based on the NI with RIO.

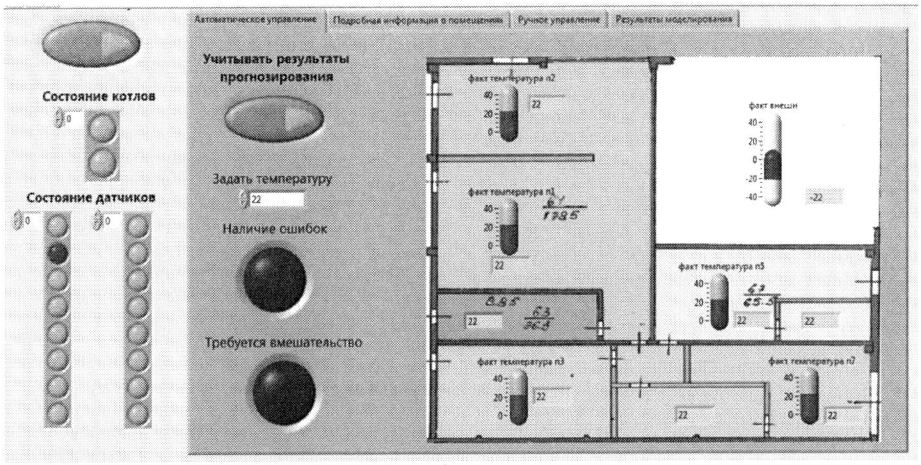

Fig.3. The screenshot of dispatching control panel.

4 Conclusion

The automated system has a number of unique features:

1. An account of the amount of heat, transmitted to the building, and comparison of actual values with the results of the simulations and modelling, which allows us to track various inclusions into the pipeline and/or change of the heating system design.

2. Prediction and forecasting of air temperature in the building on the basis of the results of computer simulations and modelling, which helps to increase accuracy of control and regulation and reduce air temperature fluctuations, as well as improve energy efficiency of functioning and work of the boiler.

3. An account, simulation and modelling of water temperature at the inlet and outlet of the heated room, assisting us in taking into account the actual (real) amount of consumed heat and correcting the cost of services, providing heating of the buildings.

The proposed method of modelling is applicable in the context of dispatching control of heating of separate buildings, as well as the district and area of heating as a whole. In the process of using it we may control the release of heat, provided not only to the private houses, dwellings and blocks of flats, but also to the public buildings, with a help of software. The use of the developed system allows us not only to make the house "Smart", but also to perform automation of boiler systems with an integration of the lower and upper levels with the functions of the manual and remote control of the heating systems of the building and boilers. Various tests and experiments show that the introduction of such a system helps us to reduce the cost of services of heating and increase the quality of control and regulation of boilers.

References

[1] Analysis and synthesis of control systems of thermal energy transfer process in the cogeneration system of central heating, Monograph, eds. PhD, Professor in Technical Sciences Susdalzev A.I. (Publishing House "Orlik & K", Orel, 2009)
[2] P. Thota, B. Krauskopf, M. Lowenberg, Nonlinear Dynamics **57**, 455(2009)
[3] C. Howcrof, B. Krauskopf, M. Lowenberg, S. Neild, SIAM Journal on applied dynamical systems **12**, 1181 (2013)
[4] A.P. Hiwarekar, International Journal of Mathematical Archive **3/3**, 1193 (2012)

[5] Automation of cogeneration heat supply systems with distributed peak loads, Monograph, eds. PhD, Professor in Technical Sciences Susdalzev A.I. (Mashinostroenie, Moscow, 2007)

[6] A. Y. Nikiforov, P. K. Skorobogatov, A. N. Egorov, D. V. Gromov, Russian Microelectronics **43**(2), 133 (2014) doi: 10.1134/S1063739714020073.

ESTIMATION OF DYNAMIC ERRORS IN LASER OPTOELECTRONIC DIMENSION GAUGES FOR GEOMETRIC MEASUREMENT OF DETAILS

Zimfir Khasanov, *Vasikh* Yasoveev, *Oleg* Khasanov and *Alexandr* Vishnevskiy*

Ufa State Aviation Technical University, 400000, Ufa, Russia

Abstract. The article reviews the capabilities and particularities of the approach to the improvement of metrological characteristics of fiber-optic pressure sensors (FOPS) based on estimation estimation of dynamic errors in laser optoelectronic dimension gauges for geometric measurement of details. It is shown that the proposed criteria render new methods for conjugation of optoelectronic converters in the dimension gauge for geometric measurements in order to reduce the speed and volume requirements for the Random Access Memory (RAM) of the video controller which process the signal. It is found that the lower relative error, the higher the interrogetion speed of the CCD array. It is shown that thus, the maximum achievable dynamic accuracy characteristics of the optoelectronic gauge are determined by the following conditions: the parameter stability of the electronic circuits in the CCD array and the microprocessor calculator; linearity of characteristics; error dynamics and noise in all electronic circuits of the CCD array and microprocessor calculator.

1 Introduction

Currently, the parts with complex curved surfaces are machined on six-coordinate machines with programmed control. The control module with optoelectronic gauges for geometric dimensions, coordinates, thickness, orientation angles and stroke speed measurements performs the required technological operations for the manufacture of parts. During the positional control of the machine's multiple connected mechanisms, it is necessary to solve the direct problem concerning the positions and geometric dimensions of the parts and the processing node in the absolute coordinate system OXYZ many times. The main obstacle to the automation of these processes is the difficulty in defining the coordinates for the operating profile points of the processed surface and measuring its actual geometric dimensions. The best way to solve this problem can be laser-optoelectronic dimension gauges for geometric measurement of details with the complex shape.

Laser-optoelectronic dimension gauge is designed for non-contact measurements of geometric dimensions simultaneously in three sections and parts position during the production process. Three interconnected optoelectronic cameras with a semiconductor laser are mounted

* Corresponding author: host_of_peace@list.ru

on a positioning stage with numerical program control. The software of the three-dimensional laser-optoelectronic gauge is based on a modified hypsometric method with the interferometry principles [1].

The measurement error of laser-optoelectronic dimension gauges for geometric measurements increases in the dynamic mode, compared to the static one due to the time lag of the measuring channel nodes and fluctuations in the signal of the measuring information [2]. At the same time, the factors leading to measurement error in the static mode, have the same effect in a dynamic one as well. Errors do not remain constant over the time due to the changes affecting the factor measurement accuracy, for example, dynamic changes in the measurement signal, and temperatures in the structural elements of the system [3].

Laser-optoelectronic dimension gauge for geometric measurements is built according to modular principle with the use of high-speed video controllers. The structure of the gauge includes sequentially connected modules of a primary transducer (MPT), an analog-to-digital converter (ADC) and a video controller (VC). The formal identification of the gauge subsystems with dynamical systems under the deterministic approach leads to mathematical models belonging to the class of differential equations in the state space [4]. The sequence of state changes caused by the test action completely characterizes the dynamic properties of the laser gauge. For example, the method of test transient processes is realized by excitation of the transient process in the converters of the laser gauge, and the converted value affects the parameters of the converters by changing their transient characteristics in the state space.

Thereby, the interaction of the laser gauge with the object being measured can be qualitatively represented based on the general systems theory models. However, it should be noted that, in contrast to the dynamical systems themselves, characterized in terms of stability, controllability, and observability, in the optoelectronic measurement processes, the main issue is the problem of representing the properties of a measurement object by numerical values and the uniqueness of these values. That is, the solution to this problem depends on how the properties of the measured object are represented. Formation of the properties' operational values is based on the processes of identification, ordering, counting, and comparison.

2 Mathematical aspects of the estimation of dynamic errors in laser optoelectronic dimension gauges for geometric measurement of details

Any optical image can be characterized by a multidimensional function. Its arguments can be the coordinates *x, y, z,* the time *t* (or the frequency ν, or the wavelength of the optical radiation λ, and in the case when the signal is described by the intensity function, the time *t*, the wavelength λ, and the polarization state ψ). The informativity of an optical signal is greater than that of an electric signal, but the conversion of such a signal is more complicated.

A fundamental feature of the converters of a laser gauge is the sequential conversion of multidimensional optical signals into one-dimensional electrical ones, and then, in the recording system, again into multidimensional signals, that are convenient for the visual perception. In addition, a large amount of naturally originated optical interference, usually accompanying the reception and processing of an optical signal should be noted. Therefore, special optical filters are introduced into the laser gauge to select a desired signal against the interference background.

In [5], a system of criteria is described that makes it possible to evaluate the optimality of conditioning the optoelectronic converter (OC) with the optical system (OS) and the video controller, in which microprocessor information processing occurs. In this system of evaluation criteria, the following coefficients are used

$$\eta_1 = \frac{I_{oc}}{I_{os}}; \eta_2 = \frac{t_{oc}}{t_{os}}; \eta_3 = \frac{\vartheta_{oc}}{\vartheta_{os}}; \eta_4 = \frac{I_{vc}}{I_{os}},$$

where I_{oc}, I_{os}, and I_{vc} is the maximum possible information capacity of the image perceived by the OC, OS, and VC, respectively; t_{os} – the time of the optical image formation in the OS and t_{oc} -- the total processing time of the image in the OC; ϑ_{oc} and ϑ_{os} – productivity of OC and OS, respectively. The criterion η_1 characterizes the speed of the OC, η_2 and η_3 – the specific speed and specific performance of the laser gauge, η_4 – the load efficiency of the video controller in the laser gauge.

The proposed criteria render new methods for conjugation of optoelectronic converters in the dimension gauge for geometric measurements in order to reduce the speed and volume requirements for the Random Access Memory (RAM) of the video controller which process the signal. In addition to the above criteria, the stability and variation range of the parameters and characteristics of the optoelectronic converter are assessed under the specified operating conditions. The sensitivity of optoelectronic converters is determined by the charge accumulated in the potential energy well of the cell and by the signal level associated with this charge. For the estimation of the photodiode arrays capabilities, it is also important to know the sensitivity non-uniformity of its individual elements and the parameters' repeatability error of electronic systems for reading signals from individual elements. In arrays, the parameters' repeatability error can reach up to 50%, which reduces their dynamic range by two or more times. Such an error clearly reduces the possibility of the optical analysis of images, which are usually estimated using the temperature difference NEΔT equivalent to the noise. The value of NEΔT increases in direct proportion to the conversion coefficients' repeatability error of the individual elements in the photodiode arrays. Thus, that the digital processing of the output signal directly in the external microcontroller allows solving the problem of implementing a high degree of sensitivity uniformity across all the elements when the ambient temperature and background illumination change.

In the optoelectronic gauge [6] operating in real time, the sensitivity of each photodetector and the output signal of each reading cells are measured and stored in a special microcontroller. In addition, such a scheme makes it possible to approximate linearly the transfer function of each element of the photodiode array without significant deterioration of the signal-to-noise ratio. When the array is operating in the accumulation mode, a linear approximation of the relationship between the optical radiation flux incident on an individual element of the photodiode array and the output signal of the readout systems provides a sufficiently high accuracy of compensation, characterized by errors in measuring the temperature difference that equivalent to noise of thousandths of Kelvin.

When the photodiode array is operating in the immediate-action mode, the dependence of the current at the output of the photodiode array reading system on the value of the input signal is nonlinear. Here, the dynamic errors in compensation systems with linear approximation are greater than in systems with accumulation. To reduce these errors, for each element of the photodiode array, it is necessary to calculate and memorize not two but three parameters, which considerably complicates the compensation system and increases the dynamic error of calculation in a special microcontroller.

The main node for control and data processing in a laser optoelectronic dimension gauge for geometric measurements is the video controller with the RAM containing the corresponding programs for information processing and measurement. A variety of means that provides communication with a powerful PC can be connected to the video controller. The exposure time is calculated using the video controller timer. The reading of the photodiode array is determined by the time of the reliable resolution of the two measurement signals and can reach 10 ÷ 50 kHz.

The dynamic errors of the optoelectronic gauge are largely determined by the dynamic parameters of all the interrogators and signal processing in the photodiode array and digital processing elements on the microcontroller [7]. These errors have not been studied, and there is no methodology for calculating them. The conversion in the gauge is performed by a complex of nonlinear devices and the mathematical tool of linear systems is not applicable to its analysis [8]. In this regard, two types of dynamic errors for MPT occur: the first type, due to the inertial properties of MPT nodes, and the second type, associated with the change in the measured signal over the conversion cycle. Then, the dynamic characteristics of the MPT are determined as follows:

$$\begin{cases} |K(jw)| = \prod_{s=0}^{n} K_s(jw), \\ A(w) = \prod_{s=0}^{n} A_s(w), \\ \varphi(w) = \sum_{s=0}^{n} \varphi_s(w), \end{cases} \tag{1}$$

where $K(jw)$ – is the frequency response; $A(w)$ and $\varphi(w)$ – amplitude-frequency and phase-frequency characteristics; n – the number of analog function elements in MPT.

In such a representation, the sum of the dynamic errors of the first and second types of MPT is determined by the equation

$$\Delta U_{\mathrm{D}}^{1}(t_i) = U_{\mathrm{out}}^{1}(t_i + \Delta t_i) - U_{\mathrm{ma}}^{1}(t_i), \tag{2}$$

and by the inaccuracy of the algorithm being implemented for its processing.

However, considering that the actual value of the measured signal $U_{\mathrm{ma}}^{1}(t_i)$ at moment when the received response relates is unknown, as well as the time $t_i + \Delta t_i$ is unknown when the measurement was actually obtained, it is impossible to use the expression (2) to determine the dynamic error. Thus, various estimations for the magnitude or parameters of the dynamic error are adopted. From (2) it is seen that the value of this error depends on the type $U_{\mathrm{out}}^{1}(t_i)$ and function of its variation with time. Therefore, the initial data, which is necessary for obtaining the estimation of this error, should include the information on the parameters of the function being evaluated and data on the process being measured.

MPT is a collection of a large number of elements that interact in a certain way and it possesses the main features and characteristics of discrete-analog systems, in a broad sense of the term. The processes of their functioning are characterized by a sequential change in the state of the MPT elements in time and can be described by a number of states

$$Q = \{q_1, q_2, \ldots, q_i, \ldots\} \tag{3}$$

and by number of stay durations in each of them

$$T = \{\tau_1, \tau_2, \ldots, \tau_i, \ldots\}. \tag{4}$$

Then, in accordance with the assigned task, for the MPT consisting of a photodiode array, an amplifier, a double-correlated processing scheme and an active filter, we determine the stay duration of the measurement signal at a discrete moment τ , i.e. elements of a set T .

The scheme of combining the CCD (charge-coupled device) array with the electronic circuit is shown in Figure 1. When reading the optical information, each photosensitive element (5) is connected to the signal reading information bus through the corresponding signal key switches. The shutter of each key switch is controlled by a shift scan register. After reset, the potential at the amplifier input is measured when the key switch is off (9). This algorithm provides a double-correlated sampling mode that eliminates reset noise, which is the largest over all other noise in CCD arrays [9].

Through the input (6) an injection pulse is applied to the reading line, causing the disappearing of the potential well and the current flow, which ensures the recombination of the carriers accumulated in the well. The voltage difference between the sampled values is determined by the value of the charge injected from the potential well. During the interval between the reading processes, the key switches (3) are switched on and the reference voltage is set on all photodetectors.

For the given optical information, the conversion dynamic error in the CCD array can be described by a functional as follows:

$$\Delta_D = f\left(\frac{dx}{dt}, \Delta t, t_{\text{int}}\right), \tag{5}$$

where $x(t)$ is optical information; Δt – conversion time; t_{int} – time of interrogation.

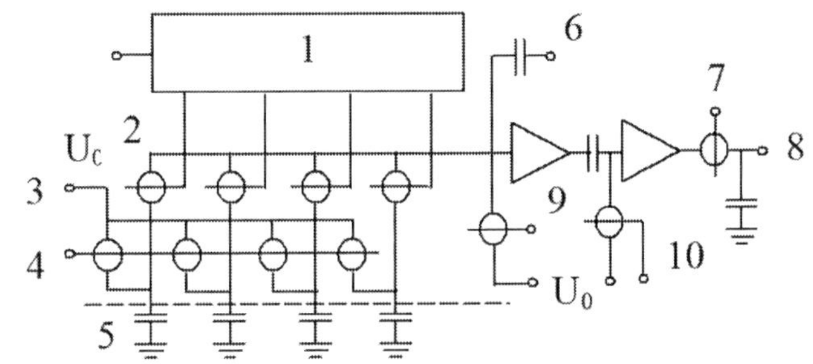

Fig. 1. The scheme of combining the CCD array with the electronic circuit:1 – scan registers; 2 – information bus key switches; 3 – addressing key switches; 4 – parallel reset key switches; 5 – photodetectors (represented in the form of capacity); 6–- injection pulse; 7 – sampling;8 – output; 9 –reset; 10 – first sampling.

Assuming the measurement information $x(t)$ is continuous during the interrogation time $t_{\text{опр}}$, it can be represented as follows

$$x(t_i) = x(t_{\text{int}}) + x'(t_{\text{int}}) \cdot \Delta t_i + \frac{x''(t_{\text{int}})}{2}(\Delta t_i)^2 + \ldots + \frac{x^{(n)}(t_{\text{int}})}{n!}(\Delta t_i)^n. \tag{6}$$

The function of the interrogation speed of the CCD array during the time $[0; T_{\text{int}}]$ is represented in the following form

$$V_{\text{CCD}}(\omega_1) = \int_0^{T_{\text{int}}} V_{\text{CCD}}(\omega_1 t)dt = \int_0^{T_{\text{int}}} \left(V_{\text{CCD(N)}} + \Delta V_{CCD(A)} \sin(\omega_1(t))\right)dt, \tag{7}$$

and the function of changing the measurement information in the time interval $[0; T_i]$ is represented as follows

$$V_{\text{mi}}(\omega_2)=\int_0^{T_i}V_{\text{mi}}(\omega_2 t)dt=\int_0^{T_i}\left(V_{\text{mi(N)}}+\Delta V_{\text{mi}(A)}\sin(\omega_2(t))\right)dt, \tag{8}$$

where $V_{\text{CCD(N)}}$ and $V_{\text{mi(N)}}$ are nominal values of speeds; $V_{CCD(A)}$ and $V_{mi(A)}$ - the amplitude values of the interrogation speed of the CCD array and speed of change in the measurement information.

Assuming $\int_0^{T_1}V_{\text{CCD}}(\omega_2 t)dt=\int_0^{T_1}\Delta V_{mi(A)}\sin(\omega_2(t))dt$, then the change in the measurement information in the time intervals $[0;T_1]$ and $[0;T_2]$ can be represented as follows

$$V_{\text{mi}}(\omega_2 T_1)=\int_0^{T_1}\left(V_{\text{mi(N)}}+\Delta V_{\text{mi}(A)}\sin(\omega_2(t))\right)dt,$$

$$V_{\text{mi}}(\omega_2 T_2)=\int_0^{T_2}\left(V_{\text{mi(N)}}+\Delta V_{\text{mi}(A)}\sin(\omega_2(t))\right)dt.$$

Integrating these expressions and solving them with respect to T_1 and T_2, the following equation is obtained

$$\Delta_{\text{mi}}(T_2)\approx\frac{T_1}{T_2}=\frac{V_{\text{mi}}(\omega_2 T_1)-2\dfrac{\Delta V_{\text{mi}(A)}}{\omega_2}\sin^2\dfrac{\omega_2\cdot V_{\text{mi}}(\omega_2 T_1)}{2V_{\text{mi(N)}}}}{V_{\text{mi}}(\omega_2 T_2)-2\dfrac{\Delta V_{\text{mi}(A)}}{\omega_2}\sin^2\dfrac{\omega_2\cdot V_{\text{mi}}(\omega_2 T_2)}{2V_{\text{mi(N)}}}}. \tag{9}$$

For any T_i, this equation takes the form as follows (вместо we get)

$$\Delta_{\text{mi}}(T_i)\approx\frac{T_1}{T_i}=\frac{V_{\text{mi}}(\omega_2 T_1)-2\dfrac{\Delta V_{\text{mi}(A)}}{\omega_2}\sin^2\dfrac{\omega_2\cdot V_{\text{mi}}(\omega_2 T_1)}{2V_{\text{ms(N)}}}}{V_{\text{mi}}(\omega_2 T_i)-2\dfrac{\Delta V_{\text{mi}(A)}}{\omega_2}\sin^2\dfrac{\omega_2\cdot V_{\text{mi}}(\omega_2 T_i)}{2V_{\text{ms(N)}}}}. \tag{10}$$

Due to the change speed irregularity in the measurement information, the relative error $\Delta_{\text{ms}}(T_i)$ is determined as follows

$$\delta_{\Delta_{\text{mi}}(T_i)}=\frac{2\Delta V_{\text{mi}(A)}}{\omega_2}\left(\frac{\sin^2\dfrac{\omega_2\cdot V_{\text{mi}}(\omega_2 T_i)}{2V_{\text{ms(N)}}}}{V_{\text{mi}}(\omega_2 T_i)}-\frac{\sin^2\dfrac{\omega_2\cdot V_{\text{mi}}(\omega_2 T_1)}{2V_{\text{ms(N)}}}}{V_{\text{mi}}(\omega_2 T_1)}\right). \tag{11}$$

At small values ω_2

$$\delta_{\Delta_{\text{mi}}(T_i)}=\frac{2\omega_2\Delta V_{\text{mi}(A)}}{V_{\text{mi(N)}}}\left(V_{\text{mi}}(\omega_2 T_i)-V_{\text{mi}}(\omega_2 T_1)\right). \tag{12}$$

The lower relative error, the higher the interrogetion speed of the CCD array.

The laser gauge must have the required dynamic accuracy which means that the error for all given measurement conversions and perturbation influences should not exceed a certain permissible value. Complete information on dynamic accuracy is given by the law of time error variation. For the deterministic change in the input effect, the error consists of two components

$$\Phi(t_i) = \Phi_{st}(t_i + \Delta t_i) - \Phi_{tr}(t_i + \Delta t_i), \quad (13)$$

where $\Phi_{st}(\cdot)$ and $\Phi_{tr}(\cdot)$ – stimulated and transient errors, respectively, which depend on external influences and on the change of all parameters of the interconnected converters of the laser gauge. To reduce these errors, various methods of dynamic correction are used. However, for the solution, the structure and parameters of measurement circuits with correcting elements in numerical form should be set, while for the dynamic synthesis case, the choice of the structure and parameters of the correcting systems is the main task. Because of that, the methods based on solving differential equations, at the stages of synthesis of structures and parameters cannot achieve the desired results.

At the stage of the synthesis of the laser gauge, it is rational to apply frequency methods that allow a direct and visual connection of the dynamic properties of functional subsystems with metrological parameters and characteristics. The main correction tools are the correcting elements, the inclusion of which to the measuring circuits provides the required improvement in characteristics. The task is even more complicated if the dynamic characteristics in the analogue part of the information conversion do not remain constant but change in time randomly.

One of the types of such correction is dynamic reflex transformation [10]. When this method is used at the output of the CCD array, which has unstable dynamic properties, a chain of correcting elements is switched on, dynamic characteristics of which are automatically maintained in accordance with the current state of the photoconverters dynamic parameters. To simplify the recording, the photodetector circuit of the CCD array consists of two elements: an inertial linear one, the static conversion function of which is stable, and the inertialess linear one with time-dependent conversion function. Let us assume that the inertial element of the photodetector circuit of the CCD array is described by the equation

$$L(K_i^*) = W(p)L(\varepsilon_i), \quad (14)$$

where K_i^* is the variable corresponding to the output signal of the inertial element. Then the conversion function of the inertialess linear element is determined by

$$K_i = f_0(t) + f_1(t)K_i^*, \quad (15)$$

where $f_0(t)$ and $f_1(t)$ is a function of time. In turn, the function of the correcting element is defined as follows

$$\varepsilon_{k(i)} = a_0 + a_1 K_i. \quad (16)$$

To determine K_i^*, the inverse Laplace transformation is used:

$$K_i^* = L^{-1}[W(p)L(\varepsilon_i)]. \quad (17)$$

Then K_i takes the form as follows

$$K_i = f_0(t) + f_1(t)L^{-1}[W(p)L(\varepsilon_i)]. \quad (18)$$

By selecting parameters for $f_0(t)$, $f_1(t)$, $W(p)$ and $L(\varepsilon_i)$ the required value K_i is achieved.

3 Conclusions

Direct parallel or sequential correction devices in electronic circuits of the CCD array with a microprocessor calculator provide the maximum dynamic quality of the optical signal conversion with a wide range of perturbation influences such as ambient temperature and background illumination. The requirement to provide an independent choice of parameters and quality criteria regarding the effective influences and to each of the perturbation influences leads to the necessity to organize its own combined measurement contour for each influence. From the point of view of achieving high dynamic measurement accuracy with a wide range of changes in the environment and the variety of background illumination, the multichannel structure of digital processing of CCD array sensor signals is the most rational.

Thus, the maximum achievable dynamic accuracy characteristics of the optoelectronic gauge are determined by the following conditions: the parameter stability of the electronic circuits in the CCD array and the microprocessor calculator; linearity of characteristics; error dynamics and noise in all electronic circuits of the CCD array and microprocessor calculator. The following factors determine the achievement of high dynamic characteristics of a laser optoelectronic dimension gauge for geometric measurements: nonlinearity, dynamics of the errors, intrinsic noise, deviations of parameters due to changes in ambient conditions, mutual cross-links generated by the practical implementation specifics of the electronic circuits in the CCD array and a microprocessor calculator.

Based on the obtained theoretical and experimental results, the recommendations for the parameter selection for laser optoelectronic dimension gauge for the industry application are proposed. The results for the dynamic error calculation of the developed dimension gauge are well consistent with the results of real metrological tests with an accuracy of 3 ÷ 5%.

4 Conflict of interests

The authors declare that there is no conflict of interest regarding the publication of this paper.

References

[1] Z.M. Khasanov, O.Z. Khasanov, Testing. Diagnostics, 6, 45 (2006)
[2] A.A. Vishnevskiy, Caspian Journal: Management and High Technologies **2(30)**, 193 (2015)
[3] A.A. Vishnevskiy, V. Kh Yasoveev, Proceedings of the third international conference: "Intelligent technologies for information processing and management", 65 (2015)
[4] A.A. Vishnevskiy, V.Kh Yasoveev, Caspian Journal: Management and High Technologies **3(31)**, 158 (2015)
[5] A.L. Shestakov, *Methods of the theory of automatic control in dynamic measurements* (Publishing Centre: SUSU, Chelyabinsk, Russia, 2013)
[6] Z.M. Khasanov, Testing. Diagnostics, 59 (2006)
[7] Z.M. Khasanov, O.Z. Khasanov, Testing. Diagnostics, 62 (2007)
[8] A.A.Vishnevskiy, V.Kh. Yasoveev, Z.M. Khasanov, 2nd International Ural Conference on Measurements (UralCon), Chelyabinsk, Russia, 319, (2017), doi: 10.1109/URALCON.2017.8120730
[9] Z.M. Khasanov, Mekhatronika, Avtomatizatsiya, Upravlenie **4**, 37 (2006)
[10] A.A. Vishnevskiy, V.Kh. Yasoveev, Z.M. Khasanov, International Conference on Industrial Engineering, Applications and Manufacturing (ICIEAM), St. Petersburg, Russia, 1 (2017), doi: 10.1109/ICIEAM.2017.8076120

NEW MODELING OF STEADY-STATE MODES OF COMPLEX ELECTRICAL GRIDS OF POWER SYSTEMS

Arman Akhmetbayev[1], *Dauren* Akhmetbayev[a2], *Serik* Zhumazhanov[2], *Bauyrzhan* Zhakishev[2]

[1]Directorate of Information Systems of "Kaztelecom" JSC, Almaty, Kazakhstan
[2]Department of "Electrical Power Supply" KATU n.a. S. Seifulullin, Astana, Kazakhstan

Abstract. Classical methods for modeling the steady-state modes of complex electrical networks and systems are based on the application of nonlinear node equations. Nonlinear equations are solved by iterative methods, which are connected by known difficulties. To a certain extent, these difficulties can be weakened by applying topological methods. In this paper, we outline the theoretical foundations for the formation of the inverse form of nodal stress equations based on the topology of electrical networks and systems. A new topological method for calculating the distribution coefficients of node currents is proposed based on all possible trees of a directed graph of a complex electrical network. A complex program for calculating current distribution coefficients and forming steady-state parameters in the MATLAB environment has been developed.

1 Introduction

When analyzing the functioning and management of power system regimes, it becomes necessary to perform a large amount of work on the calculation of steady-state regimes [1]. Equations of steady-state regimes of complex energy systems are nonlinear and can be solved only by iterative methods. The difficulties of obtaining real solutions of non-linear equations of steady-state regimes can be overcome if one approaches from the point of application of the inverse form of nodal equations [2]. There exist several indirect methods for determining the matrix of nodal resistances, which in the general case did not lead to the desired result with respect to complex systems. In [3], topological methods were developed for calculating electric power systems in which the steady-state regime was directly formalized from the topology of the circuit, which turned out to be of little use for a complex system. Below we propose an improved topological method for modeling the steady-state regimes of complex electric power systems.

[a]Corresponding author: Axmetbaev46@mail.ru

2 The transformation of the node equation

The inverse form of the node equations is as follows:

$$\dot{U}_{\Delta} = \underline{Z}\dot{\underline{J}}, \tag{1}$$

where $\underline{Z} = \underline{Y}^{-1}$ –node impedance matrix; $\underline{Y}^{-1}$ –square matrix of nodal conductivities;

$\dot{\underline{J}} = -\frac{\hat{\underline{S}}}{\sqrt{3}\hat{\underline{U}}}$ –column matrix of given current.

The analytical relationship between the given currents and the distribution of currents in the passive circuit is determined by the expressions:

$$M\dot{\underline{I}} = \dot{\underline{J}}; \tag{2}$$

$$\dot{\underline{I}} = \underline{C}\dot{\underline{J}}, \tag{3}$$

where $\underline{C}$ –rectangular matrix of current distribution coefficients; M –the first incidence matrix.

Expression (1) with provision for (2), (3) can be written down as follows:

$$\dot{\underline{U}}_{\Delta} = \underline{C}^{t}\underline{Z}_{b}\underline{C}\dot{\underline{J}}, \tag{4}$$

where $\underline{C}^{t} = \underline{Z}M\underline{Z}_{b}^{-1}$ –transposed matrix of current distribution coefficients.

The obtained equation (5) allows us to write the identity as follows:

$$\underline{Z} = \underline{C}^{t}\underline{Z}_{b}\underline{C} = \underline{Y}^{-1}. \tag{5}$$

Thus, the problem of calculating the matrix of nodal resistances is reduced to determining the matrix of the distribution coefficients of the given currents, which substantially reduces the amount of computing work.

3 Current distribution coefficients and electrical network topology

The topology of electrical networks makes it possible to establish an analytical relationship between the current distribution coefficients and the structure of the linear graph of the network. A significant place in the network topology is occupied by such subgraphs as trees and 2-trees. From the values of trees and 2-trees of the graph numerators and denominators of the distribution coefficients of the driving currents of electric networks are formed [3].

In the works [4, 5], a new approach to the determination of the numerators of the current distribution coefficients in the function of the network parameters was proposed without searching for and determining the value of the 2-trees of the graph of a complex network.For the closed network shown in Figure 1*a*, the first column of the matrix of the distribution coefficients of the node currents is determined by solution of the equation

$$\dot{\underline{I}}_{k} = \underline{Z}_{1} + \underline{Z}_{2} + \underline{Z}_{3} + \underline{Z}_{4} = \underline{Z}_{1}, \tag{6}$$

generated for the circuit (see Figure 1*b*).

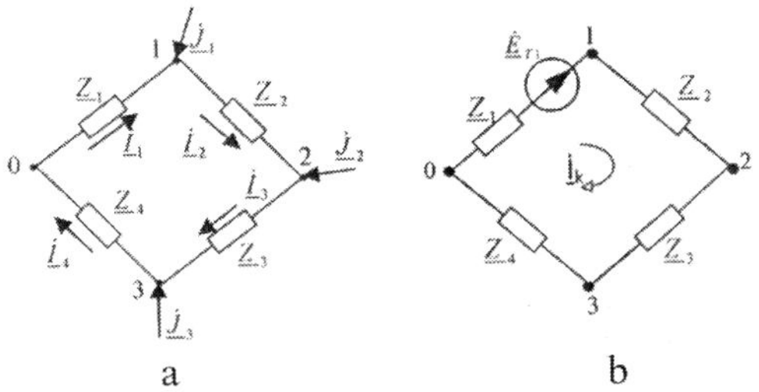

Fig. 1. The closed circuit layout: a) original circuit; b) design circuit.

$$\underline{C}=\frac{1}{\underline{Z}_k}\begin{vmatrix} -(\underline{Z}_2+\underline{Z}_3+\underline{Z}_4) & -(\underline{Z}_3+\underline{Z}_4) & -\underline{Z}_4 \\ \underline{Z}_1 & -(\underline{Z}_3+\underline{Z}_4) & -\underline{Z}_4 \\ \underline{Z}_1 & (\underline{Z}_1+\underline{Z}_2) & -\underline{Z}_4 \\ \underline{Z}_1 & (\underline{Z}_1+\underline{Z}_2) & (\underline{Z}_1+\underline{Z}_2+\underline{Z}_3) \end{vmatrix}, \quad (7)$$

where $\underline{Z}_k=\underline{Z}_1+\underline{Z}_2+\underline{Z}_3+\underline{Z}_4$ – loop resistance.

The topological properties of the current distribution coefficients (7) can be determined by formulas that can be written down after simple transformations as follows:

$$\underline{C}=\frac{1}{\underline{Y}_1\underline{Y}_2\underline{Y}_3+\underline{Y}_1\underline{Y}_2\underline{Y}_4+\underline{Y}_1\underline{Y}_3\underline{Y}_4+\underline{Y}_2\underline{Y}_3\underline{Y}_4}\times$$
$$\times\begin{vmatrix} -\underline{Y}_1(\underline{Y}_2\underline{Y}_3+\underline{Y}_2\underline{Y}_4+\underline{Y}_3\underline{Y}_4) & -\underline{Y}_1(\underline{Y}_2\underline{Y}_3+\underline{Y}_2\underline{Y}_4) & -\underline{Y}_1(\underline{Y}_2\underline{Y}_3) \\ \underline{Y}_2(\underline{Y}_3\underline{Y}_4) & -\underline{Y}_2(\underline{Y}_1\underline{Y}_3+\underline{Y}_1\underline{Y}_4) & -\underline{Y}_2(\underline{Y}_1\underline{Y}_3) \\ \underline{Y}_3(\underline{Y}_2\underline{Y}_4) & \underline{Y}_3(\underline{Y}_1\underline{Y}_4+\underline{Y}_2\underline{Y}_4) & -\underline{Y}_3(\underline{Y}_1\underline{Y}_2) \\ \underline{Y}_4(\underline{Y}_2\underline{Y}_3) & \underline{Y}_4(\underline{Y}_1\underline{Y}_3+\underline{Y}_2\underline{Y}_3) & \underline{Y}_4(\underline{Y}_2\underline{Y}_3+\underline{Y}_3\underline{Y}_4+\underline{Y}_1\underline{Y}_2) \end{vmatrix} \quad (8)$$

The obtained expressions for the elements of the current distribution matrix are the trees of the graph of the circuit as a function of the natural parameters of the electrical circuit. For all the current distribution coefficients that make up the matrix under study, the denominator is the same, which is determined by the sum of the products of the conductivities of possible trees of the graph.

Possible trees of the graphare shown in the Figure 2.

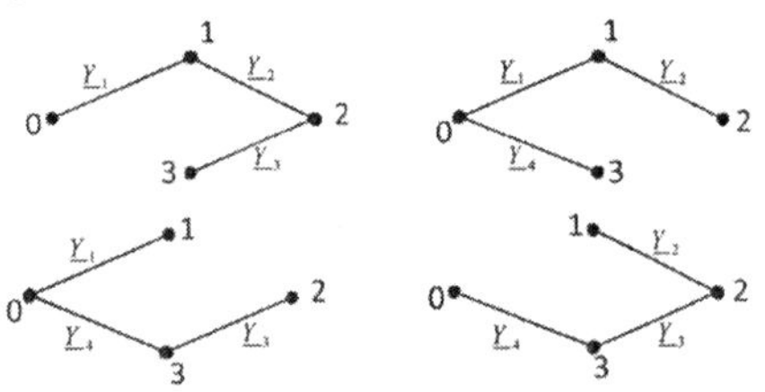

Fig. 2. Closed circuit graph trees.

The numerators of the current distribution coefficients depend on the node of the perturbation application, in the form of a single current, and the geometric image of the circuit.

For example, the numerator of the current distribution coefficient $\underline{C}_{11}$is formed by the sum of the products of the values of three trees containing the path from the first node to the basic node, through branch 1, shown in Figure 3.

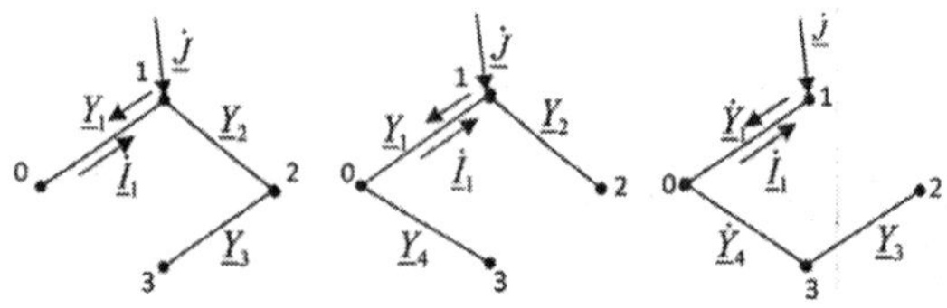

Fig. 3. Graph trees of the numerator $\underline{C}_{11}$.

The numerators of the coefficients $\underline{C}_{21}$, $\underline{C}_{31}$, $\underline{C}_{41}$.are defined likewise.

In the general case, the elements of the current distribution matrix are calculated on the basis of the topological expression

$$C_{ij} = \frac{\sum_{k=1}^{M} F_{kj}^{1}}{\sum_{k=1}^{N} F_k}, \tag{9}$$

where $\sum_{k=1}^{N} F_k$ –the total value of all possible trees of the graph;

$\sum_{k=1}^{M} F_{kj}^{1}$ –algebraic sum of the values of the specific trees of the graph of the i-th branch.

For all current distribution coefficients that are part of the network under study, the denominator is the same, i.e. is determined once by the total value of possible trees of the graph.

4 Formation of steady-state mode

The voltage at the nodes relative to the base node is determined by the expression:

$$\dot{U} = U_0 + C' Z_d C \hat{U}_d^{-1} \hat{S}, \tag{10}$$

where C –rectangular complex matrix of current distribution coefficients;

Z_d –diagonal matrix of branches resistances;

$\hat{U}_d$ – diagonal matrix of nodal conjugate voltages;

$\hat{S}$ –matrix - column of conjugate powers of node loads and generators.

5 Calculation of the steady state of a real 220 kV network

The nodalization diagram of the allocated part of the network of the real power system of the Republic of Kazakhstan is shown in Figure 4. The resistance of the branches of the replacement circuit is indicated in the Figure and the total power of the nodes are known.

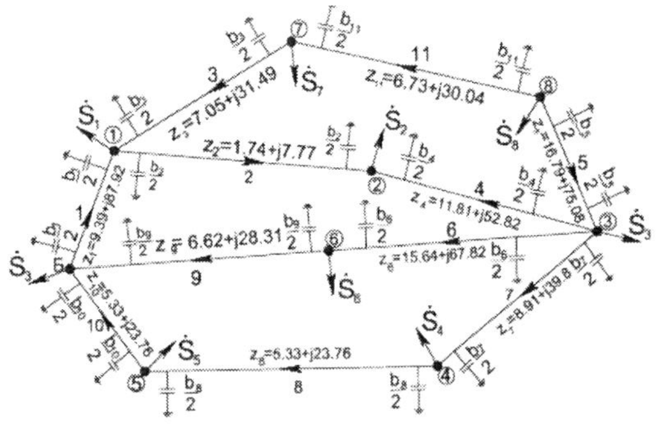

Fig. 4. Electric network nodalization diagram.

5.1 Formation of a matrix of current distribution coefficients

All possible trees of the non-directional graph of the original circuit are defined using the theory of structure numbers. To search for and identify specific trees, a special program was developed in the MATLAB environment, on the basis of which the current distribution coefficients were calculated (9).

Calculation results are shown in Tables 1a, 1b.

Table1a. The values of the current distribution coefficients for $C_{i,j=1...4}$.

-0.5043 + 0.0289i	-0.4733 + 0.0272i	-0.2631 + 0.0156i	-0.1432 + 0.0085i
0.3434 + 0.0200i	-0.5958 + 0.0188i	-0.1823 + 0.0108i	-0.0992 + 0.0059i
0.1523 + 0.0089i	0.1224 + 0.0084i	-0.0808 + 0.0048i	-0.0440 + 0.0026i
0.3434 + 0.0200i	0.4042 + 0.0188i	-0.1823 + 0.0108i	-0.0992 + 0.0059i
0.1523 + 0.0089i	0.1224 + 0.0084i	-0.0808 + 0.0048i	-0.0440 + 0.0026i
0.2357 + 0.0146i	0.2504 + 0.0139i	0.3504 + 0.0087i	0.1907 + 0.0047i
0.2600 + 0.0143i	0.2762 + 0.0133i	0.3865 + 0.0069i	-0.3339 + 0.0038i
0.2600 + 0.0143i	0.2762 + 0.0133i	0.3865 + 0.0069i	0.6661 + 0.0038i
0.2357 + 0.0146i	0.2504 + 0.0139i	0.3504 + 0.0087i	0.1907 + 0.0047i
0.2600 + 0.0143i	0.2762 + 0.0133i	0.3865 + 0.0069i	0.6661 + 0.0038i
0.1523 + 0.0089i	0.1224 + 0.0084i	-0.0808 + 0.0048i	-0.0440 + 0.0026i

Table1b. The values of the current distribution coefficients for $C_{i,j=5...8}$.

-0.0716 + 0.0043i	-0.0775 + 0.0048i	-0.4487 + 0.0259i	-0.3956 + 0.0229i
-0.0496 + 0.0029i	-0.0537 + 0.0033i	0.2222 + 0.0179i	0.1066 + 0.0159i
-0.0220 + 0.0013i	-0.0238 + 0.0015i	-0.6709 + 0.0079i	-0.5023 + 0.0070i
-0.0496 + 0.0029i	-0.0537 + 0.0033i	0.2222 + 0.0179i	0.1066 + 0.0159i
-0.0220 + 0.0013i	-0.0238 + 0.0015i	0.3291 + 0.0079i	0.4977 + 0.0070i
0.0954 + 0.0024i	-0.1914 + 0.0030i	0.2622 + 0.0133i	0.2874 + 0.0120i
-0.1669 + 0.0019i	0.1139 + 0.0018i	0.2892 + 0.0126i	0.3170 + 0.0109i
-0.1669 + 0.0019i	0.1139 + 0.0018i	0.2892 + 0.0126i	0.3170 + 0.0109i
0.0954 + 0.0024i	0.8086 + 0.0030i	0.2622 + 0.0133i	0.2874 + 0.0120i
0.8331 + 0.0019i	0.1139 + 0.0018i	0.2892 + 0.0126i	0.3170 + 0.0109i
-0.0220 + 0.0013i	-0.0238 + 0.0015i	0.3291 + 0.0079i	-0.5023 + 0.0070i

5.2 Steady-state mode calculations

Algorithms for the formation of steady-state mode are realized in the MATLAB. All possible graph trees are defined on the basis of the theory of structural numbers, and specific trees are based on a specially developed for this purpose algorithm. Calculations of nodal voltage are performed with iterative methods.

A numerical experiment was performed for the design circuit of the allocated part of the network of the real power system of the Republic of Kazakhstan, depicted in Figure 4. The resistance of the branches of the replacement circuit is indicated in the Figure and the total power of the nodes is known. In the process of preparation of the initial data, the steady-state zero-iteration mode is formalized, with subsequent refinement of the node currents to the specified accuracy of the determination of the mode parameters.

Calculation results are shown in Table2.

Table 2. The results of comparative calculations of nodal voltage.

	Calculation with matrix C		Calculation according to the RASTR program		Deviations		
Nodes number	U, kV	δ, grade	U, kV	δ, grade	ΔU, kV	ΔU, %	Δδ, grade
1	241.43	-4.8315	240.65	-4.79	0.78	0.32	-0.04
2	241.53	-4.8433	240.78	-4.81	0.75	0.31	-0.03
3	240.89	-4.3086	240.36	-4.28	0.53	0.22	-0.03
4	237.38	-3.0902	237.1	-3.08	0.28	0.12	-0.01
5	235.60	-1.8466	235.45	-1.84	0.15	0.06	-0.01
6	236.27	-1.9230	236.12	-1.92	0.15	0.06	0.00
7	239.60	-6.2297	238.33	-6.17	1.27	0.53	-0.06
8	239.12	-6.5359	238.03	-6.49	1.09	0.46	-0.05

The allocated section of a real electrical network, in fact, is not very loaded, which is confirmed by the calculations. As can be seen from Table 2, the results obtained by different methods are giving good agreements. It should be noted that the results obtained using the matrix C are close to real. This is explained by the fact that in the "RASTR" program the reactive powers generated by the line during the calculation are taken into account by constant values and are corrected for the steady-state voltage, whereas in the proposed method, the reactive power of the line is refined at each step of the iteration.

References

[1] V.A. Stroyev, I.S. Rokotyan, *Methods of mathematical optimization in power supply problems* (MPEI, Moscow, 1993)

[2] B.Z. Manussov, A.V. Lykin, Yu.M. Sidorkin, News of Higher Educational Institutions of the USSR, Energetics **9**, 3 (1974)

[3] O.T. Geraskin, News of Academy of Sciences of the USSR, Energy and Transport **5**, 106 (1967)

[4] D. Ahmetbayev, Results in Physics **7**, 1644 (2017)

[5] D. Akhmetbayev, A. Akhmetbayev, A. Suleymenov, M. Kolcun, Proceedings of the 9th International Scientific Symposium Elektroenergetika,StaráLesná, Slovak Republic **9**,12(2017)

MODELING OF ACOUSTIC PROCESSES IN SOLIDS BASED ON PARTICLE INTERACTION

Angela Kuzovova[1], *Timur* Muksunov

Tomsk State University, 634050, Tomsk, Russia

Abstract. An algorithm for the numerical propagation of acoustic waves in solids has been developed. The algorithm is based on the particles interaction model of in a tetrahedral crystal lattice. Numerical integration of Newton's equations in the calculation of particle trajectories is applied. The possibility of numerical modeling of wave diffraction processes on heterogeneities is shown. The resonant phenomena are observed in simulation of acoustic oscillations for an ultrasonic waveguide by means of proposed algorithm.

1 Introduction

Acoustic methods are widely used in physical research, in particular, they allow studying various phenomena in solids. The high sensitivity of ultrasonic waves to the presence of mechanical heterogeneities of the medium contributed to the development of ultrasonic flaw detection. Ultrasonic methods allow determining the presence and type of defects in materials that are "opaque" for optical and radio wave methods. Hence, the development of methods for numerical modeling of acoustic processes that take into account all possible wave effects and nonlinear processes is relevant. In the linear approximation, acoustic processes in a homogeneous medium are described by the wave equation. The solution of this equation makes it possible to calculate the field distribution in the medium by specified sources. To account for heterogeneities in the density and speed of sound in a medium, the wave equation is solved in the approximation of small perturbations or in the single-scattering approximation. Far from all effects can be taken into account in the linear approximation for a homogeneous background medium. To take into account nonlinear effects, the propagation of waves in heterogeneous media is necessary the application of numerical methods. Widely applied the method of finite differences in the time domain [1]. This method is based on the iterative calculation of the subsequent state on the basis of the previous, in the process of numerical integration of the wave differential equation. The finite element method is applies for the analysis of stationary fields in the frequency domain and requires the solution of large systems of linear algebraic equations and is applicable in the linear approximation, but allows to take into account the spatial inhomogeneity of the medium. The most common approach is the simulation of interacting particles of matter on the basis of Newton's equations [2, 3]. Together with this, similar approaches are used to model acoustic processes [4].

[1] Corresponding author: Ang_kuz93@mail

2 Particle model of solid media

In this paper, we propose a method for numerical simulation of the propagation of acoustic waves in solids based on an array of particles in a tetrahedral equidistant grid and Newton's equation of motion. Through the partitioning of the entire set of particles into groups and the parallel calculation of the equations of motion by means of OpenCL, is provided the accelerated modeling of acoustic processes in a solid. The properties of matter are determined by the dependence of the force interaction of the particles on the distance between them and the mass of the particles. For the modeling of a solid body, the spatial arrangement of the particles presented in Figure 1 is considered, where the distance between all nearest particles is the same and equal to d. We assume that the particles interact only with the nearest neighbors. For the lattice under consideration, each particle will have 12 neighbors.

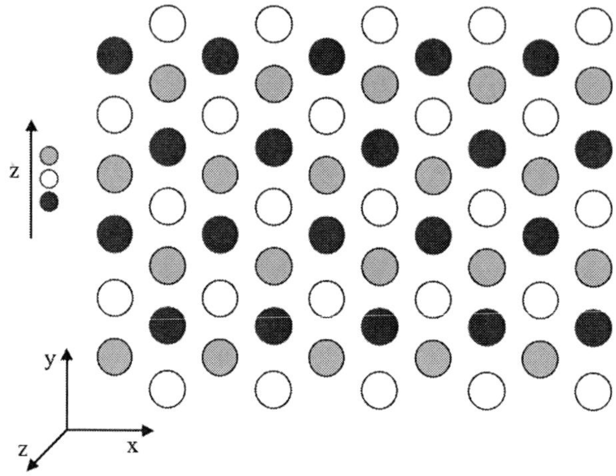

Fig. 1. Structure of the crystal lattice under consideration.

The interaction strength between the particles is given by the graph shown in Figure 2 and is determined by the formula (1):

$$F(x) = (1 - d/x)(d\sqrt{2} - x)^2 (x < d\sqrt{2}) \quad (1)$$

where is d – the distance between the particles, $F(x)$ - the force of interaction between particles, x – particle coordinate (in Figure 2 x is measured in values of d).

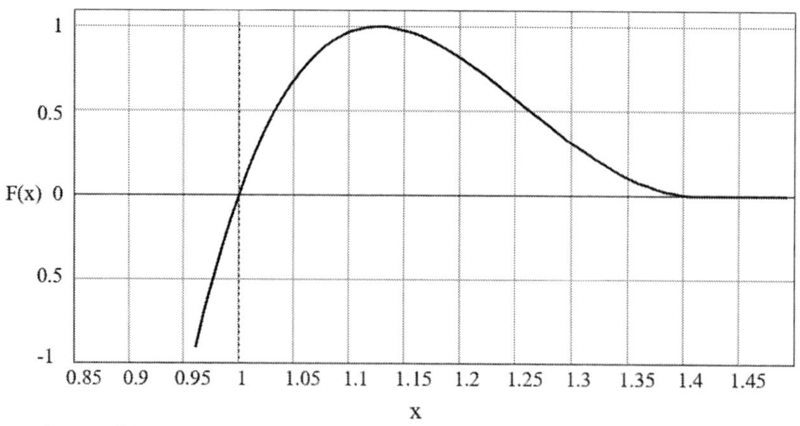

Fig. 2. Dependence of the attractive force between particles from distance.

It is seen that for a distance d (when $x = 1$) between the particles, the interaction force is zero, and so the unperturbed structure will be stationary. Near the equilibrium point, the attractive force depends linearly on the distance, which allows modeling of linear acoustic processes and in the linear approximation to estimate the propagation velocity of longitudinal waves.

We consider a one-dimensional periodic chain of particles. Let the period of this chain be d. Then, in the state of equilibrium, the coordinate of the n-th particle of the chain is $x = nd$. When the particle is shifted from the equilibrium state, in the linear approximation, a force proportional to the magnitude of the shift appears (Hooke's law). We denote the shift of the n-th particle from the equilibrium state by X_n. Hence the force acting on the particle on the left side can be written as $F_1 = k(X_{n-1} - X_n)$, the force acting on the side of the right particle will be written as $F_2 = k(X_{n+1} - X_n)$. For a central particle of mass m, we can write Newton's equation in the form:

$$m\frac{\partial^2 X_n}{\partial t^2} = F_1 + F_2 = k(X_{n+1} - 2X_n + X_{n-1}) \qquad (2)$$

Let us consider the case when the step d is much smaller than the characteristic distance, on which there is a significant change in the deviation of the particles (or much less than the wavelength). Then the quantity $(X_{n+1} - 2X_n + X_{n-1}) \approx d^2 \frac{\partial^2 X}{\partial x^2}$, therefore

$$m\frac{\partial^2 X_n}{\partial t^2} = kd^2 \frac{\partial^2 X}{\partial x^2}, \text{ or } \frac{\partial^2 X}{\partial x^2} - \frac{m}{kd^2}\frac{\partial^2 X_n}{\partial t^2} = 0 \qquad (3)$$

– this is the canonical form of the wave equation, from which it follows that the speed of sound is determined by the expression: $\frac{1}{c^2} = \frac{m}{kd^2}$, therefore , the velocity of sound $c = d\sqrt{k/m}$.

At a distance $d\sqrt{2}$ is the next nearest particle, so at this distance the interaction force should be zero. In total, each particle has a maximum of 12 nearest neighbors, which must be taken into account when calculating the sum of the interaction forces, this accelerates the computation in comparison with a rectangular grid where it is necessary to process 26 neighbors.

3 Numerical modeling

Let us consider a numerical model of the propagation of acoustic waves in a solid in which the scattering object (heterogeneity) represents the plate at an angle, and the background medium is the equivalent of brass (sound velocity 4430 m/c, density 8485 kg/m3). When carrying out a numerical experiment in the area with dimensions of 128 mm by 55 mm. In the case of a solid medium, with the propagation of a perturbation in the medium, two types of waves will already be propagated: longitudinal and transverse. This is due to the fact that a shear elasticity exists in a solid medium, in contrast to a liquid one. Figure 3 shows the result of modeling a wave emanating from a particle from above. The red indicates the velocity of the particle along the X axis, green along the Y axis, blue along the Z axis. The initial perturbation was directed along the X axis. It can be seen that the longitudinal wave is ahead of the transverse wave, the waves are diffracted on heterogeneity, and wave reflection, occurs according to Snellius law (the angle of incidence equal to angle of reflection).

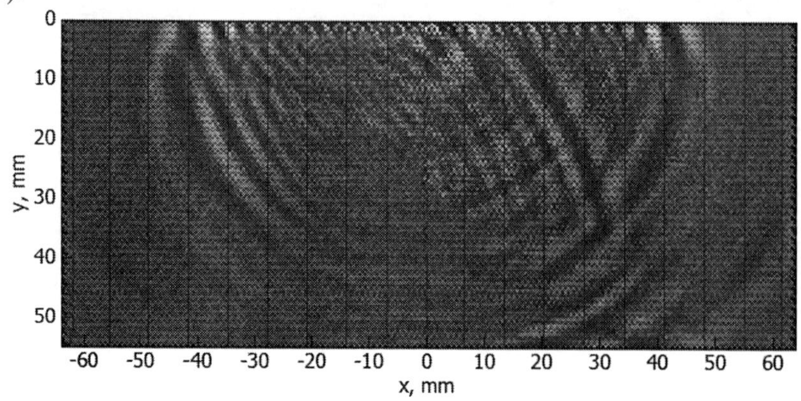

Fig. 3. The result of simulation of oscillations in a lattice after excitation of one particle by a monopolar pulse.

In addition, a numerical simulation of a structure composed of three different solid media was carried out. This structure is an ultrasonic waveguide (Figure 4) made of titanium connected to a piezoelectric element and a reflector made of steel. To specify materials with different properties, the values of the particle mass and the slope of the curve of the dependence of the force on the distance near the equilibrium point were regulated.

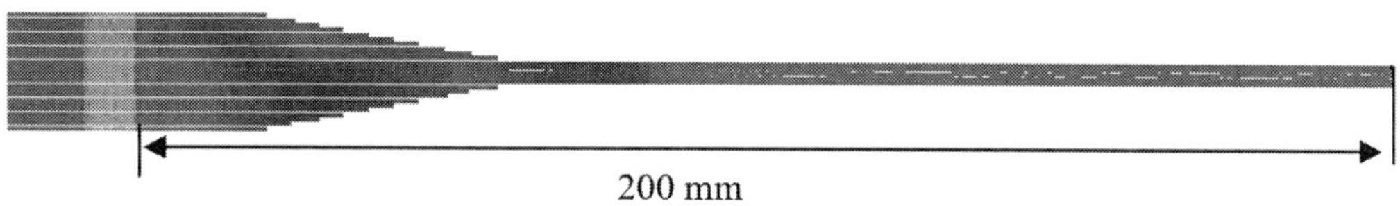

Fig. 4. Model of ultrasonic waveguide.

In this case, the piezoelement is a source of oscillations and its entire volume is subject to external disturbing impact. The model provides additional compressing and stretching of the piezoelement depending on the time, but it interacts with other materials and conducts sound.

As a result of numerical simulation, the values of the resonant frequencies for the given structure were obtained (Figure 5).

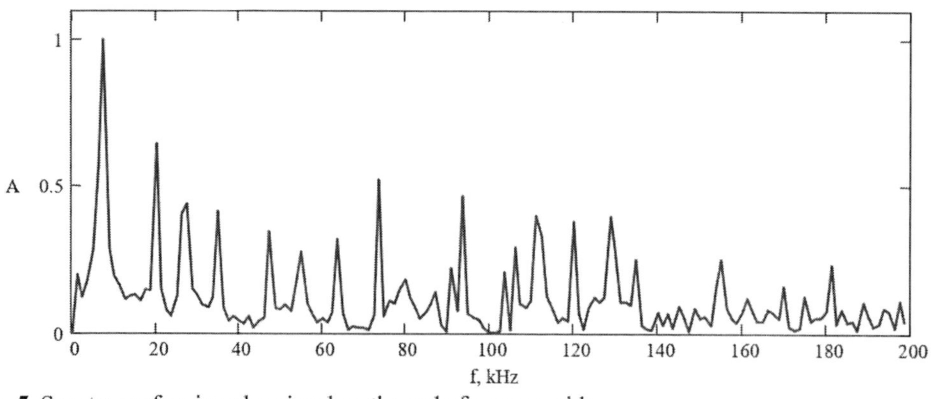

Fig. 5. Spectrum of an impulse signal on the end of a waveguide.

At a frequency of 20 kHz, the amplitude of the oscillations waveguide behaves as shown in Figure 6. It can be seen that when the excitation frequency coincides with the resonant frequency an almost linear increase in the amplitude of the vibrations on the end of the waveguide occurs.

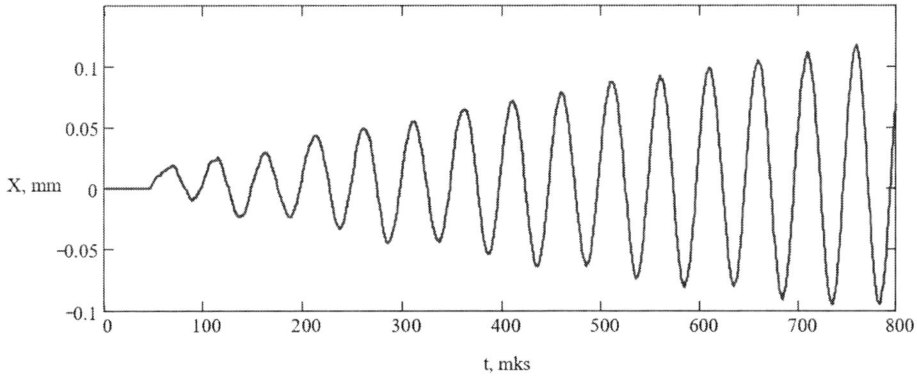

Fig. 6. Harmonic signal at a frequency of 20 kHz on the end of the waveguide.

From the above examples of numerical simulation, one can observe a number of acoustic effects: diffraction, reflection, interference, resonance phenomena in waveguides. In the linear approximation, this model reduces to the wave equation. The peculiarity of the model is that it makes it possible to significantly reduce the calculation time by reducing the number of processed particles. The model allows taking into account nonlinear effects in the medium, by assigning the force dependence graph.

4 Conclusion

As a result of the work, a computer model has been created that makes it possible to calculate the process of propagation of an acoustic wave in a solid body in the presence of heterogeneity. Here was shown a possibly of prediction of the diffraction and reflection effects of waves from an object of arbitrary shape.

Acknowledgments

The work was supported by the Ministry of Education and Science of Russia within the framework of the project "Creation of a high-tech complex of ultrasound surgery" (Unique identifier of the project RFMEF157517X0163).

References

[1] V.A. Barkhatov, Russian Journal of Nondestructive Testing **43**, 605 (2007) doi: 10.1134/S1061830907090069
[2] R.W. Hockney, J.W.Eastwood, *Computer Simulation Using Particles* (N. J.: A. Hilger. 1988)
[3] A.M. Krivtsov, N.V. Krivtsova, The Far East. mat. **f. 3**, 254 (2002)
[4] D.A. Sukhoruchkin, O.A. Sapozhnikov, Proc. of the school-seminar "Waves-2016" **22** (2016)

DEVELOPMENT OF A HARDWARE AND SOFTWARE SYSTEM FOR METEOROLOGICAL OBSERVATIONS

Mikhail Shcherbakov*, *Alexey* Borisov

Altai State Technical University, 656000, Barnaul, Russia

Abstract. Article is devoted to development of a hardware and software system for monitoring of weather conditions. The hardware of this complex is the Arduino platform, which are connected sensors of temperature, humidity, atmospheric pressure, speed and direction of wind, level of ultra-violet radiation, which allow to watch environment parameters. Data are transferred to the remote server which is constructed on the basis of the single board computer Raspberry Pi, by means of a radio channel. For ensuring functioning of a complex, the software was written and it consisting of three programs. The first program was written to on Si similar language for operation of the Arduino controller. The second from programs was written for functioning of the server, it consists of MySQL DBMS, a script for data record in a basis and the PhpMyAdmin servers for administration of a DB. The third was written in the C# programming language and represents the application Windows forms. The made experiment showed that the complex allows to carry continuously meteorological observations, at the same time data remain in the database and the consumer software allows to carry their processing.

1 Introduction

The existing meteorological stations have changing for more than 100 years. At the beginning of the 20th century they represented laboratories, with devices for meteorological observations which were controlled by the person. Now it is almost completely the automated complexes which are collecting data on the environment practically without intervention of a person and giving results of observations already in the processed look.

Automatic meteorological stations with, partial cycle of observations (only collection of information, storage and/or it transfer) are widespread. They are used for expansion of a zone of observation of metrological stations or control of weather conditions in certain territories, for example, the airports, production sites, etc. [1].

The theory and practice of application of similar hardware and software systems showed their effectiveness. They usually are autonomous, simple in using, suitable for carrying out the researches which aren't demanding excessive accuracy, but conducted a long time.

* Corresponding author: shermih1996@mail.ru

2 Relevance

Nowadays almost each person learns temperature or watching a weather forecast per day before an exit from the house that put on according to weather conditions. That is why the problem of obtaining of current information on state of environment haven't an equal relevance.

Presently the state and commercial systems of monitoring weather conditions and their forecasting are developed, but their accuracy in various territories differ and in some places the error of measurements and forecasts is high. It is connected with the fact that now the shortage of posts of meteorological observations is observed in Western Siberia and in some other regions of Russia that leads to a lack of information for creation of the exact forecast of weather conditions. For example, in Altai Krai there are only about 30 posts for meteorological observations which are widely spaced and located unevenly [2]. In this regard in Russia automatic meteorological stations receive the increasing popularity. They are used by the commercial organizations, groups of researchers or ordinary people [3].

They differ in usability, mobility, a set of complete sets. Their functionality depends on the price. Meteorological stations with broad functionality differ in high cost. It does their acquisition difficult by enterprises with has small capital, some higher educational institutions, or research groups. This fact does development of an automatic meteorological station with the low cost and broad functionality very relevant.

3 Work algorithm

The automatic meteorological station is used for meteorological observations without participation of people. It provides the systematized data on the made observations and it usually is partially or completely autonomous.

The meteorological station consists of two principal elements: data acquisition unit and server.

The data acquisition unit consists of a set of sensors, for collection of meteodata, the controller of control, the interface of connection and power supply unit. All elements are in the tight cases. Treat sensors:

1. Barometer. The meteorological station measures atmospheric pressure by means of this device.
2. Thermometer. It takes air temperature.
3. Hygrometer. The device defines humidity of air.
4. Weather vane. The device shows the direction of wind.
5. Anemometer. The device for fixing of force and speed of streams of wind.
6. Sensor of level of ultra-violet radiation. It measures quantity of the solar radiation reaching the Earth's surface.
7. Sensor of an amount of precipitation. It measures the level of precipitations.

The controller controls of sensors. It realizes data collection, their transfer on the server and power management controls sensors.

All parameters from sensors are transferred to the server. The server stores measurements in the database, reports results in the form of tables, diagrams, etc. General algorithm of work of a hardware and software system:

1. The block of data collection located on the street by means of sensors, conducts monitoring of conditions of the environment;
2. The controller polls all sensors and collects from them information one time during a certain time frame:
3. The controller transfers data to the server on wireless connection by means of a radio channel:

4. The server writes data in the database.

5. A user connects the personal computer to the server by means of the Wi-Fi network. By means of the special software it is connected to the database of the server, and it performs the data operations provided by a software functionality.

The scheme of action of system is submitted in the Figure 1.

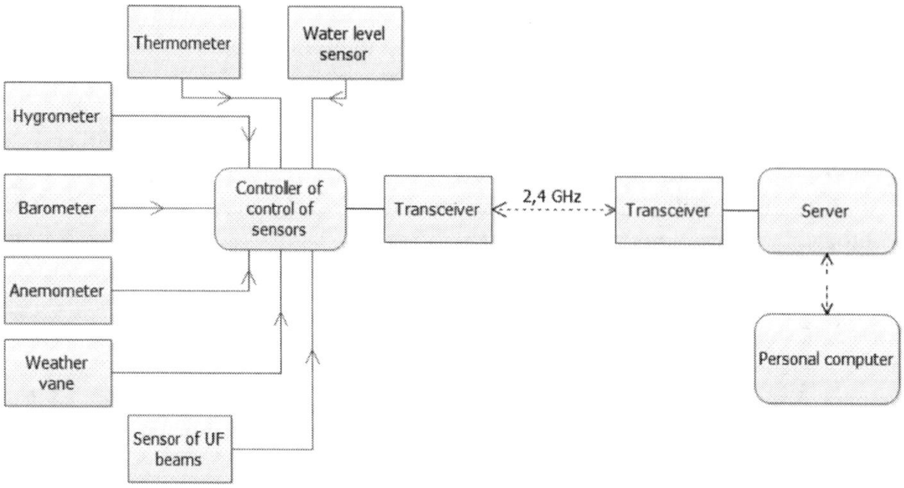

Fig. 1. Scheme of action of system.

4 Hardware-software providing

As it was already provided in the previous point, for monitoring the following parameters of the environment are selected: temperature, air humidity, atmospheric pressure, level of an ultraviolet light, direction and wind speed, level of precipitates.

The acquisition unit keeps their monitoring of these parameters. The appropriate sensors are used for monitoring of these parameters. By means of real-time clocks exact time of data collection is marked. A supply of the unit is carried out from an external source and/or the built-in power supply (the accumulator and a solar panel). All modules are run by the controller [4].

For the block of data collection the following components are used:

– microcontroller Arduino Mega 2560. Clock rate of the processor 16 MHz, the RAM of 8 KB, ROM of 4 KB, a working tension 7-12B, the consumed current up to 90 mA;

– sensor of temperature and humidity DHT22, range of temperatures is -40-125 (±0.2) °C, humidity of air 0%–100 of % (±2%), max. current 2.5 мA, measurement up to 0.5 Hz (times in 2 seconds), working tension 3–5 V;

– atmospheric pressure sensor (barometer) BME280, operating range: 300-1100 GPa, an error of 0.2 Pa, an error of sensitivity of ±0,25%, the maximum consumed current 3.5 mkA;

– the sensor of ultra-violet (UF) radiation GUVA-S12SD, the range of length of the found waves of 240–370 nanometers, viewing angle of 130 degrees, the consumed current of 1 mA;

– sensor of quality of air MQ-135, type: electrochemical, supply voltage: 5V, the consumed current: 150 mA, measurement range: 0.001 - 1% (10—1000 ppm);

– water level sensor. Working current 20 mA, tension 3–5V;

– real-time clocks DS3231, values of the report: second, minute, hour, a day of the week, day of month, month, year, high-speed performance of the interface up to 400 kHz, are supplied with the independent power supply;

– microSD memory card module, the memory card of the MicroSD card format, memory size is 4 GB;
– power control unit, tension 9V, maximum current 1A;
– solar panel, the given tension 7.5 B, the max. given power of 5.5W;
– accumulator, 5000 Ach;
– module of the wireless connection NRF24L01, transmit frequency is 2.4 GHz, range is up to 1100 meters, in urban development up to 520 m, transmission rate up to 2 Mbit/s, 126 channels for data transfer, a maximum current (in a transmission mode) 13.5 mA, in a sleep mode 25mkA, a working tension from 1.9 to 3.6V.

The server includes the following components:
– single board computer Raspberry Pi 3 B+, processor ARMv8 4x1.2 GHz, RAM 1GB, Wi-Fi, Blutooth 4.1, Ethernet port, 4 USB ports;
– module of the wireless connection NRF24L01 (the description is given above).

The flowchart of the hardware of the complex is provided in the Figure 2.

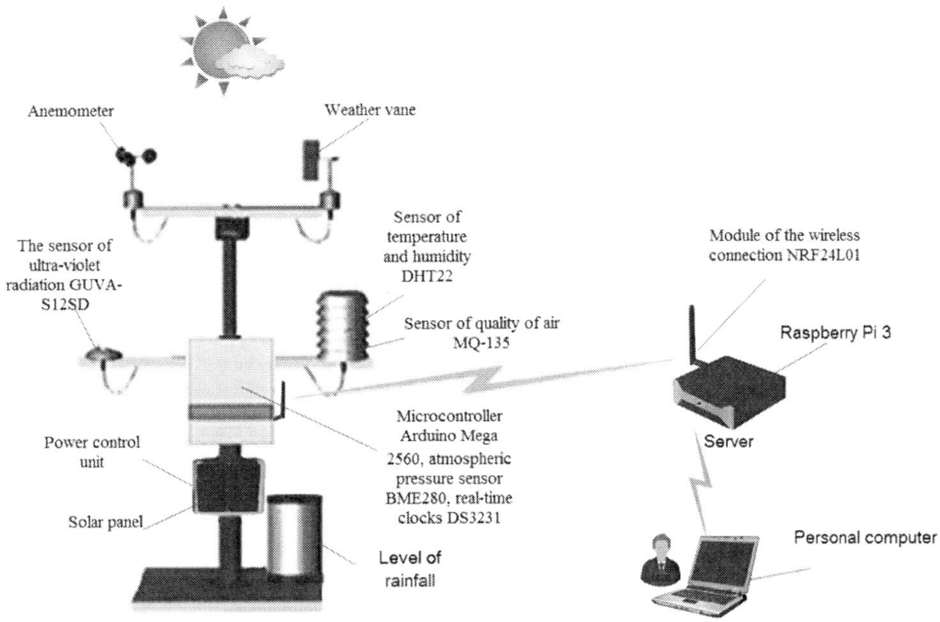

Fig. 2. Flowchart of the hardware of the complex.

The general algorithm of functioning of the hardware complex consists in the following:

1. Arduino collects environment state-of-health data from the sensors which is built in data acquisition unit;
2. Arduino adds to data the exact time taken from real-time clocks, arranges it in the form of a line and sends it on the radio channel to the server.
3. Raspberry Pi writes data in the MySQL database.
4. If it is necessary, data are sented on the third-party device on the Internet network, by means of Wi-Fi or via the RJ-45 interface.

If a user needs to get data access, it will take the following actions:

1. He will connect the personal computer to the server by means of Wi-Fi or a cable;
2. He will get access to the database, having used the special software;
3. User will perform necessary operations which are provided by the software functionality.

The software consists of three programs: the first is written for functioning of the Arduino microcontroller, the second for creation of the server on the Raspberry Pi computer, the third for working on the personal computer.

The program for the controller is written to C the similar programming language which is specially adapted for Arduino. It will execute the following functions:

– Data collection from sensors.
– Transfer of received data to a line and sending to the server.
– Data collection from management system a supply (a charge of the accumulator, a given current and tension).
– Transfer of all received data to the server.

The program for Raspberry Pi represents MySQL DBMS, the script for data record in a basis and the PhpMyAdmin servers for administration of a DB. The software performs the following operations:

– Data acquisition from data acquisition unit and their recording in the database.
– Saving data in the file in necessary format.
– Provisions of access to the database for the client.

The program for the computer is written in the form of a window Windows application in the C# programming language. Performs the following functions:

– Receiving of access to the database of the server.
– System status request (accumulator charge, operation of sensors).
– Execution of manipulations with information from the database according with the software functionality (an output of the current indications of sensors, statistics for a certain period, etc.).

The interface of the software is provided in a Figure 3.

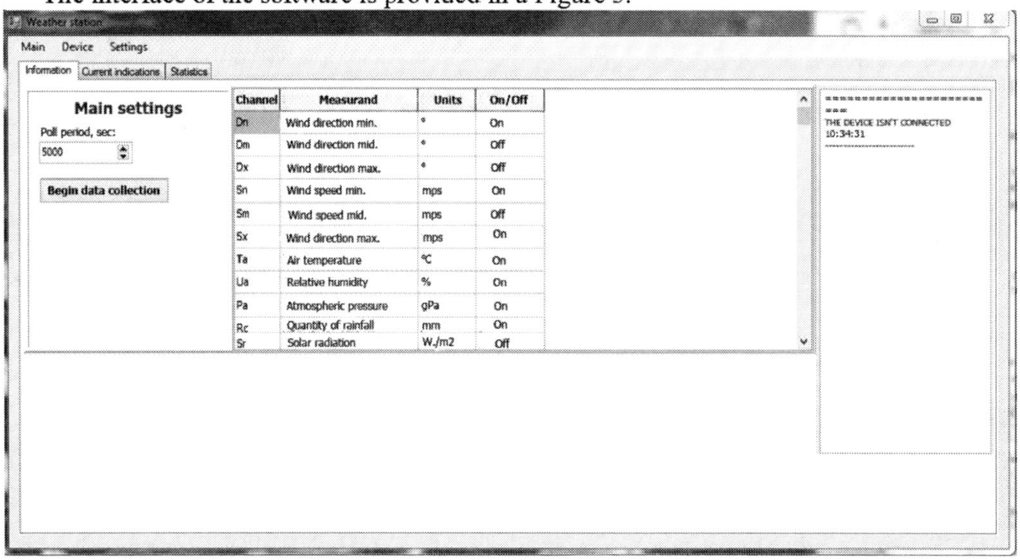

Fig. 3. Software interface.

The interface contains the following elements:

The main menu contains three points: "Principal", "Device" and "Settings".

In the “Setup” point the user has an opportunity to make system operation adjustment: the list of sensors from which data, an inquiry interval, etc.

In the “Device” point the interface removes state-of-health data of the hardware complex such as: connection of the server, data acquisition unit, status of sensors, accumulator charge, etc.

In the "Main" point there are three subparagraphs. In the subparagraph "Information" the user has an opportunity to get acquainted with a current status of system, that is whether the PC is connected to the server what functions of system are switched on or switched off, etc. The view of a window is provided in a Figure 3.

In the subparagraph "Current indications" the dialog of a request of indications from all sensors of the unit of observation is located. The user has an opportunity to examine dataful monitoring, taken directly from the unit of observation. The single request, or information up-dating is possible each several seconds (the period of a request is specified in textbox). In a Figure 4 the interface of a dialog of a request and result of a request is provided.

Fig. 4. Conclusion of the current indications of sensors.

The information output dialog from the database of the server in the form of the table or the diagram is posted in the section "Statistics". In the table information from all sensors for the period of measurements set by the user is output. In the form of the diagram the statistics of indications of temperature, relative air humidity, an atmospheric pressure, wind speed, the index of solar activity, impurity of CO2 air, for the set time frame can be requested (24 hours, 48 hours, 72 hours and so on). In a Figure 5 the diagram of air temperature, from the data obtained during meteorological observations an interval of record of indications of a basis of 60 minutes, the period of observations of 24 hours is provided.

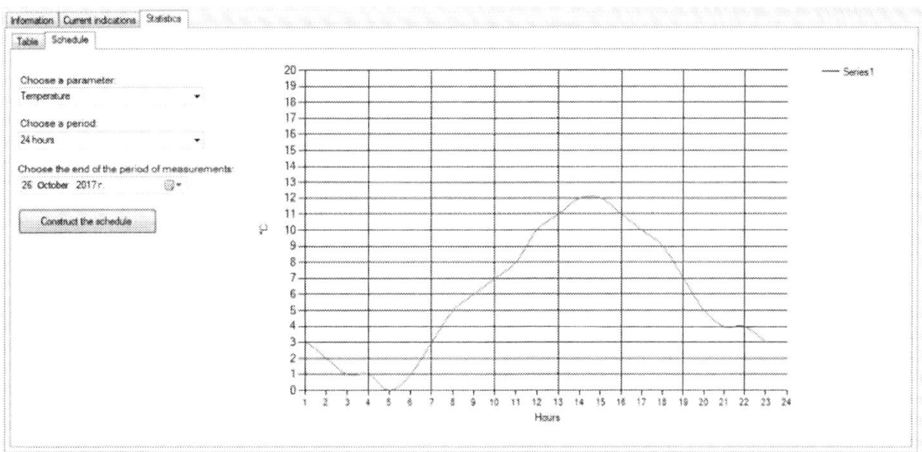

Fig. 5. The conclusion of the schedule of temperature for the set period.

As it was told earlier, the data acquisition unit and the server communicate with each other by means of a radio channel. The radio wave in the course of distribution in space occupies volume in the form of an ellipsoid of revolution with the maximum radius in the middle of flight which is called a Fresnel zone. Natural (the earth, hills, trees) and artificial (buildings, poles) barriers getting to this space weaken a signal. For optimum operation of a meteorological station and a steady signal between modules calculation of a Fresnel zone for a radio channel was made. Radius of the first Fresnel zone is calculated with the help of the following formula

$$R = 17{,}3\sqrt{\frac{1}{f}\cdot\frac{SD}{S+D}} \quad (1)$$

where R – the radius of a Fresnel zone (m); S, D – distance from antennas to the highest point of an estimated hindrance (km); f - frequency (GHz) [5].

During researches the distance between antennas was equal about 200 m, distance to a hindrance of 20 meters. Result of computation:

$$R = 17.3\sqrt{\frac{1}{2.437}\cdot\frac{20\cdot180}{20+180}} = 26.8\ (m) \quad (2)$$

Therefore, that attenuation of a signal was minimum, devices have been placed so that the distance to an obstacle didn't come into Fresnel's zone with a radius of 26.8 m.

5 Conclusion

The carried-out laboratory and field tests with use of a hardware and software system for meteorological observations showed satisfactory results. The complex allows to control over the selected environment parameters in real time, and also to carry out data analysis, collected for a certain time frame.

References

[1] M. Shcherbakov, A. Borisov, Collection of scientific articles of the III International scientific-practical **1**, 185 (2017)
[2] Official site of Hydrometeorological center of Russia URL: http://www.meteoinfo.ru
[3] C. K. Wright, K. M. de Beurs, Z. K. Akhmadieva, P. Y. Groisman and G M Henebry Environ. Res. Lett. **4**, 045020 (2009) doi:10.1088/1748-9326/4/4/045020
[4] M. Shcherbakov, A. Borisov, Materials of the regional youth scientific and practical conference "Program Technical Supply of Automated Systems" **1**, 111 (2017)
[5] The relation "signal noise" in digital communication systems. Creation of antenna-feeder paths and radio systems with external antennas URL: http://www.intuit.ru/studies/courses/1004/202/lecture/2796

24-HOUR FETAL/MATERNAL MONITORING SYSTEM BASED ON PHONOCARDIOGRAM ANALYSIS

Dmitry Zhdanov[a1,2], *Artem* Bureev[1,2], *Yana* Kosteley[1,2]

[1]Limited Liability Company Diagnostica+, 634055, Tomsk, Russia
[2]National Research Tomsk State University, 634050, Tomsk, Russia,

Abstract. The article describes an engineering prototype of the Device for 24-hour monitoring over the functional state of the fetus and mother's cardiovascular system in the antenatal period of pregnancy. The Device is a hardware and software system that ensures the registration and analysis of fetus and mother's phonocardiograms with further processing and interpretation. The interpretation process is designed to timely detect conditions that constitute a threat to a fetus. The Device operates as part of a telemedicine network that allows obstetricians-gynecologists managing pregnant women to remotely control their state. The article describes the Device design concept and features of its specific components. The engineering prototype of the Device was used to carry out biomedical tests that confirmed its suitability for 24-hour monitoring over the functional state of fetus and mother's cardiovascular system in home settings.

1 Introduction

Long-term control over the state of fetus in the 2nd and 3rd trimesters of pregnancy has been relevant for more than 100 years [1]. Despite all medical, scientific and technical achievements, the problem of maintaining fetus' life and health in normal conditions and in case of various pregnancy pathologies has not been solved completely yet [2–4]. The analysis of the most recent data (2016) shows that more than a half of the intrauterine fetal death (IUFD) episodes take place in home settings [2, 5].

The share of antenatal mortality in the stillbirth rate reaches 77.0%. Very many of these cases happen at the late stages of pregnancy when emergency delivery is indicated. About 90% of the antenatal death episodes happen in home settings or rural areas where it is hard to receive emergency medical care. At the same time, antenatal death is preceded by long-term intrauterine fetal agony. The primary causes of fetal death at the late stages of pregnancy have been determined, and the most important causes are timely undiagnosed intrauterine hypoxia and acute gestation course disorders [2, 6].

The experts from the World Health Organization consider 24-hour monitoring over pregnant woman's state of health and fetus development to be the most effective method of

[a] Corresponding author: D_S_Zhdanov@mail.ru

reducing the antenatal mortality rate [2,7]. However, now the state of fetus is objectively controlled only episodically by means of cardiotocography (CTG). At the same time, there are several problems connected with the use of CTG. The most serious issues are the following [2,7]:

– ultrasound signals capture not only the cardiac structures of a fetus, but also large blood vessels. The amplitude of Doppler signals reflected from them is sometimes not lower than the amplitude of signals reflected from cardiac structures, and it is impossible to distinguish between these signals via simple methods;

– exposure to energy is undesirable for a developing fetus: there are still no unambiguous data on its impact on a fetus;

– the prognostic value of CTG does not exceed 10-14 days, which is absolutely insufficient in case of the actual threat of antenatal fetal death and requires regular updates of test findings;

– CTG is performed at clinical laboratories of functional diagnostics, which can be inaccessible for many pregnant women and requires hospitalization.

An alternative solution is to detect parameters similar to the CTG indicators on the basis of fetal electrocardiograms (FECG). However, singling out low-frequency 10–30 mcV FECG signals against the background of external noise and interference, pregnant woman's movements and respiratory activity is a quite sophisticated problem. Digital signal processing methods require considerable computation capacities [8]. Therefore registering FECG and building CTG on its basis are usually performed at clinical laboratories, which is often challenging.

Previously, the authors provided an analytical review of the contemporary state of this problem and described the existing approaches and methods that allow assessing the state of fetus in a mother's womb. This review can be found in [9].

To solve the problem of 24-hour maternal/fetal monitoring, in 2014-2016, the development team of LLC Diagnostica+, including the authors of this article, decided to develop a hardware and software system that would ensure the continuous monitoring of the cardiovascular system (CVS) of both the pregnant woman and her fetus in the antenatal period of gestation. At the same time, the system under development had to be suitable for use outside medical institutions (in home settings), had to have a feature of remote monitoring over physician's examinations and test findings and be able to detect conditions constituting a threat to a fetus. It would allow providing timely medical care to save a fetus and maintain mother's health.

2 Materials and Methods

As a result, the development team created an engineering prototype of the Device that ensures 24-hour monitoring over the functional state of a fetus and mother's cardiovascular system. The Device is a hardware and software system that registers fetal and maternal phonocardiograms (PCG) from the surface of a pregnant woman's body and sends them to specialized software. The Device software allows performing the preliminary processing of phonocardiograms and using them to build fetal and maternal cardiointervalogramms (CIG). These cardiointervalogramms are analyzed to assess the fetal and maternal CVS and detect conditions constituting a threat to a fetus.

The Device hardware operates in the following way (Figure 1). The acoustic sensors located in the mother's abdominal and cardiac areas register the acoustic signals of fetal and maternal heart beats and convert them into electric signals. These signals are transferred to an analog unit where they are amplified and filtered to reduce the impact of interferences and unwanted noises on the device.

Then a microcontroller (microcontroller unit) uses an analog-to-digital converter to transform analog signals into a digital code. These data are saved in a built-in memory and shared with a software component by means of a wireless data transfer unit (Bluetooth).

The control and indication elements are used to display the status and operation mode of the prototype Device hardware, notify of dangerous conditions and operate the hardware component of the Device.

The power unit provides power supply within a necessary voltage range and currents of the component blocks, as input voltage changes due to the discharge of the battery. The charge level control unit is used to monitor the voltage level of the battery.

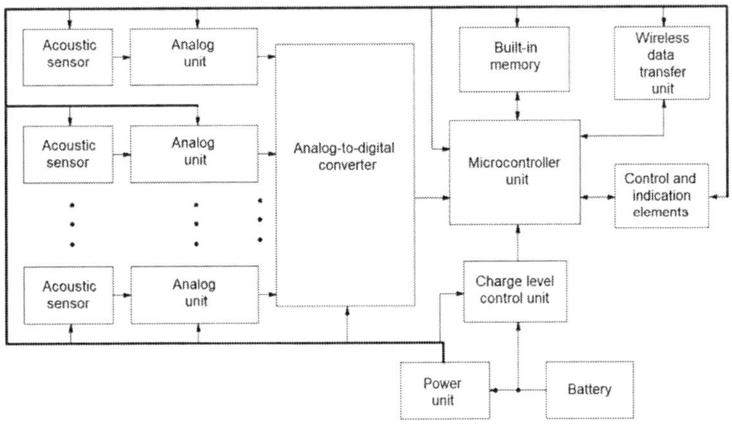

Fig. 1. Functional scheme of the Device hardware.

The specially developed piezoceramic sensors put into a soundproof body are used as acoustic sensors. These sensors allow achieving an average signal/noise ratio of 23.5-35 dB depending on the quality of contact between the sensor and pregnant woman's skin.

The Device software consists of several interconnected components responsible for different tasks:

1) Mobile application (Patient MWS) installed into a pregnant woman's smartphone. It is designed to receive PCGs from the Device hardware, perform preliminary processing (remove the sounds of rubbing against the patient's skin, detect dangerous conditions) and transfer PCGs and analysis findings to a remote server (Application Server). If conditions constituting a threat to a fetus are detected, the application sends commands to the prototype Device hardware informing it of a need for sound and visual signaling and sends alarm messages (in addition to independent sound and visual signaling) to the specialist who manages this patient and her relatives informing them of a need to provide emergency medical care;

2) Server software (Application Server). It is meant to receive, store and provide (upon the request from Physician AWS) patient data and test findings;

3) Software performing the functions of an obstetrician-gynecologist automated work station (Physician AWS). It is meant to keep electronic patient records, attach specific samples of the prototype Device hardware components to them, add (enter to the balance sheet) new samples of hardware components and review test findings on the functional state of the fetal and maternal CVS in both delayed and real-time modes. In addition, this software can send alarm messages via internal mail distribution;

4) Mobile application (Physician MWS) installed into the smartphone of the obstetrician-gynecologist who manages this patient. This is a mobile version of the Physician AWS that allows the obstetrician-gynecologist to review the findings of ongoing or previous tests and check internal mail regardless whether they are at their work place now or not.

3 Results and Discussion

The applicability of the prototype Device (Figure 2) was assessed in 2016 during biomedical trials involving both the development team and experienced obstetricians-gynecologists. The trials were carried out in strict compliance with the developed biomedical trial program. 50 patients took part in these trials. The patients were informed of the goal of these trials and action principle of the prototype Device and gave their informed consent. The results received from the Device were compared with reference equipment, i.e. the UNICOS-01 and MAK-2Ch fetal monitors.

Fig. 2. External design of the prototype Device.

Right before the trial, the obstetricians-gynecologists used the Physician AWS interface to register the electronic records of the patients who took part in the trial and to attach the Device hardware sample to them. Then the sensors of reference equipment and the Device hardware component were attached to the surface of pregnant women's body, and the process of registering data on the fetal CVS was launched under constant surveillance by the obstetrician-gynecologist (Figure 3).

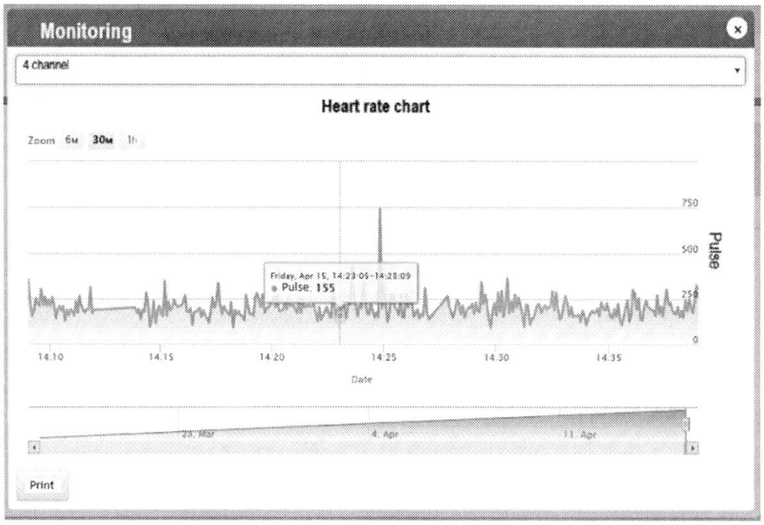

Fig. 3. Physician AWS interface during remote monitoring.

No episodes of fetal state deterioration were detected during the trials. The PCG analysis findings obtained via the prototype Device were characterized by a margin of error amounting to ±2 BPMs vs. the reference devices. Overall, the trials confirmed the applicability of the Device for 24-hour fetal/maternal monitoring.

4 Conclusion

The work deliverables show that phonocardiography allows assessing the state of the fetal CVS with an accuracy rate comparable to that of the commonly used CTG method. At the same time, the accuracy of CIG construction is characterized by the margin of error amounting to ±2 BPMs. Figure 4 demonstrates an example of a CIG created as the result of processing a fetal PCG registered on the pregnant women's abdominal surface.

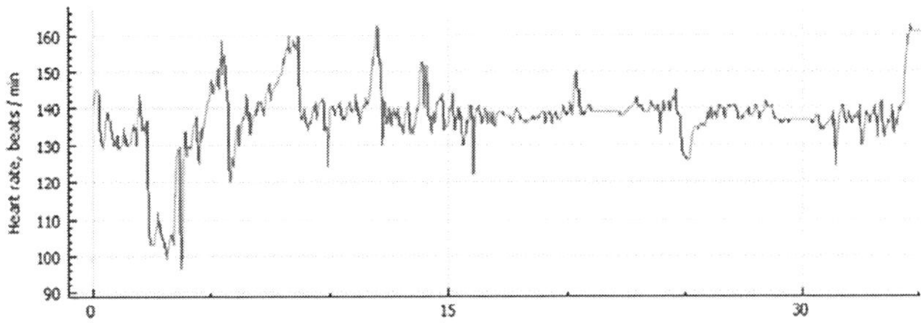

Fig. 4. Example of a CIG created by the Device on the basis of fetal PCG analysis findings.

Using the method of registering and analyzing fetal and maternal phonocardiograms allows assessing the state of the fetal CVS not only at medical institutions but also at any other place, as this method requires no special skills and additional equipment and means.

5 Conflicts of Interests

The authors declare that there is no conflict of interest regarding the publication of this paper.

These findings have been obtained within the framework of implementing Agreement No.14.579.21.0019 (unique project identifier – RFMEFI57914X0019) on the subject "Developing a device for 24-hour fetal and maternal health monitoring during pregnancy via control over cardiovascular parameters based on acoustic data" concluded between LLC Diagnostica+ and the Ministry of Education and Science of the Russian Federation.

References

[1] J. W. Ballantyne, Brit. Med. J. **3222**, 583 (1922) doi: https://doi.org/10.1016/S0002-9378(25)90967-2

[2] P.R. Melinte, G.S. Dragoi, P. Paitici, I. Dinca, E. Patrascul, Rom. J. Leg. Med. **23**, 69 (2015) doi: 10.4323/rjlm.2015.69

[3] R.A. Pilliod, J.M. Page, R.M. Burwick, A.J. Kaimal, Y.W. Cheng, A.B. Caughey, Am. J. Obstet. Gynecol. **213,** 410.e1 (2015) doi: 10.1016/j.ajog.2015.05.022

[4] A.P. South, K.M. Stutey, J. Meinzen-Derr, Am. J. Obstet. Gynecol. **209,** 114.e1 (2013) doi: 10.1016/j.ajog.2013.04.032

[5] A.S. McLennan, C. Gyamfi-Bannerman, C.V. Ananth, J.D. Wright, Z. Siddiq, M.E. D'Alton, F.M. Friedman, Am. J. Obstet. Gynecol. **217**, 80.e1 (2017) doi: 10.1016/j.ajog.2017.03.002
[6] J. Adam, Rev. Obstet. Gynecol **5**, e132 (2012)
[7] Levels & Trends In Child Mortality (United Nations Children's Fund, New York, 2015)
[8] R.M. Rangayyan, *Biomedical Signal Analysis* (Wiley-IEEE Press, New-York, 2015)
[9] A.Sh. Bureev, D.S. Zhdanov, N.N. Zilberman, E.Yu. Kiseleva, S.Yu. Yuriev, Biosci. Biotech. Res. Asia **12**, 1743 (2015)

CONTROL OF VALVE CONVERTERS WITH NATURAL CURRENT CHARACTERISTICS

Sergazy Mendybayev, *Yermek* Sarsikeyev[a], *Dameli* Kapanova

Kazakh Agro Technical University named after S. Seifullin, 010000, Astana, Kazakhstan

Abstract. The article considers the valve converter with a natural current characteristic, which is called the current of parametric, for which the primary is not the source of the electric force but a parametric source of current, to the output of which connected to the counter-parallel thyristor sets assembled on the bridge scheme. The control of thyristors in the current converter is carried out taking into account the characteristic features of the parametric source, to avoid overvoltage on its reactive elements, analyzed algorithmic and modulating functions of the system borne temporary diagrams.

1 Introduction

One of the most important tasks of modern production is energy saving, which is largely determined by the quality of electrical energy. The quality of electrical energy depends essentially on the type of harmonic composition of the current consumed. The harmonic composition of the network current is essentially dependent on the semiconductor transducers, which is a non-linear load for the supply network, which leads to significant distortion of the form of the network current and, as a consequence, deterioration of quality indicators electricity [1–4].

2 Main part

One of the ways to solve this problem is the application of the current parametric converters, which have a natural current characteristic. In the current of parametric, in contrast to traditional, primary is the source of voltage, and from the parametric source (PS), to which output the controlled valve sets (VS) are connected. The scheme of one of the variants of the bridge controlled CPC is shown in Figure 1.

[a] Corresponding author: sarsikeyev.ermek@yandex.ru

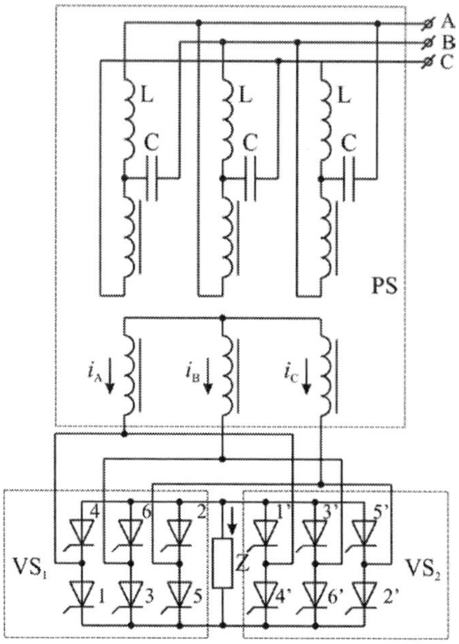

Fig. 1. Power scheme of CPC.

The control of thyristors in the CPC should take into account the following characteristics of the PSC [5, 6]:

-Necessity of presence of conducting valves at every moment in all phases of CPC, in order to avoid occurrence of surges on reactive elements of the current source;

-When unlocking the valves in a circuit with a voltage other than zero, there may be undesirable modes in which the thyristors are discharged capacitors, in circuits with almost zero resistance.

The main functions of the current converter control system Highlands The peculiarities of its operation. The control systems of the CPC should convert the set of input control and auxiliary signals into the control signals of thyristors.

The control signals can be digital and analogue. The control signals synchronized by the clock generator shall ensure the unlocking of the valves in the sequence necessary for the formation of the output current, the voltage of the required form-algorithmic function

The valve unlocking should occur at the moment of natural switching in order to avoid unfavorable transition processes. At the same time, the control signals should be such as to ensure the reliable inclusion of the valves in the case of breaking currents, having a minimum average control capacity ($R_u = R_{u\ min}$).

The algorithmic function of the control system is implemented in a logical device, and it can be performed in the form of a machine or combinational logic schemes. The logical device, depending on input information coming to its input, sets the following modes of operation of the converter:

- Positive load current ($I_н$+);
- Negative current ($I_н$-);
- Pause at Open load ($I_н = 0$);

Modulating function of the control system of the CCI is realized in the device number-pulse control and performs modulation of temporal parameters of control effects:

-duration of pulses of control on thyristors of the current of parametric synchronized with the moment of natural unlocking of valves;

-output frequency converter;

-regulation of the output current by changing the whole number of cycles of pause.

The generalized structure diagram of the device the number-pulse control is shown in Figure 2.

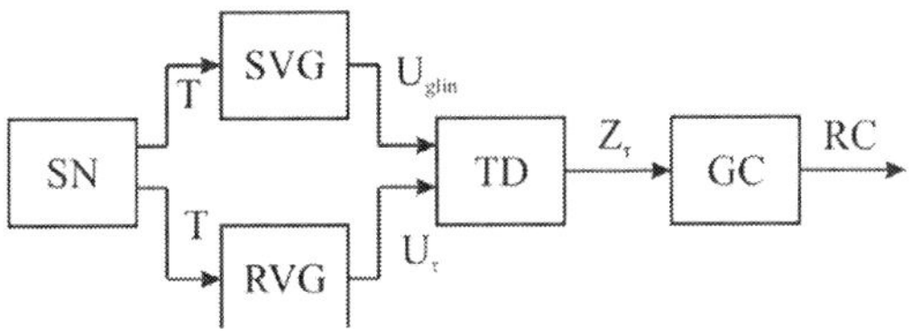

Fig. 2. Structural scheme of the device number-pulse control.

Sawtooth voltage is formed in the block of the sawtooth voltage generator (SVG). This synchronized voltage is compared with the support, coming from the output of the reference voltage generator (RVG), in the threshold device (TD) and at the moment of comparison at the output of the code generator (GC) the mode of operation of the converter changes.

In the threshold device, at the moment of comparison of these signals, the impulse which influencing on the code generator changes the mode of operation of the converter.

Analytical output signal device number-pulse regulation Z(t) is described:

$$Z(t) = m, \quad at\ t \leq t_{response}$$
$$Z(t) = n, \quad at\ t \geq t_{response}, \tag{1}$$

where m is the code *Z (t)* before the threshold device is triggered; *n*– code *Z (t)* after the threshold device is triggered; $t_{response}$. –- response time of the threshold device.

The synchronization node is a sensor of moments of natural unlocking of valves. As such can be the current sensors included in the output phases of the parametric source, voltage sensors, controlling the moment of transition of the voltage "anode-cathode" valves through zero. Let us consider modulation of temporal parameters for nonreversible and reversible CPC.

1. Nonreversible current parametric converters. In Figure 3 there are temporary diagrams explaining the work of DNIC, where it is indicated: τ_1 –the duration of current flow in the load; τ_2- duration of a noncurrent pause; $U_{\tau2}$– the value of the reference voltage, which specifies the duration of the τ_2; T_2 – period of the output current period.

Up to the time of t_3, the load has a straightened current, corresponding to the combination of code words at the output of the code generator – 10 ($Z_{\tau1}$).

At the moment of time t_3 voltage $U_{glav} = U_{\tau2}$, as a result there is a response of TD and on output of the code generator there is a signal $Z_{\tau2}$ corresponding to code 11. This code determines the duration of the noncurrent pause. With change of $U_{\tau2}$ changes $Z_{\tau2}$, thus the average value of a load current is regulated.

The following variants of work are possible:

a) $\tau_1 + \tau_2 = T_2 = const$. With the change of the reference voltage U_τ, simultaneously change the duration of signals $Z_{\tau1}$and $Z_{\tau2}$

The period of the output current remains constant and is set by the period of quantization of the sawtooth voltage generator (Figure 3). The device number-pulse control has one channel containing a sawtooth generator and a reference voltage with one threshold device.

b) $\tau_1 + \tau_2 = T_2 = var$. In this case, in turn, the following options are possible:

$\tau_1 = const,\ \tau_2 = var$. The duration of the straightened current τ1 remains constant and is set by the reference voltage $U_{\tau1} = const$. The duration of the noncurrent pause $\tau_2$2 is a function of time and is determined by the reference voltage $U_{\tau2} = var$, т.е. $\tau_2 = f(U_{\tau2})$. $\tau_1 = var$, $\tau_2 = const$. The duration of the straightened current $\tau_1(t) = f(U_{\tau1})$, $U_{\tau1} = var$. The duration of the noncurrent pause remains constant and is set by the reference voltage $U_{\tau2} = const$. $\tau_1 = var,\ \tau_2 = var$. The duration of the straightened current and the non-current pause are the functions of time and vary according to the change in the support voltage $U_{\tau 1}$ и $U_{\tau 2} = var$, $\tau_1 (t) = f(U_{\tau 1})$, $\tau_2 = f(U\tau_2)$.

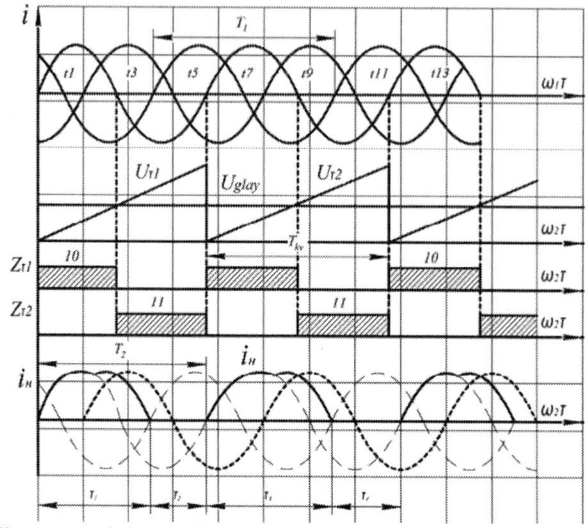

Fig. 3. Temporary diagrams of the device operation number-pulse regulation of the nonreversible CPC.

The device number – pulse control in the second variant has two channels, each of which contains a sawtooth voltage generator, a reference voltage generator and a threshold device. 2. Reversible currents parametric converters. In reversible current parametric converters are possible four time intervals on the period of the output current.

The device's temporary diagrams of the number-pulse control are shown in Figure 4, where indicated:

τ_1 – the duration of the positive current (I_H +);
τ_3 – the duration of negative current (I_H -);
τ_2 – the duration of the noncurrent pause at transition from I_H+ to I_H-;
τ_4 – the duration of the noncurrent pause during the transition from I_H- к I_H+.
10 – $Z_{\tau 1}$ signal code, specified by in + mode I_H+;
11 – signal code $Z_{\tau 1 and}$ и $Z_{\tau 2}$, that specify in $I_H = 0$;
01 – signal code $Z_{\tau 3}$ setting mode I_H-.

At the same time duration of intervals $\tau_1 \div \tau_4$ two variants are possible:
a) dependent on each other;
b) independent of each other.

In the first variant the modulator has one generator of sawtooth voltage and two generators of reference voltages ($U_{\tau 1}$, $U_{\tau 2}$), with corresponding threshold devices.

The duration of the signals $Z\tau_1$ are determined by the values of the reference voltage $U\tau_1$. Analytically this can be expressed as follows:

$$Z_{\tau 1} = f(U_{\tau 1}),$$
$$Z_{\tau 2} = Z_{\tau 4} = f(U_{\tau 3} - U_{\tau 4}), \tag{2}$$
$$Z_{\tau 3} = f(U_{\tau 3}).$$

In the second case the modulator has four independent channels, each of which contains generators of sawtooth and reference voltages with a threshold device.

$$Z_{\tau 1} = f(U_{\tau 1}),\ Z_{\tau 2} = f(U_{\tau 2}),\ Z_{\tau 3} = f(U_{\tau 3}),\ Z_{\tau 4} = f(U_{\tau 4}). \tag{3}$$

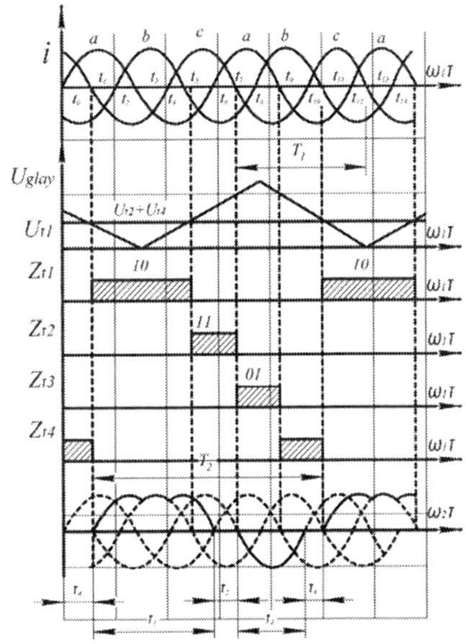

Fig. 4. Temporary diagrams of the device operation number-pulse regulation of reverse chamber with independent regulation.

3 Conclusion

Thus, the considered basic functions of the control system of CPC can be used at practical design of valve converters with a natural current characteristic for the most energy-intensive consumers, which allows to improve the quality of electric power and reduce the negative impact of non-linear load on the electric network.

References

[1] B. Lukutin, Y. Sarsikeyev, M. Surkov, D. Lyapunov, Proceedings of 14th International Conference on Environment and Electrical Engineering, EEEIC, 6835913(2014)

[2] A. Petrusev, V. Rulevskiy, Ye. Sarsikeyev, D. Lyapunov, Proceedings of 2nd International Conference on Industrial Engineering, Applications and Manufacturing, ICIEAM, 7911461 (2016)

[3] N. V. Chicherina, V. Kazakov, O. Yu. Korneva, E. A. Titenko, The European Proceedings of Social & Behavioural Sciences (EpSBS) **19**, 349 (2017) doi:10.15405/epsbs.2017.01.47

[4] S.G. Obukhov, Ye.Zh. Sarsikeyev, I.I. Sholokhova, Proceedings of 2nd International Conference on Industrial Engineering, Applications and Manufacturing, ICIEAM, 7911454 (2016)

[5] K. Suker *Power Electronics. Developer's Guide (*Moscow, 2007).

[6] N. V. Chicherina, V. A. Tolkacheva, European Proceedings of Social & Behavioural Sciences (EpSBS) **19**, 683 (2017) doi:10.15405/epsbs.2017.01.90

CALCULATION OF THE MAGNETIC FIELD OF THE STATOR WINDINGS OF THE SYNCHRONOUS MACHINE

Kuanyshbay Zhumagulov, *Yermek* Sarsikeyev[a], *Dameli* Kapanova

S.Seifullin Kazakh Agro Technical University, 010000, Astana, Kazakhstan

Abstract. This article discusses the possibility of applying a cylindrical coordinate system when examining the magnetic field of a synchronous machine.

1 Introduction

For the study of the magnetic field (MF) in the air gap created by the stator winding is not clearly a pole synchronous machine (SM) and the parameters of the machine, caused by this field, the most convenient is the application of analytical method of study, as the distribution of the magnetic field on the circumference of the air gap has a sharply pronounced peak character and in the study of such field numerical methods arise difficulties and inaccuracies [1]. In addition, when using numerical methods calculation is performed for each set of parameter values and the overall picture of the magnetic field is obtained as a result of a large number of calculations.

From the current analytical methods of the solution of the field equation, the method of direct solution is applied; there is a solution of the equation Laplace by the method of dividing variables. It is known [2–6] that the method of direct solution of the equations of the field is reduced to finding a potential function that satisfies the equations of Laplace or Poisson, as well as the specified boundary and other conditions defining the field in a particular investigated area of the space where the electric machine is located.

When analyzing a field, it is particularly important to select a coordinate system in which a constant value of one or two coordinates describes the boundary configuration between regions with different values of magnetic RDP, which allows finding a solution standard way.

2 Main part

The method of dividing variables is usually used for analytical calculation of electromagnetic fields in areas whose boundaries are coordinate surfaces. At the current time, Cartesian and cylindrical coordinate systems are mainly used for the analytical study of the magnetic field of electric machines. Cartesian coordinates are convenient for analyzing fields

[a] Corresponding author: sarsikeyev.ermek@yandex.ru

with rectangular boundaries, and a cylindrical system for defining fields in areas whose boundaries in the plane are concentric circles and radii.

As this article aim is to investigate the magnetic field of standard electric machines of alternating current, in which the boundaries between areas with different magnetic properties are basically cylindrical forms, then for convenience of accounting Boundary conditions connecting gradients of potential functions, the cylindrical coordinate system is applied. In the analytical solution of the field equation, it is especially important to consider the amount of space splitting where the electric machine is located. The method of calculating the electromagnetic field of the stator winding does not revolving the synchronous machine by solving the Laplace equation first for the point conductor with current i, the method of dividing the variables when presenting the space of the machine location let's imagine the five areas [7] (Figure 1):

1. External area $\rho > d$ with index 0 and magnetic permeability $\mu_0 = 4\pi 10^{-7}$ gn/M;
2. Stator area with $d > \rho > c$ with index 1 and permeability μ_1;
3. Air gap $c > \rho > b$ with δ index and permeability μ_0;
4. The area of the rotor $b > \rho > a$ with index 2 and permeability μ_2;
5. Inner area $a > \rho >$ with index 3 and permeability μ_0.

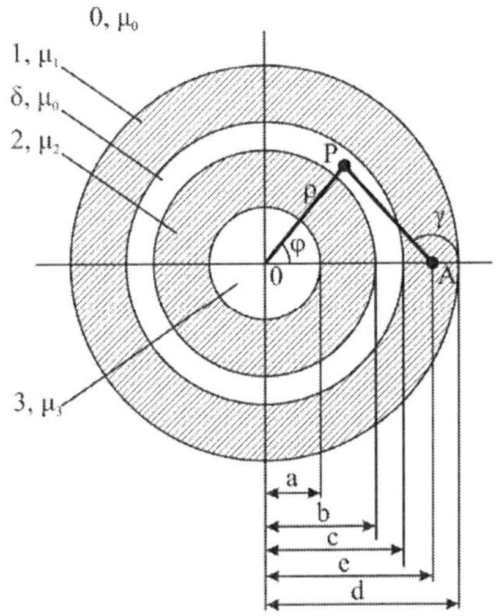

Fig. 1. Distribution of areas.

Due to the absence in this 2d model of volumetric distribution of current density, the magnetic field can be characterized by scalar magnetic potential V and in the cylindrical system of coordinates to be considered as satisfying differential equation Laplace in private derivatives,

$$\frac{\partial}{\partial\rho}\left(\rho\frac{\partial V}{\partial\rho}\right)+\frac{1}{\rho}\frac{\partial^2 V}{\partial\varphi^2}=0\,, \tag{1}$$

where ρ-distance from the investigated point of the space to the center of the machine; φ-polar angle.

The solution of the equation (1) for the case of the point conductor location at the stator disposition for the five areas of space is

$$\mathrm{V}_0=\frac{\mathrm{i}}{2\pi}\varphi+\sum_{n=1}^{\infty}\mathrm{D}_{\mathrm{on}}\rho^{-n}\sin n\varphi\,, \tag{2}$$

$$V_1 = \frac{i}{2\pi}\gamma + \sum_{n=1}^{\infty}\left(C_{1n}\rho^{n} + D_{1n}\rho^{-n}\right)\sin n\varphi, \tag{3}$$

$$V_\delta = \sum_{n=1}^{\infty}\left(C_{\delta n}\rho^{n} + D_{\delta n}\rho^{-n}\right)\sin n\varphi, \tag{4}$$

$$V_2 = \sum_{n=1}^{\infty}\left(C_{2n}\rho^{n} + D_{2n}\rho^{-n}\right)\sin n\varphi, \tag{5}$$

$$V_3 = \sum_{n=1}^{\infty} C_{3n}\rho^{n}\sin n\varphi, \tag{6}$$

The constants in formulas (2) – (6) for harmonic *n*-th order, found from conditions on the boundaries of the section of environments with different magnetic permeabilities have the following form

$$D_{0n} = C_{1n}d^{2n} + D_{1n}, \tag{7}$$

$$C_{1n} = \frac{\mu_1 - \mu_0}{\mu_1 + \mu_0}d^{-2n}D_{1n}, \tag{8}$$

$$D_{1n} = \frac{c^{2n}}{1+m_1}C_{\delta n} + \frac{1}{1+m_1}D_{\delta n} + \frac{i}{\pi n} + \frac{1}{1+m_1}c^{n}, \tag{9}$$

$$C_{\delta n} = \mu_1\frac{i}{\pi n}\frac{c^{n}m_{12}}{c^{2n}(\mu_0 + \mu_1 m_{12}) + (\mu_0 - \mu_1 m_{12})\frac{1-m_{21}}{1+m_{21}}b^{2n}}, \tag{10}$$

$$D_{\delta n} = -b^{2n}\frac{1-m_{21}}{1+m_{21}}C_{\delta n}, \tag{11}$$

$$C_{2n} = \frac{\mu_0}{\mu_2}\frac{1}{1-m_2}\left(C_{\delta n} - b^{-2n}D_{\delta n}\right), \tag{12}$$

$$D_{2n} = \frac{\mu_0}{\mu_2}\frac{m_2}{1-m_2}\left(b^{2n}C_{\delta n} - D_{\delta n}\right), \tag{13}$$

$$C_{3n} = C_{2n} + D_{2n}a^{-2n}. \tag{14}$$

From the formulas (2) – (6) are determine the radial and tangential components of the field strength in all five regions of space.

Practically for calculation of reactivities caused by components of a field of an air backlash, the radial component of a arrears magnetic field, defined by the expression is of special interest.

$$H_{\delta\rho} = \sum_{n=1}^{\infty} n\left[D_{\delta n}\rho^{-(n+1)} - C_{\delta n}\rho^{(n-1)}\right]\sin n\varphi. \tag{15}$$

For a current layer of width equal to the width of the groove slot and located on the arc of the circumference of the stator's sharpening with an internal spatial angle equal to 2 α we have

$$H_{\delta\rho сл} = \sum_{n=1}^{\infty}\left[D_{\delta n}\rho^{-(n+1)} - C_{\delta n}\rho^{(n-1)}\right]\frac{\sin n\alpha}{n\alpha}\sin n\varphi. \tag{16}$$

Further in indices of values of notation δ and ρ will be omitted, having in view, that the one or another value characterizing a field refers to a radial component of a field of an air backlash of the machine.

Then for the intensity of the field of air gap created by the stator winding coil, consisting of $w_{\kappa 1}$ turns, the sides of which stand apart on the arc of the circumference of the stator sharpening to the internal spatial angle of the β when combining the origin with the axis of this coil expression will have the appearance.

$$H_{\kappa} = 2w_{\kappa 1}\sum_{n=1}^{\infty} K_n \sin n\frac{\beta}{2}\frac{\sin n\alpha}{n\alpha}\cos n\varphi . \tag{17}$$

where $K_n = n\left[D_{\delta n}\rho^{-(n+1)} - C_{\delta n}\rho^{(n-1)}\right]$.

For a group of coils of winding with number of grooves on a pole and a phase q_1 and an internal spatial angle α_{z1} between axes of two neighboring coils we have

$$H_q = 2w_{\kappa 1}q_1\sum_{n=1}^{\infty} K_n \kappa_{\text{об} n}\cos n\varphi , \tag{18}$$

where $\kappa_{win\,n} = \sin n\frac{\beta}{2}\frac{\sin n\alpha}{n\alpha}\frac{\sin\left(nq_1\frac{\alpha_{z1}}{2}\right)}{q_1\sin\left(n\frac{\alpha_{z1}}{2}\right)}$ – wrapping coefficient for harmonic n-th order. The expression for the tension of the magnetic field in the air gap created by the phase double winding of the stator with the whole q_1 at flow on it a sinusoidal symmetrical three-phase current frequency f_1 will have a kind

$$H_T = 4pw_{\kappa 1}q_1\sum_{n=1}^{\infty} K_n\kappa_{win\,n}\kappa_{pq\,n}\left[\sin n\left(\varphi - \frac{2p-1}{p}\frac{\pi}{2}\right)\sin\left(\omega_1 t + \frac{2\pi}{3}\right) + \right.$$
$$\left. + \sin\left(\varphi - \frac{2p-1}{p}\frac{\pi}{2} - \frac{2\pi}{3p}\right)\sin\left(\omega_1 t - \frac{2\pi}{3}\right) + \sin n\left(\varphi - \frac{2p-1}{p}\frac{\pi}{2} - \frac{4\pi}{3p}\right)\sin\omega_1 t\right], \tag{19}$$

where $\omega_1 = 2\pi f_1$ – angular frequency; t – time.

In Figure 2 are given the curves of distribution on the circumference of the stator of the magnetic inductions of the individual components of the result field of the air gap created by the double winding of the stator generator type TBB–200–2, as well as the distribution of the sides coils for moments of time t = 1/600 and 1/300 with within one pole division.
In the study of the magnetic field in the air gap created by a winding not revolving synchronous machine alternating current. The resulting field is divided into three components:

1. Main field with magnetic induction B_p;
2. Field of scattering on crowns of teeth with induction $B_{\kappa z}$;
3. Field of Groove scattering with induction $B_{n\alpha}$.

The main field is a field with an order equal to the number of pairs of poles of the machine, and the field of scattering on the crowns of the teeth-the sum of all spatial harmonic fields of the air gap with the order, starting from the n_{kz} and above. The n_{kz} value can be found from the expression $n_{nz} = (Z_1/2) - p$, where Z1 is the number of stator grooves. This expression is found from the symmetry condition of the magnetic induction distribution curve of the scattering field, the crowns of the $B_{\kappa z}$ teeth and the circumference of the stator distraction in relation to the abscissa axis.

As seen in Figure 2, the curve of a field of scattering on crowns of teeth is practically symmetrical concerning an axis abscissa, and each its pair of poles is located symmetrically concerning an axis of the corresponding groove with current. The value of magnetic induction of this field under other identical conditions is proportional to the value of the full current of a groove, and magnetic power lines of one pair of poles cover only one groove with an electric shock, and magnetic poles settle down symmetrically concerning an axis each groove with a total current different from zero. With the expiration of time, that is, as the star turns the vectors of the phase currents the axis of each pair of poles remains motionless relative to the winding, coinciding with the axis of the corresponding groove, changes only the value of magnetic induction practically proportionally the size of the full groove current. The sum of all the remaining spatial harmonic fields of the air gap to a subsonic order, with the exception of the ground one, is a scattering field. The sum of the gap and crowned teeth of the components of the air gap field is the field of the differential dispersion of the winding.

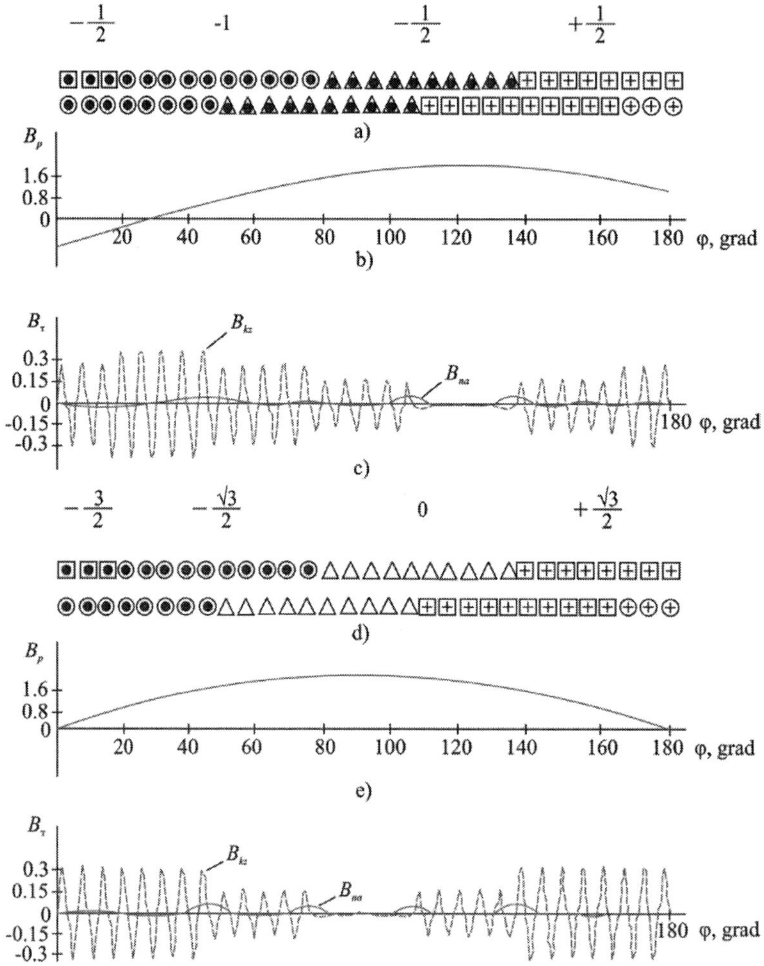

Fig. 2. Results of calculations.

The field of scattering on crowns of prongs of a winding of an anchor leaving from a sharpening of a stator practically vertically its surface, that is in a radial direction, passes through an air backlash, both in radial, and in tangential directions.

The part of this field, reaching to a surface of a core of a rotor, penetrates its body and can put in its circuits EMF with the frequency different from the frequency of working current circuits in a rotor. The component of the field of differential scattering of the anchor winding due to shortening of the pitch, distribution of coils on the grooves several times, and sometimes to order and more can be less scattering on crowns of teeth. When calculating the field, the equivalent values of magnetic RDP stator μ 1 and rotor μ 2, found separately for each component of the air gap field, which correspond to the actual operating mode of the machine, should be used. Although for all three components of the field of Air gap created by the winding, magnetic is common, but the magnetic circuits for each of them different. Therefore, the values of equivalent magnetic RDP for each component of the field are different.

3 Conclusion

External magnetic fields of the electric machine, depending on electromagnetic fields in the air gap, can have a significant impact on the electronic control equipment [6– 11], processors and computers located [12] near the machine.

References

[1] A. Alekseev, M. Kostenko, *Turbogenerators* (Moscow, 1939).
[2] P. Alger, The Calculation of the Armature Reactance of Synchronou Machines **47**, 493 (1928)
[3] I V Plotnikova, N V Chicherina, S S Bays, R G Bildanov, O Stary, IOP Conference Series: Materials Science and Engineering **289**, 012029 (2017) doi:10.1088/1757-899X/289/1/012029
[4] E. V. Beyerleyn, P.V. Tyuteva, Proceedings of 15th International Conference of Young Specialists on Micro/Nanotechnologies and Electron Devices (EDM-2014), (2014)
[5] N. V. Chicherina, V. A. Tolkacheva, European Proceedings of Social & Behavioural Sciences (EpSBS) **19**, 683 (2017) doi:10.15405/epsbs.2017.01.90
[6] B. Tsybikov, E. V. Beyerlein (Beierlein), P. V. Tyuteva, MATEC Web of Conferences **91**, 1(2017)
[7] G. Odilov. *The regenerating character of the anchor field in the air gap of the multiphase alternating current machine* (Tashkent, 1996)
[8] N. V. Chicherina, V. Kazakov, O. Yu. Korneva, E. A. Titenko, The European Proceedings of Social & Behavioural Sciences (EpSBS) **19**, 349 (2017) doi:10.15405/epsbs.2017.01.47
[9] B.V. Lukutin, Y.Zh. Sarsikeyev, M.A. Surkov, D.Yu. Lyapunov, Proceedings of 2014 14th International Conference on Environment and Electrical Engineering, EEEIC, 6835913 (2014)
[10] A.S. Petrusev, V.M. Rulevskiy, Ye.Zh. Sarsikeyev, D.Yu. Lyapunov, Proceedings of 2nd International Conference on Industrial Engineering, Applications and Manufacturing ICIEAM, 7911461 (2016)
[11] S.G. Obukhov, I.A. Plotnikov, M.A. Surkov, L.P. Sumarokova, IOP Conference Series: Materials Science and Engineering **177**, 012143 (2017)
[12] D.S. Akhmetbayev, D.A. Aubakir, Y.Z. Sarsikeyev, Results in Physics **7**, 1644 (2017)

https://doi.org/10.1051/matecconf/201815501049

DECREASE RADAR CROSS-SECTION OBJECTIVES USING THE METAMERIAL

Aleksandr Mironchev[1], *Vladimir* Yakubov

National Research Tomsk State University, 634050, Tomsk, Russia

Abstract. In this work influence of an electromagnetic field on lattices from close located ring conductors is shown. RCS is also calculated and shown that at $0 \le n^2 < 1$ there is the considerable decrease of an object.

1 Introduction

Efficient scattering area in a radar-location — the area of some dummy even surface located normally to the direction of an incident plane wave and which is the ideal and isotropic reradiate which, being placed in a purpose arrangement point, creates in the location of the antenna of radar station the same fluency of power, as the actual purpose [1].

2 Experimental study

RCS is the quantitative measure of property of an object to disseminate an electromagnetic wave. Along with energy potential of a send-receive path and intensification coefficient of antennas of radar station, RCS of an object enters the equation of range of a radar-location and determines the range at which an object can be found by the radar. The increased RCS value means larger radar visibility of an object, decrease in RCS complicates detection. Decreases of RCS of the purposes it is possible with use of composition metamaterial of a radio-frequency range.

On the basis of numerical model operations on interaction of an electromagnetic radiation with structure from elemental annuluses [2, 3], the experimental model of metamaterial is created. The exemplar was executed in the form of matrix structure from the ring conductors located in frothed polystyrene. Distances between parallel ring conductors were made by 0.5 cm (Figure 1).

[1]Corresponding author: mironchev42@mail.ru

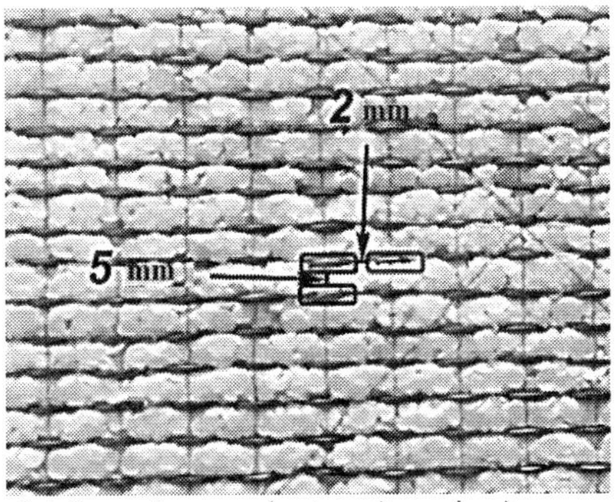

Fig. 1. The metamaterial model from ring conductors (top view).

A series of the pilot studies was made:

1. The composite from ring conductors one layer, thickness is 1 cm.
2. The composite from ring conductors two layers, thickness is 2 cm.
3. A composite from ring conductors three layers, thickness of 3 cm.
4. The composite from ring conductors four layers, thickness is 4 cm.
5. Plexiglass, the thickness of 1.8 cm.

Carrying out an experiment was carried out by method is based on measurement of electrophysical properties in the free space [4].

In the Figure 2 dependences of a phase (1) and amplitude (2) on the frequency of the signal from 0.05 GHz up to 22 GHz which passed through a single-layer lattice from elemental annuluses are presented. Lines (3) and (4) represented the linear regression for a phase (3) and amplitude (4) respectively. Point noted mean values and fiducially intervals. In a superwide frequency band the curve of a phase has linearly the decreasing character that is characteristic of metamaterials.

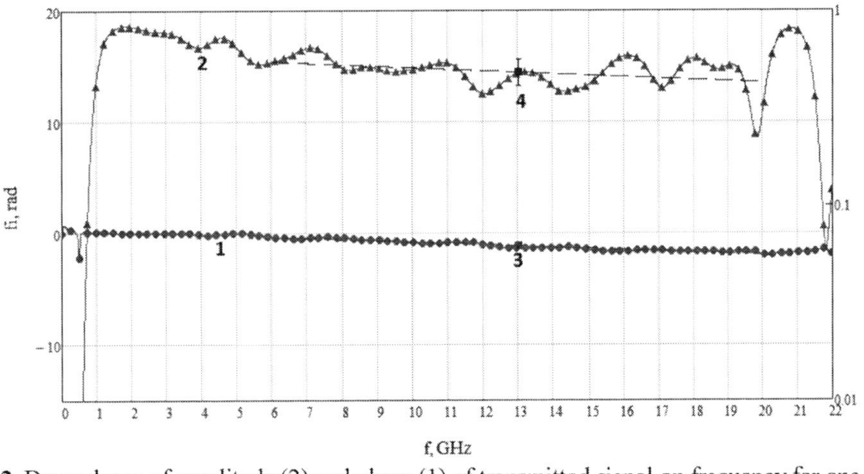

Fig. 2. Dependence of amplitude (2) and phase (1) of transmitted signal on frequency for one layer of rings 1 cm thick, linear regression of phase (3), linear regression of amplitude (4).

Complex refractive index in the frequency range from 6 GHz to 20 GHz n=0.46±0.02+i0.09±0.01. The result obtained indicates that a single-layered lattice of closed ring elements is capable of reducing the complex refractive index due to the backwash of the wave. As discussed in the previous chapters, this becomes possible due to the appearance of a near field near a spatial lattice in which there is no phase increase, as occurs in the far radiation zone.

Consider the results of an experiment carried out with two layers of Figure 3. You can see, as in the previous case with one layer, the phase of the transmission coefficient is linear. Throughout the entire frequency interval, a negative slope of the phase curve is observed, which is characteristic of a metamaterial.

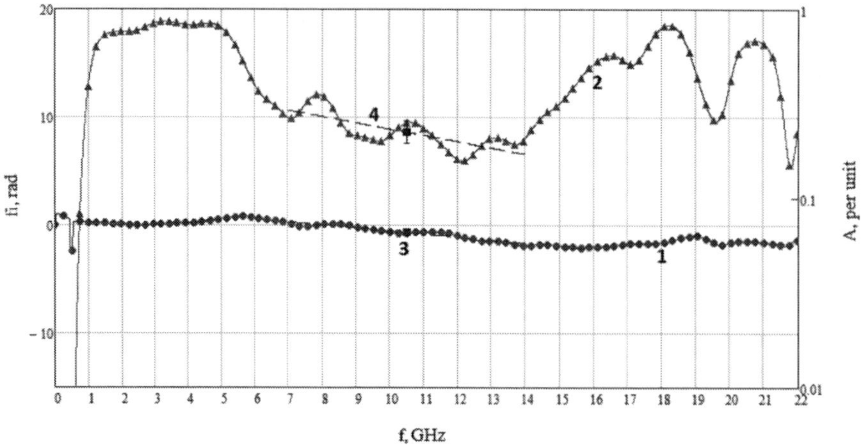

Fig. 3. Dependence of amplitude (2) and phase (1) of transmitted signal on frequency for one layer of rings 2 cm thick, linear regression of phase (3), linear regression of amplitude (4).

An experiment carried out with two layers showed the result n=0.35±0.05+i0.14±0.03 in the 7÷14 GHz band. Such values of the real part of the refractive index indicate an increase in the properties of the metamaterial in a lattice made up of annular conductors. At the same time, the imaginary part of the refractive index begins to increase with the increase in the number of layers.

Next, Figure 4 shows the dependence of the amplitude (2) and phase (1) of the transmitted signal on the frequency for three layers of rings 3 cm thick. The behavior of the curves is similar to the curves given for two layers with a total thickness of 2 cm. Also, there is a frequency interval on which the phase foray has a negative course.

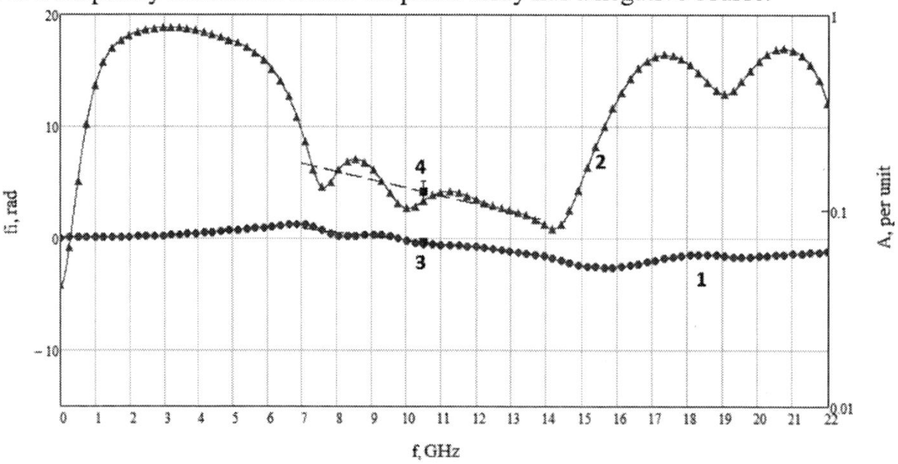

Fig. 4. Dependence of amplitude (2) and phase (1) of transmitted signal on frequency for one layer of rings 3 cm thick, linear regression of phase (3), linear regression of amplitude (4).

Consider an interval from 7 GHz to 14 GHz. In this frequency range, there is a smooth decrease in the amplitude of the signal transmitted through the three-layered lattice from the closed conductors. It can be seen that the phase shift decreases with increasing frequency. The complex refractive index for a lattice of three layers of rings is n= (0.44±0.05)+i (0.15±0.05). It is worth noting that by increasing the number of layers, the real part of the refractive index increased in absolute value, and the imaginary part decreased and became close in value for the two layers.

Figure 5 shows graphs showing the dependence of the amplitude and phase of the signal passed through a four-layered grid of closed conductors with a total thickness of 4 cm. From 8 GHz to 15 GHz, a strong amplitude sagging associated with a large concentration of ring conductors is seen. At this interval, the refractive index is (n =0.68±0.05+i (0.07 ± 0.04)).

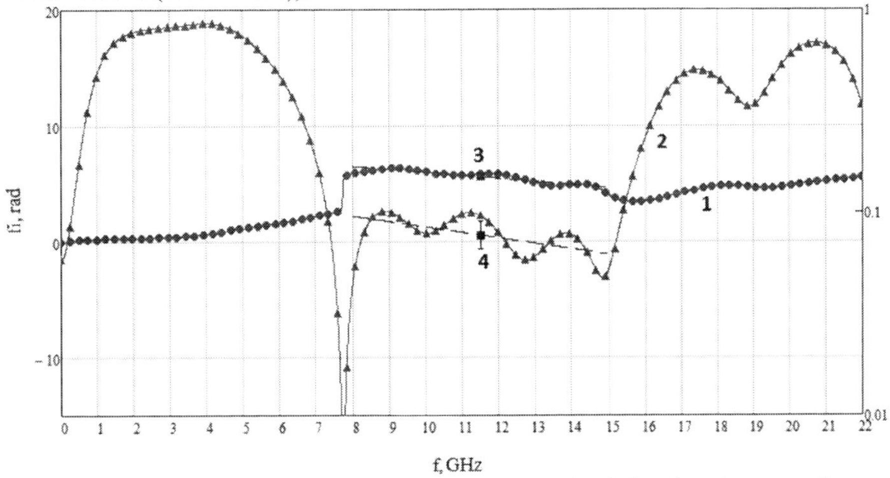

Fig. 5. Dependence of amplitude (2) and phase (1) of transmitted signal on frequency for one layer of rings 4 cm thick, linear regression of phase (3), linear regression of amplitude (4).

As a result of the experimental study, it can be concluded that such a lattice of annular conductors has a refractive index $n \in [0,1]$.

3 Radar cross section

Possible applications of a metamaterial with a negative refractive index have already been discussed. Consider a possible application of a metamaterial with a refractive index $n \in (0, 1)$ to reduce radar cross section (RCS) of objects.

The use of a composite material to reduce the RCS targets is illustrated by the example of a coated ball. If the diameter of the ball R , and the thickness of the dielectric coating h, then for the impact distance p , at which the beam of the radio wave, refracted at the boundary, will touch the surface of the metal under the condition (Figure 6) $p = (R+h)\sin\alpha = nR\sin\beta$.

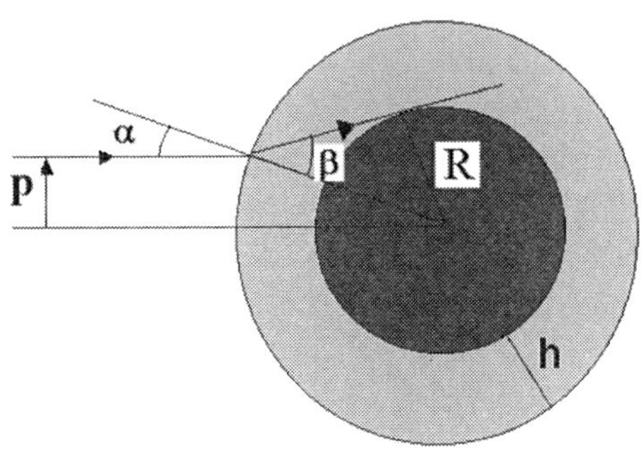

Fig. 6. To the determination of the RSC of a coated ball.

In this case, the extremely large impact distance, at which the refracted beam will still touch the ball, corresponds to the condition $\sin\beta = R/(\mathrm{R+h})$. Thus, the greatest sighting distance of the visibility of a coated ball is determined by the condition $p = Rn\dfrac{R}{R+h}$.

Since for an EPR ball is equal to the area of a circle with a radius of an extremely large impact distance, we have for it the relation $S = \pi p^2 = S_0\left(n\dfrac{R}{R+h}\right)^2$, where $S_0 = \pi R^2$ - the cross-section area of the uncoated ball.

The magnitude of the decrease in the RCS of a ball due to a dielectric coating is determined by $G = \dfrac{S_0}{S} = \left(n\dfrac{R}{R+h}\right)^2$. Effective reduction is achieved when $0 \le n^2 \le 1$, that provides the material considered. This will be observed in a sufficiently wide frequency band. When $n^2 = 0$ any object will be invisible. When using conventional dielectric coatings ($1 \le n$) or metamaterials ($n \le -1$) there will be an increase in RSC.

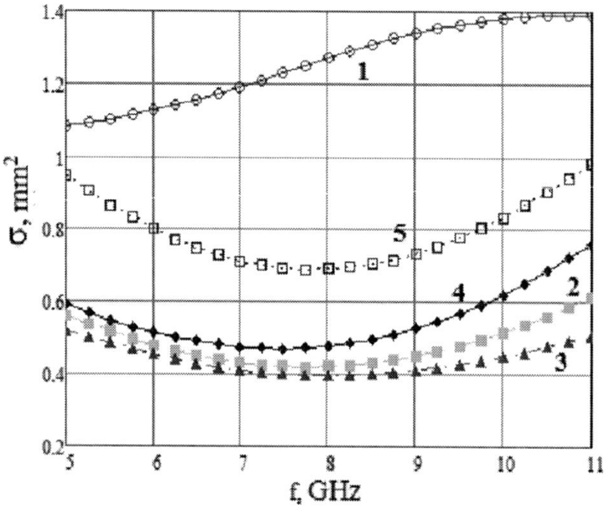

Fig. 7. RSC of the sphere covered with material with an index of refraction:
1) $n=-1+0.1i$ 2) $n=0.5+0.1i$ 3) $n=0.7+0.1i$ 4) $n=-0.5+0.1i$ 5) $n=0.3+0.1i$.

Here are the results of modeling in the software package "CST Microwave studio". As a model, we take a metal ball covered with a material with a different refractive index. The simulation was carried out on the frequency range 5–11 GHz. The RCS values were taken with a frequency step of 0.25 GHz. Figure 7 shows the five RCS dependencies (σ, мм2) of the frequency, for various refractive indices of the coating material, all the results were normalized to the RCS of a coated ball having a refractive index n = 1. It can be noted that when $n=-1+0.1i$ RCS increases in comparison with the metal ball, which is not covered, while the coating of materials $0 \le n^2 < 1$ reduces the effective scattering area.

4 Conclusion

This study shows that the fields scattered by the object covered with a metamaterial with a refractive index $n = 0.3 + 0.1i$ Similar in structure to fields with scattered spheres with a shell of material $n = -1 + 0.1i$, but the RCS of the object is reduced. At the same time, the effect of decreasing RCS is more significant when using a coating with a refractive index $0 \leq n < 1$, than at $-1 < n \leq 0$.

References

[1] M.I. Finkelstein, *Fundamentals of radar. Textbook. for universities. 2 ed.* (Radio and Communication, Moscow, 1983)
[2] V.P. Yakubov, A.S. Mironchev, A.G. Andreitsov, I.O. Ponomareva, Artificial Metamaterials and Radio Tomography **53,** 859 (2011)
[3] A. Mironchev, A. Gorst, MATEC Web of Conferences **79**, 01064 (2016)
[4] A.S. Mironchev, A.V. Klokov, A.V. Gorst, Conference Proceedings of CriMiCo 2014 24th International Crimean Conference Microwave and Telecommunication Technology **1,** 702 (2014)

MATEC Web of Conferences **155**, 01050 (2018) https://doi.org/10.1051/matecconf/201815501050
IME&T 2017

PRINCIPLES AND MECHANISMS OF BRAIN FUNCTIONING

Vladimir Shumilov *

National Research Tomsk State University, 634050, Tomsk, Russia

Abstract. Principles of brain functioning and a mechanism for their realization are formulated in the paper. An absolutely local mechanism provides independent of any external control formation of traces of events in the form of connections between neurons (accumulation of experience), storage of these traces (fading memory), interpretation of traces (response to the impact of the environment in accordance with accumulated experience). Traces of events influence formation of the pathways of signals from the receptors to the effectors and, consequently, the organism's reaction. Retraining is provided by formation of new and therefore stronger links.

1 Introduction

For over 60 years, researchers have been working on the task of creating artificial intelligence (AI), but this task is still far from the acceptable solution. The information capacity of the brain, where the human intellect is located, is so large (at least 300 terabytes), that hundreds of thousands and millions of years of labor of highly skilled specialists are needed in order to fill in such capacity by information [1]. This circumstance makes creation of AI by today's methods economically inexpedient, although, there are no principal barriers to creation of AI that is not inferior to human intelligence. Required computer power will be achieved in the near future.

However, creation of AI will contribute almost nothing to understanding functioning of the brain of humans and animals. In addition, for the constant development of the created intelligence, it is necessary to ensure its openness, the ability to perceive the flow of information with no less speed than a person does.

2 The main properties of the natural carrier of intelligence as an information processing system

The nervous system (NS) and the brain are a natural self-organizing information processing system, which is the system that converts signals about dangerous (harmful) environmental influences on the body generated in receptors into signals that control the mechanisms of evading these impacts on the organism.

* Corresponding author: vnshumilov@rambler.ru

Historically, the nervous system of biological organisms with receptors, a signal processing system (actually the brain) and effectors (manipulators) has emerged as a means of organisms' survival by means of avoiding dangerous effects. This ensured the survival and expansion of organisms with the NS.

The paths of the signals along the NS include a multitude of neurons and the connections between them in the retransmission mode without generating new signals independently. Neurons provide undamped progress of signals through inter-neural connections. This advance is ensured by excitation of the neuron when a total input signal arriving at its input exceeds a certain excitation threshold. The total input signal represents the common impact on the neuron of previous neurons (or receptors, which are special input neurons) delivering signals to the neuron's input through all of its multiple input inter-neural connections. The neuron is excited by a relatively small input signal. Upon excitation, the neuron generates a powerful signal at the output that participates in the excitation of subsequent neurons through their input connections. This ensures undamped progress of signals through the brain structures (from receptors to effectors).

The energy of a powerful signal is extracted by a neuron from food products delivered to it through the blood. In addition, the functioning of the neuron provides many other functions.

3 The pace of information processing by the human brain

It has been proved, that the experience of biological beings is accumulated in the connections between neurons in the corresponding zones of the brain [2]. This occurs with an average rate of forming of connections between neurons in the human brain of about 30,000 bonds per second during a lifetime (the number of possible connections / lifetime) [1]. It is clear, that no centralized mechanism can provide such speed of forming connections in absolutely certain places. Where could it come from? The impossibility of existence of a centralized mechanism to control formation of inter-neural connections in the brain becomes evident if we recall that the development of the organism begins with a single mother cell, which has no place for additional information. This cell may contain only information about the self-duplication of this cell, about possible variations in the generation of new copies of cells depending on the presence of various components (building materials) in the neighborhood of the cell for new cells and the mechanism of cell functioning. On the other hand, even if the central mechanism existed, how would it manage to get a huge amount of information about the place, in which the next inter-neural connection should be established?

4 A local mechanism that ensures functioning of the brain

The problem of forming a connection between neurons at the required speed and in the right place is easily solved by the local mechanism proposed by us. We note that the connections between neurons are paired, and they are formed only between two contiguous neurons that are excited at the same time. If this were not true, all possible connections would quickly be formed even between unexcited neurons. Moreover, the connections formed would not correspond to anything. In fact, observations show that the connections between brain's neurons of animals are intensively formed precisely in those areas of the brain that are responsible for the type of activity that the animal is being taught [2].

What mechanism could provide the formation between the simultaneously excited neighboring neurons? Such a mechanism is provided by a neuron, a model of which we propose. In contrast to the classical McCulloch-Pitts perceptron model [3], with which we can illustrate only some aspects of the functioning of artificial neural systems, we propose a neuron model, which explains in fact all the basic processes, occurring in the brain. Figure 1 represents a scheme of this model.

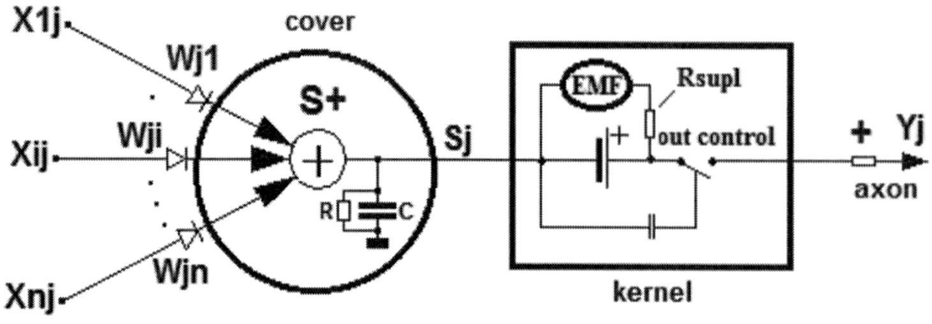

Fig. 1. Model of the neuron, which ensures functioning of the brain.

The key at the neuron output (at the beginning of the axon) can be manipulated in different ways. It can be a thyristor or a field effect transistor, or even just a mechanical shutdown of the ion channel. The answer to this question is not so fundamental and can be obtained later.

The model proposed differs significantly from the classical model because it clearly indicates the source of energy for the formation of a pulse. This impulse is transmitted further by the neuron when it is excited and when the level of the total input signal exceeds the excitation threshold.

The analysis of the above circuit gives an idea of its response to the total input signal. This response is shown in Figure 2*a*, 2*b* with the response of the **H(Sj)** McCulloch-Pitts perceptron.

Let us explain the Figure. The perceptron forms the response potential as a Heaviside function as soon as the total input signal **Sj** exceeds a certain threshold **Ut**. The output potential in this model is zero when the input signal drops below the Ut level.

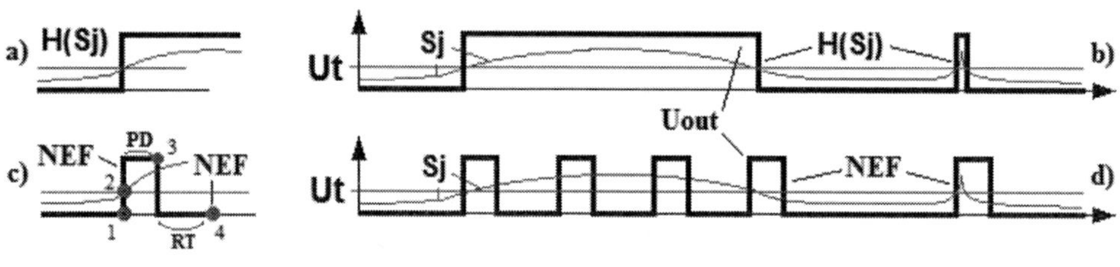

Fig. 2. Excitation function of the McCulloch-Pitts perceptron and the proposed model of the neuron.

The response of our model is significantly different. It is presented in Figure 2*d*. Here, on the vertical axis, the potential difference between the point of the total input signal **Sj** and the axon of the neuron is reflected.

We call the function depicted in Figure 2*c* as a Neuron Excitation Function (**NEF**). Let us consider it. The characteristics of the function are determined by the parameters of the neuron and its current state. The properties of this function are the folowing:

1) For a prolonged **Sj <Ut NeF = 0** (to point 1 on the left). That is, the neuron is not excited (state of rest).

2) When the conditions **(Sj ≥ Ut) & (^ RT)** are satisfied, the neuron is excited (points 1 and 2 in Figure 2*c*.) A positive pulse with a duration of **PD** (Pulse Duration) begins, which continues to point 3.

3) At point 3, the impulse ends because of expending of accumulated energy.

4) At point 3, restoration of the neuron's operability begins, excitation of which has just ended. Recovery lasts for a "dead" time **RT** (Recovery Time). **RT** ends at point 4, at the moment of restoring the neuron's ability to get excited. The excitation cycle is complete on this. The neuron is ready for the next excitation. If the value of the input signal **Sj>Ut** is longer than the duration of one cycle (for example, for 4 cycles), at the neuron output, we will observe not one long pulse, as for the standard Heaviside function, but a series of 4-cycle pulses generated very simple because of presence of "dead" time. The pulse duration **PD** and the "dead" time **RT** are determined by the parameters of the neuron (the capacity of the energy store and the time of its charging).

Given that the neuron is electrically isolated rather well but there are leakage currents, we come to a conclusion that the potentials of the inputs, the total input and output of the unexcited neuron are close to 0 ("ground"). In addition, when the neuron is excited, because of the closure of the key, the total input potential **Sj** switches in the potential difference U2 between the total input and the axon. In electrical terms, a neuron is "suspended" on its inputs and outputs. Therefore, as a result of the appearance of a potential difference between the total input and the axon, the axon potential becomes positive when the neuron is excited, and the potential of the total input becomes negative.

5 A mechanism of formation of connections between neurons

From the most general point of view, all natural processes produce some changes. The propagation of signals in the brain from the receptors to the effectors also produces some changes. In the energy sense, the process of signal propagation produces the greatest change in the effector, which is a place of transformation of weak signals into mechanical or chemical transformations. In addition, the propagation of signals through the brain leaves traces of events in the form of connections between neurons formed during the propagation of signals.

After considering the appearance of potentials at the output and input of an excited neuron, it becomes clear how the bonds are formed, and only between excited neurons. Connections between neurons can be formed only when the potential difference between neighboring points of two excited neurons is higher than some bond potential (BP) for formation of a bond. We see that the proposed scheme easily ensures fulfillment of this requirement. Formation of connections between neurons is explained in Figure 3*a*.

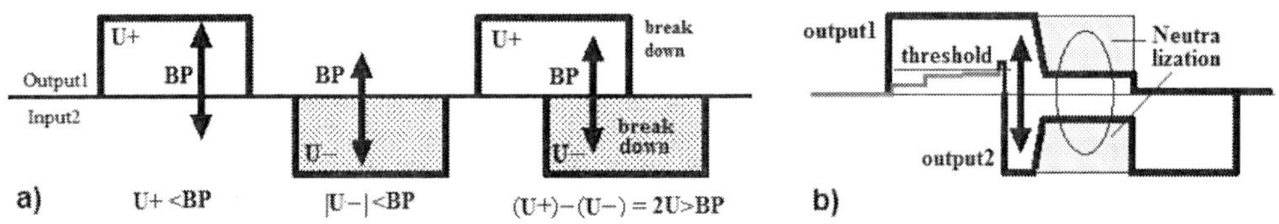

Fig. 3. Formation of a new connection made by the potential difference, which is greater than the bond formation potential, and compensation of the potentials of excited neurons.

The potentials U + and U- at the output (axon) and at the total input of the excited neuron are individually smaller in modulus than BP, and | U + and U- | are larger than BP. Therefore, a galvanic bond can be formed between adjacent parts of two excited neurons, and communication can be formed only between excited neurons. For example, as a result of electrical breakdown (or other irreversible phenomenon), a long-term conducting channel can be formed, which is an inter-neural connection with some initial conductivity (logical weight). Moreover, the conductivity of this channel (logical connection weight) may decrease with time. For example, it happens due to diffusion. This reduction in the weight of links represents a process of forgetting over time.

That is, the neuron model proposed easily provides formation of traces of events affecting the organism that is associative connections between excited neurons, which are a reflection of the organism's experience. Subsequently, formed traces of events or inter-neuronal connections (experience) are involved in formation of the reaction, that is, in formation of the paths of signals propagating from receptors to manipulators. Namely, because of the presence of a connection between neurons excited approximately at the same time with the appearance of one stimulus, a chain of neurons associated with another stimulus will immediately be excited through the previously formed bond. As a result of this induced excitation through the previously formed bond, the organism will react to the second, not yet come, stimulus, as if foreseeing its appearance. Since in nature not random events, but the chains of causally related events occur, thanks to this one-part forecast an organism managed even by only two or three neurons can survive with greater probability than an organism that does not have the ability to fix traces of events that have affected organism. Thus, because of formation of traces of past events in the brain, the past experience influences the body's response to the current impact of the environment on the body. That is the key useful feature of the brain necessary for the survival of the organism and expansion of its species.

Thus, by changing the structure of the brain because of formation of traces of events (new pathways of signals following) the model provides training (fixation of information, storing and interpreting it). Interpretation of the fixed information at the local level looks like signal passing through different paths (the inter-neural connections). From the point of view of technical implementation, this ensures the full functioning of the information processing system. That is, our model ensures fixation, storage and interpretation of the fixed information.

6 Analysis of the mechanism of neuron interaction

The input and output circuits of neurons have a finite load capacity. This leads to the fact that the potentials of the inputs and outputs of the excited neurons are mutually compensated, as shown in Figure 3*b*. As a result of compensation of potentials, the first excited neuron prevents the excitation of other neurons because of the reduction (compensation) of the potentials that had caused its excitation. Because of the compensation of potentials, a situation occurs, which in the theory of neural networks is viewed as a requirement and as a principle "The first receives everything". This requirement in our model is a consequence and prevents the self-excitation of the processing system. Self-excitation will occur if the average multiplication factor of signals (K-factor) is greater than 1. In this case, after each next step in the propagation of signals, their number will grow as K-factor^N. K-factor is also regulated within the species evolutionarily, from generation to generation because of the more likely survival of organisms with a suitable neuron parameter set.

Figure 4*a* shows a scheme for determining the point, at which a connection is formed between the output of an excited neuron (at the top of the Figure) and the input of another excited neuron (in the right part of Figure 4*a*). However, in nature there are no regular flat rectangular intersections. The outputs and inputs of neurons are located in a 3-dimensional volume, which provides even more opportunities for formation of connections (Figure 4*b*).

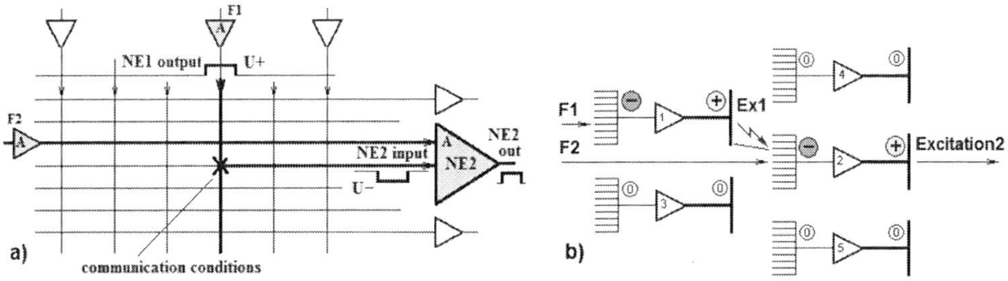

Fig. 4. A scheme for selecting the point of formation of a connection between simultaneously excited neurons.

We assume that the connections between neurons are formed because of the electrical breakdown of the isolation separating the excited neurons. Instead of a breakdown, any other process irreversibly forming a link between excited neurons can act. Moreover, the connections formed must first have the greatest logical weight (the conductivity of the channel arising as a result of electric breakdown). Over time, the conductivity of the channel (connection) naturally decreases because of diffusion-type processes and the influence of the connection decreases. Fading of memory (forgetting) occurs.

The neuron model proposed provides not only propagation of signals and formation of new connections, but also selection of the best ways of the signal propagation by comparing the signal configurations and the weights of the connections of different neurons. It happens as follows. The first to be excited is the neuron, in which the configuration of the coupling weights corresponds best to the configuration of the input signals. By compensating its input and output signals, neuron obstructs excitation of other neurons (which have approximately the same input signal configuration as the newly excited neuron). Therefore, the signal will propagate only along the optimal path, carrying out the optimal reaction corresponding to the configuration of stimuli and to the organism experience, that is, to the weights (forces) of connections between the neurons.

Let us consider what will happen when the situation changes, when the old reaction ceases to be effective. It means that for a long time an irritating signal arrives and continuously excites the neurons of the chain generating an old and ineffective response. As a result, neurons do not have time to restore their working capacity in full. Neurons "get tired". The threshold of their excitation increases. A similar picture of "getting fatigue" is observed with the continuous stimulation of "electric" fish. The potential of the discharges and the magnitude of the amperage decreases each time. As a result of the increase in the threshold of neurons excitation in the chain of the old reaction, it becomes possible to excite neurons adjacent to the old chain. When they are excited, new connections are formed. So the signals move along new paths and a new reaction is formed. It means that because of "getting fatigue", a chain of a new reaction can be formed in the brain. However, this new reaction may also turn out to be ineffective. In this case, the procedure for formation of a new reaction will be repeated. The process will continue until a successful reaction will appear. So neurons "fatigue" allows the brain to change reactions and to be retrained.

Subsequently, if a similar configuration of the input signals occurs, first, a new reaction will be activated, not the old one. This will happen for the reason that the inter-neural connections of the new chain are stronger than the connections of the old chain. The neurons of a new chain win a competition for the exciting signal because the weight of connections is continuously decreasing over time. Therefore, the new connections have the maximum weight (conductivity). It means that forgetting is not a purely negative characteristic.

Thus, in a natural way, the problem of suppression of the old reaction with appearance of a new reaction (reflex) is resolved because of the decrease in the weight of all inter-neural connections over time. Ivan P. Pavlov, a physiologist, a Nobel Prize winner, called this suppression as "the damned question of physiology" [4]. Our model easily explains the suppression of old reflexes when new ones appear. Moreover, it does not need to introduce into consideration additional entities (inhibitory neurons and connections). Appearance of inhibitory structures would cause the necessity to centralize control of all neurons with a centralized comparison of the results. Our model also provides the same result because of the existence of competition for the signal between the new stronger and old weaker inter-neuronal connections at the local level.

Thus, our local model allows us to explain all the most important features of biological systems of information processing.

The results stated (the possibility of independent training and retraining) are confirmed in practice by the functioning of the created electronic model of the simplest nervous system composed of 3 electronic neural-like elements. The model created has shown the possibility of self-training and retraining [5].

7 The emergence of thinking

How does the thinking process happen in the brain? The more stimuli are taken into account by the brain, the higher are the chances of individual organisms to survive. Therefore, from generation to generation, the number of neuron chains reacting to individual stimuli will increase. These chains will form reactions that are even more complex. The number of connections between neurons will grow, the chains will get longer, will intersect and join each other. Predictions for the arrival of stimuli will become increasingly multi-tiered and an increasing number of factors will be taken into account. Because of the increasing length of the chains and the increase in the number of intersections of the chains, the pathways of the signals from the receptors to the effector become more complex. A signal generated by some receptor can move for a long time through the structures of the brain receiving additional stimuli from signals from other stimuli. Such motion of a signal along the brain structures can be called as the flow of thought through the structures of an intelligence carrier. This carrier can be natural (in humans and animals) or a device, which is an artificial intelligence carrier (AIC). The source carrier itself does not have intelligence. It must be trained. Similarly, a computer becomes useful only after installing the software. Both animals and people train and educate their progeny. After passing the course of primary tuition (education), the bearer of the intellect (a young representative of a man or an animal) receives intellect of a certain level. This level depends on the capabilities of an intelligence carrier and on the quality of instruction. It is especially high among people because of the developed system of training and information transfer from primitive drawings on rocks and cuneiform tablets to modern systems of information transmission and visualization. A man gets the mind.

8 Conclusion

In the brain, in contrast to the currently used technical information processing systems, information is fixed not by changing the state of existing elements of the information processing system, but by formation of new elements, which are traces of events. This is how natural memory works, where the organism's experience is fixed. The processing of information in such a system consists in signal passing through already existing brain structures (where the organism's experience is) to effectors (manipulators). Signals are generated in receptors under environmental exposure on the body. These signals enter the brain, pass through its structures in accordance with the experience of the organism and come to the effector, which provides an effective reaction of the organism controlled by the brain to the effect of the environment. Signals can fade along the path and not reach the effectors.

When passing through the structures of the brain signals make traces, it means that new connections are formed between the neurons. This is the way how the organism's experience is accumulated.

Such a system for processing input information based on the experience of the organism (the previous information fixed in the brain) is so simple that it is self-organized (self-formed) without an external controlling (directive) influence in each organism with an appropriate carrier based on neurons with the structure and functions proposed in this paper.

The proposed model of a neuron, which is the main structural component of the brain, and simple natural requirements allow us to explain practically all the main features of the brain, which is a natural carrier of the intellect.

Because of the bipolarity of signals at the input and output of the neuron, training (formation of new connections) is provided.

Because of the mutual compensation of bipolar signals, selection of optimal path of signals is ensured, that is an optimal response to configuration of stimuli in accordance with the past experience of the organism.

Because of the "fatigue" of neurons, retraining is provided, that is formation of new links under the conditions of change.

Because of the decrease in the weight of connections over time, forgetting and coming of new reactions to the forefront occurs, that is "inhibition" of old reactions.

Consideration and understanding of the features of functioning of the natural intelligence carrier enables to create an artificial intelligence carrier. Our model shows that this is quite possible with the current level of circuit technologies [6, 7]. To do this, it is sufficient to create a device with a sufficient number of microelectronic neural-like elements (NEs) and to conduct initial training of the artificial intelligence carrier created. In microelectronic structures, it is inconvenient to create large electrical capacitances. Therefore, to ensure formation of bipolar potentials, it seems more convenient to use a circuit with two supply voltages U + and U-. The pulse output time PD and the restoration time RT are simulated using mono-vibrators.

Training always occurs based on already existing connections between neurons (knowledge). Therefore, when creating AIC, it is necessary to establish initial connections between neurons in it. These connections are equivalent to the unconditioned reflexes of natural organisms formed in the species of organisms over many generations.

Acknowledgements

The research presented in this paper was supported by the Russian Foundation for Basic Research (grant No 16-29-04388/17). The authors are grateful to Tatiana B. Rumyantseva from Tomsk State University for English language editing.

References

[1] V.N. Shumilov, *Principles of the brain functioning. Engineer's look.* (Ekaterinburg, 2008).
[2] G. Yang, F. Pan, W.-B. Gan, *Nature* (2009)
[3] G. McCulloch, W. Pitts, Automatic, 363 (1956)
[4] I. P. Pavlov. Brochure of the Academy of Science of the USSR **XXVIII, 4** (1932)

[5] V.N. Voronin, V.N. Shumilov, Yu.A. Shumilov, A.R. Tsoy, A.V. Kiselev, S.M. Sharshov. Report at the 11th International Congress "Neuroscience for Medicine and Psychology", Sudak, (2015).
[6] G.M. Mladenov, V.M. Spivak, E.G. Koleva, A.V. Bogdan, Nanoelectronics URL: http://me.kpi.ua/downloads/Nanoelectronics_1.pdf
[7] V.N. Shumilov, V.I. Syryamkin, M. V. Syryamkin, AIP Conference Proceedings, 1688, 040007 (2015); doi: 10.1063/1.4936040

LOCAL QUASIOPTICAL RESONATOR DIAGNOSTICS OF SEMICONDUCTOR WAFERS

Igor Dorofeev[a], *Grigorii* Dunaevskii, *Alexander* Badin, *Kirill* Dorozhkin, *Vitaly* Bessonov, *Sergey* Khodovitskiy

National Research Tomsk State University, 634050, Tomsk, Russia

Abstract. The method of measurement of reflection coefficient behind an orifice in one of reflectors of the quasioptical open resonator is considered. According to the applied mathematical model, the method has the restriction which is consist in use of a short-wave approximation for the orifice size (it dimension is much more than a wavelength). For evaluation of limits of applicability of this method the experimental setup in the three-centimetric range of wavelengths consist of the quasioptical open resonator and the system of positioning of the measuring probe is created. By the experimental study of distribution of fields behind an orifice in the reflector of the open resonator it is found that the method, except restriction of mathematical model, has also restrictions, due to the wave's interference, reflected from edges of this orifice. It causes a deviation from the uniform amplitude distribution of the field in an opening that can lead to increase in an error of a reflection coefficient measuring. It is also shown that by decrease of the sizes of an orifice formation of a bunch with small divergence, but with significantly more expressed distribution of the field localized about a bunch axis in an aperture is possible. It can be used for diagnostics of inhomogeneity of samples with higher locality.

1 Introduction

Now intensive development of element base of millimetric range of wavelengths on the basis of semiconductor technologies is observed. At the same time especially significant demands are placed on a prime material of which scheme elements are made. For millimetric and submillimetric (terahertz) ranges it is possible to single out from perspective materials GaN and materials on its basis [1], in particular, applied to creation of HEMT transistors [2, 3]. For successful realization of final devices the great value gets an opportunity of preliminary diagnostics of wafers with high space resolution.

It is essential that such diagnostics can and has to be carried out at those frequencies at which future semiconductor elements will be used. Millimeter waves to fall into to the range in which it is effectually to use quasi-optics elements instead of some traditional waveguide and resonator elements. Some of quasioptical approaches were realized also in the centimeter range therefore it is the most reasonably to expand the range of their use, having used already tried-and-true techniques. On more low frequencies for a research of

[a] Corresponding author:idorofeev@mail.tsu.ru

plane-parallel samples which size in transversal directions significantly exceeds a wavelength usually use a method of the free space which analog in quasioptics is the measurements in a lens waveguide. However the sensitivity of such methods is small. Significantly resonator techniques have much higher sensitivity. The quasioptical resonator is the open resonator which is often called the Fabri-Perot resonator. At the same time intra resonator methods of diagnostics of sheets have the locality limited to beam aperture and are intended only for materials with low losses [4].

2 The method of measurements

In [5] the quasioptical resonator technique of measurement of local values of the module and a phase of a reflection coefficient from plane-parallel samples through an orifice in one of reflectors of the open resonator is presented. In this case the locality of measurement can be almost 10 times more, than in case of intra resonator location of sample. Besides, there is an opportunity to investigate value distributions of a reflection coefficient on the surface of material and also to work with samples which size does not allow carrying out measurements in the free space. The developed method was successfully applied to a research the plane-layered materials with significant losses.

It should be noted that for this method the locality is limited to the orifice size which in turn, determines beam divergence outside of the resonator. Two contradictory requirements are imposed to the size of an orifice. On the one hand the size of the orifice has to be larger in comparison with a wavelength to provide the flat phase front and small divergence of a wave behind it, on the other side of its size has to be small in comparison with diameter of a beam on a mirror of the open resonator if to consider inhomogeneity of the field in the directions, perpendicular resonator axes. In practice it is rather difficult to meet these two requirements therefore often it is necessary to be limited to compromise option. Within the mathematical model realized for a method it is impossible to carry out any estimates of border of influence of the sizes of an orifice therefore it is reasonable to carry out the experimental studies of correctness of the theoretical assumptions.

First of all it is necessary to estimate distribution of the field behind an orifice. The experimental setup on the basis of the open resonator was created for this purpose, in one of reflectors of which the orifice for interaction of the field of resonance oscillations with the studied object was located. Considering scalability of the methods of diagnostics with change of a wavelength based on use of open resonators, researches can be carried out in the most convenient sites of the frequency range. The three-centimeter range of wavelengths was chosen for the purpose of simplification of the device of scanning of the measuring probe and increases its resolution capability.

3 Experimental setup

The open resonator consisted of the flat and concave mirrors (Figure 1). Its geometrical sizes corresponded to a quasioptical case, that is there were much more wavelengths and made: diameter of a plane mirror is 30 cm, concave mirror is 40 cm and distance between them 53 cm. Radius of curvature of a concave mirror – 120 cm.

Fig. 1. Open resonator for measurements.

In the center of a concave mirror the orifice of square shape with the side of 38 mm was located. On a plane mirror elements of communication of the open resonator with external circuits (Figure 2) was created. These elements represented the pieces of waveguides converging on a narrow wall to narrow (about 100 microns) splits.

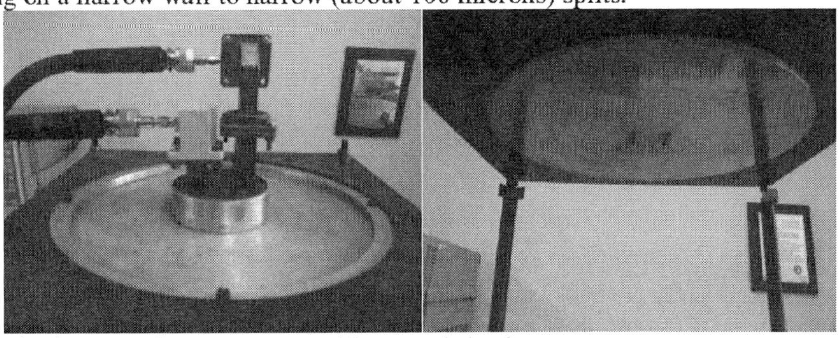

Fig. 2. Couple elements of open resonator with external circuits.

Oscillations with a vector of an electric field which was orthogonal to splits were excited in the open resonator. Thereby polarization degeneration was removed. Two couple elements were used so that there was an opportunity to define value of a transmission coefficient via the open resonator. The sides of the orifice on the concave mirror were located strictly in parallel and perpendicular to splits.

Below a concave mirror the table for samples which had a possibility of movement in the direction of an axis of the open resonator (Figure 3) was located. The plane of a table is made from polished brass and served also for calibration (the reflection coefficient module from it was accepted to unit, a phase -- for 180^0).

Fig. 3. Table for calibration and samples with the movement mechanism.

The surface of the concave mirror was covered with the absorber for decrease of influence of the resonances resulting from formation of standing waves between a sample and a mirror.

For measurement of field's distributions with a probe method it is necessary to choose probes which sizes there is much less wavelength, at the same time it is better to use the absorbing probes than scattering. However in this case there is no need to use the pointwise probe. As the orifice is square, that electric field to one of its sides will be orthogonal, relative others – in parallel. As the tangential component of the field on the relevant sides will tend to zero, deviations from the uniform distribution will be much more in the direction, perpendicular to the direction of a vector of an electric field. Therefore the long threadlike probe from the thin absorbing carbon fiber was used (Figure 4).

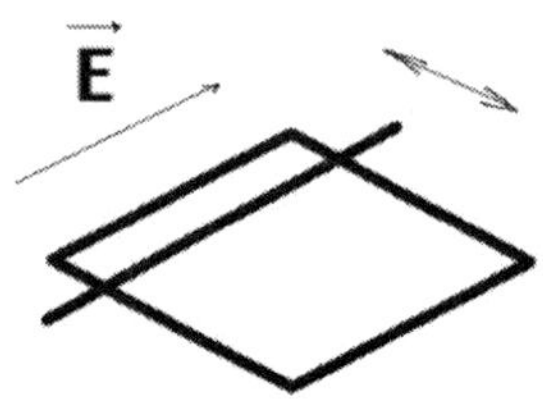

Fig.4. Probe location relative to electrical field vector.

The probe was fixed on a frame and moved in the direction, perpendicular to a vector of an electric field by means of the mechanism given by the step engine with a step of 158 microns. At each position of the probe the value of a transmission coefficient of the open resonator on a resonant frequency of the corresponding mode which change was in proportion to the power absorbed by the probe was fixed. Its value can be estimated as follows $P_p = \frac{1}{2}\sigma\int_{-a}^{a} E^2 dl$. Here σ – conductivity of material of the absorbing thread. Integration is carried out along the probe. As the field along it is orthogonal to the sides of an orifice, it can be considered the homogeneous in this direction in case the sizes of an orifice are rather big in comparison with a wavelength. So $P_p = a\sigma E^2$. Thus, change of a transmission coefficient of the open resonator on a resonant frequency is proportional to the value of an electric field square in an orifice. The probe moved to the planes, the most close to a mirror taking into account its thickness and thickness of the absorber.

4 The results of measurements

The network analyzer Agilent E8363B was used to measurements of values of transmission coefficient of the open resonator on a resonant frequency. At measurements in a frequency band which size was determined by the pieces of waveguides forming coupling elements 8 main fundamental modes of the open resonator were used. Observed dates of distribution of three of them are given in graphs on Figure 5. Not to encumber graphs the most reference dependences were chosen. The type of the others is similar to given.

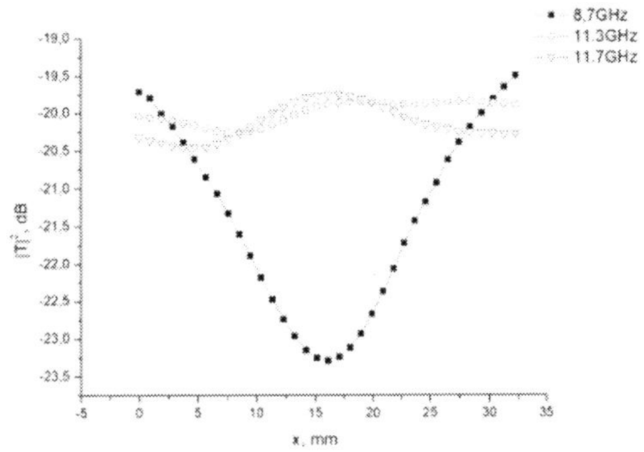

Fig.5. Field distribution near the measurement orifice in the direction, perpendicular to electrical vector.

As appears from the received results, most the uniform distribution of the field was observed for mode with a resonant frequency of 11.3 GHz, but not predictably, on the most high frequency. For it the field has oscillations. The interference of the fields scattering by the opposite sides of a square orifice is the reason of it. For the most low-frequency mode it is possible to observe rather non-uniform distribution similar created by quasioptical beam. Thus, the assumption which is that this orifice cuts out from a resonance beam the site with the uniform amplitude distribution of the field is not carried out not only for orifice insufficiently larger in comparison with a wavelength, but also and at rather larger. However it is possible to achieve enough homogeneous distributions of the field practically on all area of an orifice by selection of it size.

Deviations of distribution of the field from the uniform will lead to increase of an error of definition of a reflection coefficient from the sample corresponding to value of a reflection coefficient for a plane wave. Therefore it is necessary to use an orifice of the variable sizes for measurements in a wide frequency band.

However in some cases, for problems of diagnostics of uniformity of exemplars an absolute value cannot matter, resolution capability is much more important. Its high value will able to find out inhomogeneity of more low scale. And in this case, as appears from Figure 5, decrease of the sizes of an orifice leads to increase in locality of interaction of the electromagnetic beam created by a measuring orifice and a sample.

For evaluation of influence of such non-uniformity of distribution of the field to beam divergence behind an orifice, dependences of transmission coefficient of the open resonator on the location of a measuring table were measured. These dependences are given in Figure 6.

If to take for criterion of divergence of a beam the value of decrease of a maximum of a transmission coefficient at increase in distance between a table and a mirror, then, apparently from these graphs, these values are quite comparable among themselves.

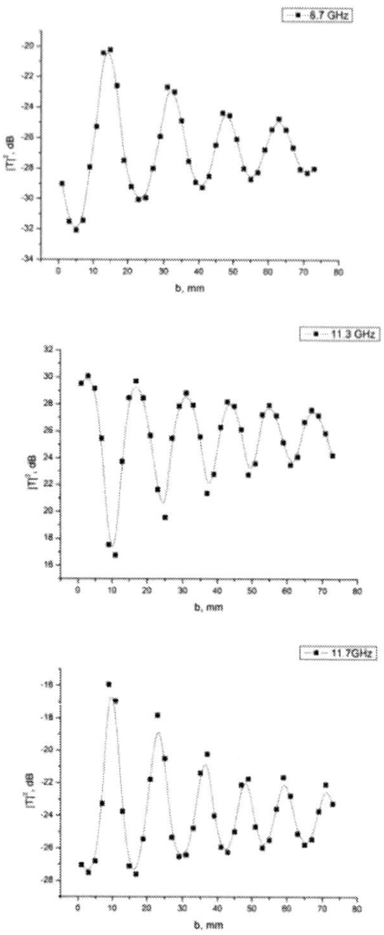

Fig.6. The dependence of transmission coefficient of open resonator on resonant frequency of corresponding mode from the distance between the mirror and the table.

5 Conclusion

Thus, as a result of the experimental study of the field distribution it is established that the method of measurement of a reflectivity behind a measuring orifice in the reflector of the open quasioptical resonator has restrictions not only a short-wave approximation (the size of an orifice have to be big in comparison with a wavelength), but also long-wave because of an interference of the fields reflected from edges of an orifice.

If this method to use for diagnostics of the relative values of inhomogeneity of samples, by means of a small orifice it is possible to create a beam which has a small divergence, but higher locality of the field.

Use of the open resonator rather just allows expanding a method in a high part of a gigahertz frequency range and also in terahertz. The method does not demand exemplars of the larger sizes, works in the wide range of change of their parameters, provides high locality, provides a possibility of scanning of samples. In the presence of a source of electromagnetic oscillations it can be realized as the self-contained unit.

Acknowledgements

This research was conducted as part of the project № 14.578.21.0240 (RFMEFI57817X240) Ministry of Education and Science of the Russian Federation.

References

[1] K. Ahi, Opt. Eng. **56**, 1. (2017)
[2] T. Otsuji, M. Shur, IEEE Microwave Magazine **15**, 43. (2014)
[3] W. Knap, D. Coquillat, N. Dyakonova, F. Teppe, O. Klimenko, H. Videlier, S. Nadar, J. Łusakowski, G. Valusis, F. Schuster, B. Giffard, T. Skotnicki, C. Gaquière, A. El Fatimy, Comptes Rendus Physique,**11**, 433. (2010)
[4] B. Yang, S. Katz, K. Willis, M. Weber, I. Knezevic, S. Hagness, J. Booske, IEEE Trans. on Terahertz Science and Techn. **2**, 449. (2012)
[5] G. Dunaevskii, *An Open Quasioptical Transformers*. (Izdatelstvo NTL, Tomsk, 2006)

https://doi.org/10.1051/matecconf/201815501052

WAYS OF EFFICIENCY IMPROVEMENT OF THE ORGANIZATION OF STUDENTS' INDEPENDENT WORK WITH COMPUTER TECHNOLOGIES USAGE

Inna Plotnikova[1,*], *Oksana* Efremova[1], *Nataliya* Chicherina[1], *Victor* Dalinger[2]

[1]Tomsk Polytechnic University, Institute of Non-Destructive Testing, 634050, Tomsk, Russia
[2]Omsk State Pedagogical University, 644099, Omsk, Russia

Abstract. Recently the need for the qualified engineering personnel increases in the world. High-quality preparation of such shots begins at the university already with the first courses and it is directly connected with the effective organization of students' independent work. In the article the organization of independent work when training students of technical colleges in mathematics with use of computer means is considered. The complex of the tasks developed by teachers included tasks in mathematics of four levels. It is shown that the independent solution of problems of different levels by the students of experimental group contributed to the development of mathematical abilities.

1 Introduction

At the design of working programs of subjects, courses, disciplines (modules) in higher education institution the appropriate educational technologies which allow to provide in the most effective way the achievement of the planned results of training are provided.

The intensification of educational process and new requirements to it force teachers to find new ways of increase of independent work efficiency. In this regard, the role of students 'independent work considerably raises. At the same time it is possible to use new forms of the organization and control of students 'independent work, or to improve already available traditional ones. Processes of a computerization and informatization inevitably influence both knowledge forms, and training technologies.

In the modern conditions the functions of stimulation of students'cognitive activity, optimization of their independent work, ensuring availability, maximum presentation and timely updating of a training material are the functions of electronic training.

Various elements of electronic training have strongly entered the educational process of higher educational institutions. However, the question of efficiency of use of electronic training in educational process remains open nowadays.

*Corresponding author: inna@tpu.ru

In the academic circles the question of efficiency of use of electronic training in development of abilities of students in the course of educational activity, in particular mathematical abilities of students is constantly discussed. Importance of this problem is caused by the fact that it is aware of mathematics there is a mastering not only knowledge necessary in professional activity of future expert, but also development of abilities to the analysis and synthesis which are the key to successful work in any field of activity.

Thus, the problem of creation and use of new information technologies in training (and especially in training in mathematics) becomes more and more relevant.

We will notice that use of computer technologies is not end in itself, but means of increase in efficiency of educational process and is focused on its information support.

Nowadays many higher educational institutions develop tools, including the automated training systems, for drawing up individual trajectories of training and assessment of real academic loading.

Computer technologies are used at schools and in higher education institutions in different contexts [1]. A part of researchers use computers for modeling and visualization of objects [2–5], during the work on creative projects [6]. It is necessary when training in mathematics, physics, biology and other disciplines [7].

Other researchers consider influence of the computer on pupils' behavior when training them by means of game technologies [8, 9], for estimation of knowledge [10–13], as an assistant for development of discipline on remote access, for exchange of knowledge on social networks [14, 15], when training mentally retarded pupils, students with the low level of knowledge [16,17], etc.

We want to apply computer technologies not only as the technologies that help to estimate adequately knowledge of a subject, but also contribute to the development in the studying abilities in the process of educational activity.

2 Methodology

Authors offer a complex of tasks for the organization of both classroom, and out-of-class independent work of students at the course of mathematics. The peculiarity of the developed complex is the use of the computer as a mediator between the teacher and the student, assistance in the organization of independent work of students at the individual level, implementation of regular monitoring with the subsequent correction of the training process, orientation to a motivational component of educational process.

The organization of independent work offered by authors was carried out with use of the training system which has been developed by teachers of the Tomsk Polytechnic University.

We will consider the organization of students' independent work on the example of a mathematics course.

At the initial stage there took place identification of the student (the choice of group of the list and a surname of the student in the database). After the online registration the student chose a subject and got acquainted with the methodical instructions on a subject.

Further the student passed to studying of the theory, and then started fulfilling the tasks. The student had an opportunity to change a training trajectory, pass the theory, at once to start performance of tasks. For each correct answer to the student a certain number of points according to the rating system of assessment was charged. The result depended on the level of a task complexity.

The tasks are presented by four levels. Tasks of the first level have been formulated in the form of theoretical questions or presented in the form of drawings with the choice of answers from the offered variants. Problems which students usually solve at the classroom lessons are considered as the tasks of the second level. Such tasks belong to standard tasks. Their decision allows to learn the basic concepts and properties characterizing concepts. At difficulty in the decision there is an opportunity to use "Hint" which arrives in the form of a floating window. At the same time the student sees the number of points which is removed for a task when using "Hint" on the computer screen. The student can address the theory or the reference book for what points aren't removed as well.

The third level is presented by algorithmic tasks, but at their decision there is no opportunity to address "Hint". The answer is entered in the form of the numerical answer into the specially allotted field, or the sequence of the decision is restored, or are chosen compliance between objects.

For the answer to the task connected with sequence restoration, the student at first chose the first, on his opinion, the element in the sequence and established an order of its following, further to the second, etc.

Change of an order and viewing of the received sequence can be carried out before final input of the answer.

The example of a task for compliance is presented in Figure 1.

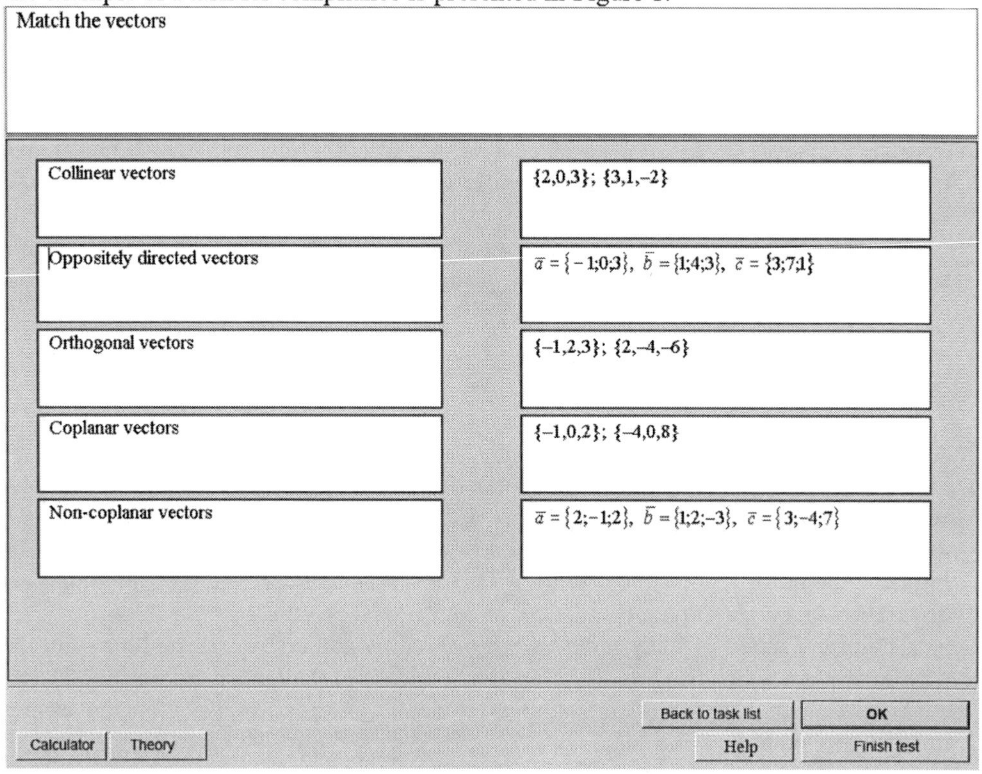

Fig.1. «Matching» task.

Tasks of the fourth level are presented by tasks at which solution the application of mathematics in practical activities is shown. The targets of the fourth level are the generally physical tasks interesting to students of technical university.

The example of a task of the fourth level is given below.

The statement of the task. From the hole having the form of the truncated cone with a big radius of R_1 and smaller radius of R_2 it is necessary to take out some mix having the changeable density of $p\ (x) = ax + p_0$. Here x is the coordinate (depth) concerning the earth level. It changes ranging from 0 to N.

It is required not only to find the spent work, but also to answer some questions.

For example. Whether it will be better to make a hole of the same volume, but in the form of the cylinder? How will work change if before taking out to mix this mixture so that its density became uniform?

The main condition is that the fourth level task has to be not from the textbook. It means that information on an object isn't necessary and sufficient for the decision (it is possible that it is absent and it is necessary to find), and it means that there is no ready answer to which it is possible to compare the result. And still it means that there is no ready template on which it is possible to solve. Therefore, the student has to show knowledge of the substantial side and existence of theoretical thinking.

Having performed a task of the fourth level, the student loads the scanned decision for check by his teacher.

After finishing the subject of a course the student passes the current control work.

3 Research results and discussion

For both groups supported identical conditions: the research was conducted in the first half of the day (11:00) in audience, duration of a research was 30 minutes investigated students carried out testing independently; the use of electronic devices was forbidden.

Comparison of the testing results separately on each task offered examinees has shown that students after training in mathematics in an experimental technique coped with tasks of the checking test of the level of synthetic and analytical skills development much more surely.

After training in mathematics in the majority of questions (in 14 of 20 tasks that makes 70% of questions) abilities of experimental group students to the analysis and synthesis were much higher, than at the students of the control group.

In five of twenty offered tasks (it has made 25% of questions) more than 80% of experimental group students have shown excellent results in comparison with an indicator of control group students. At the end of the experiment in two of twenty questions (it has made 10% of number of tasks) 90% of experimental group students have shown high level of development of abilities to the analysis and synthesis. For the comparison: the number of tasks in which 90% control group students have succeeded was only 5%.

In six of twenty tasks (30% of questions) the number of the experimental group students who have answered right has exceeded twice number of the students who have shown the correct answers to test questions prior to the experiment.

And the number of tasks which less than a half of the interviewed students have answered right has made at the first year 60% (twelve of twenty offered tasks). In 25% of tasks (five of twenty questions) the difference between the number of the examinees who have successfully coped with a task before electronic training made from 21% to 45%.

And only in three of twenty tasks (15% of questions) experimental group students after training of a mathematics course couldn't surpass the results of control group students.

Comparison of testing results to results of the academic progress of experimental group students showed that the high level of analytical and synthetic skills development allows the student to learn a training material more effectively.

All above is the evidence of efficiency of use of e-learning in mathematics in aspect of development of mathematical abilities to the analysis and synthesis.

4 Conclusions

Certainly, the introduction of mathematical tasks into the educational process depends on the level of students' mathematical competence (understanding as an opportunity to solve professional problems by mathematical methods). Therefore at an initial stage the tasks of the first, second and third levels aim at the development of mathematical abilities to the analysis and synthesis of the studied material.

The statement of educational targets of the fourth level (applied tasks) demands quite serious work of the teacher.

These tasks have to possess the whole set of characteristics:

1) in the perspective they have to be clear to students;

2) at their decision students have to illustrate the studied material and tools for the purpose of his fixing;

3) they have to be relevant.

Use of applied tasks in mathematics leads to increase in motivation of students on training, to creation of communications between the theory and practice and also updatings of students knowledge from other areas and disciplines and to the working off the skills of information search.

In the conclusion we will note that use of computer technologies in the educational process of a higher educational institution is justified and extremely expedient, and a variety of forms and methods of the independent work organization allows to make active the creative potential of students.

References

[1] M. Hadfield, M. Jopling, Innovations in Education and Teaching International **51(6),** 607(2014) doi:10.1080/14703297.2014.924747

[2] B.R.J Jansen, E. De Lange, M.J. Van der Molen, Research in Developmental Disabilities **34 (5)**, 1813 (2013) doi: 10.1016/j.ridd.2013.02.022

[3] M. Planchard, EDG J **71(2)**, 1 (2007)

[4] V.S. Sherstnyov, A.I. Sherstnyova, I.A. Botygin, D.A. Kustov, Key Engineering Materials **685**, 867 (2016) doi: 10.4028/www.scientific.net/KEM.685.867

[5] S. Obuhov, I. Plotnikov, M. Surkov, L.Sumarokova, K. Savkin, MATEC Web of Conferences **91**, 01036 (2017) doi:10.1051/matecconf/20179101036

[6] O. N. Efremova, I.V. Plotnikova, O.N. Tchaykovskaya, J Ponte **73(4)**, 158 (2017) doi: 10.21506/j.ponte.2017.4.45

[7] N. V. Chicherina, V. Kazakov, O. Yu. Korneva, E. A. Titenko, The European Proceedings of Social & Behavioural Sciences (EpSBS) **19**, 349 (2017) doi:10.15405/epsbs.2017.01.47

[8] D. Martinov, G.H. Burgess, C.M. Pomerleau, C. Marin, Computers in Human Behavior **49**, 487 (2015) doi: 10.1016/j.chb.2015.03.039

[9] N. V. Chicherina, V. A. Tolkacheva, European Proceedings of Social & Behavioural Sciences (EpSBS) **19**, 683 (2017) doi:10.15405/epsbs.2017.01.90

[10] C. Timmers, B. Veldkamp Computers and Education **56(3)**, 923 (2011) doi: 10.1016/j.compedu.2010.11.007

[11] C. Timmers, J. Braber-Van Den Broek, S. Van Den Berg Computers and Education **60(1)**, 25 (2013) doi: 10.1016/j.compedu.2012.07.007

[12] C. F. Timmers, A. Walraven, B. P. Veldkamp, Computers and Education **87**, 1 (2015) doi: 10.1016/j.compedu.2015.03.012

[13] F.M. Van Der Kleij, T.J.H.M. Eggen, C.F. Timmers, B.P. Veldkamp, Computers and Education **58(1)**, 263 (2012) doi: 10.1016/j.compedu.2011.07.020

[14] N. Behringer, K. Sassenberg, Computers in Human Behavior **48**, 290 (2015) doi: 10.1016/j.chb.2015.01.069

[15] A.E. Goldshtein, G.V. Vavilova, S.V. Mazikov, MATEC Web of Conferences **79**, 01009 (2016) doi:10.1051/matecconf/20167901009
[16] F. Lai, R. Luo, L. Zhang, X. Huang, S. Rozelle, Economics of Education Review **47**, 34 (2015) doi: 10.1016/j.econedurev.2015.03.00
[17] V.V. Redko, N.S. Starikova, L.A. Redko, G.V. Vavilova, IOP Conference Series: Materials Science and Engineering **81**, 012083 (2015) doi: 10.1088/1757-899X/81/1/012083

MODEL OF BROADBAND METAMATERIAL FOR MICROWAVES

Aleksandr Gorst[1], *Andrei* Klokov, *Andrei* Zapasnoy

National Research Tomsk State University, 634050, Tomsk, Russia

Abstract. In this work are presented, results of computer model operation of the structure executed from triangular elements with openings in the center. Results of mathematical calculation of amplitude and phase distribution of an electric field with use of CST Microwave studio are given. Also, calculation of the equivalent index of refraction when passing a plane wave through this structure is given.

1 Introduction

Metamaterials – the perspective direction of development in various spheres: Development of antenna systems [1], creation of a raincoat of the invisible being [2]. Such materials promote emergence of the new phenomena [3].

All known metamaterials are based on the resonance effect, which includes a resonant circuit. Such metamaterials give a negative value of the refractive index, but this method leads to a sharp decrease in the working frequency band, which is a negative factor. In the 21st century, with the development of ultra-wideband devices, the use of metamaterials with a 1 GHz band is not appropriate. In turn, scientists suggest the creation of tunable metamaterials, which complicates the structure of the lattice and there is a need to change the layouts from the tasks.

In the previous publications [4, 5] authors showed a possibility of creation of metamaterials from the selfcontained ring conductors focused definitely, placed in polystyrene foam. Creation of such structures processes the long-lived and laborious. The model from ring conductors is rather heavy, its weight is 500 g. The accuracy of manufacturing is not ideal, in view of manual manufacturing. Is there a need for such elements, in the manufacture of which, the human factor would have the least influence, the law minimizing the discrepancy between the parameters of the layout and the model.

2 Modeling the structure

The creation of new elements was based on the simplicity of manufacturing and the amplification of the effect of the inverse phase advance in comparison with the model of ring conductors. The structure was based on triangular elements with a hole in the center

[1] Corresponding author: gorst93@gmail.com

(Figure 1). The gap between such elements along the axis OY can be considered as a magnetic Hertz dipole, or as a slot antenna and have a significant effect on the magnetic field component. The element itself along its entire length along the OY axis can be considered as a set of electric dipoles with different lengths. Accordingly, this will have an effect on the electrical component of the field. The holes in the triangular elements protrude in the role of the rings, which are combined radiators and interact with both the electrical and magnetic field components, while ensuring the presence of negative magnetization and polarizability. To ensure the repeatability of elements and ease of their manufacture, it is decided to perform them on the basis of printed circuit boards. The same method of fabrication of structures is used by other researchers [6].

Simulation of this structure was carried out using the software CST Microwave studio. The CST Microwave Studio program uses the finite integral method (FIT) – a fairly general approach that first describes the Maxwell equations on a spatial grid, taking into account the energy conservation law, and then by them forms a system of specific differential equations, such as the wave equation or the Poisson equation. The method can be implemented both in the time domain and in the frequency domain. For our case, we use the time domain.

The structure model consisted of 1344 elements (Figure 2). The geometric parameters of the elements are as follows: an equilateral triangle with a side length of 12 mm, a hole radius of 2.5 mm, a triangle distance of 1 mm, and an element thickness of 0.1 mm. The distance between the elements along the 0X axis is 3 mm, along the 0Y axis 1 mm, along the 0Z axis 3 mm. The elements were arranged mirror-like on the dielectric layer (ε = 3.5) with a thickness of 1 mm. The material of the elements is an ideal conductor, the area around the lattice is vacuum, the boundary conditions are open space, the plane wave along the 0X axis is modeled, the electric field intensity vector is directed along the 0Y axis, the magnetic field vector is directed along the 0Z axis in the form of a wide-range pulse signal frequencies of 1–10 GHz.

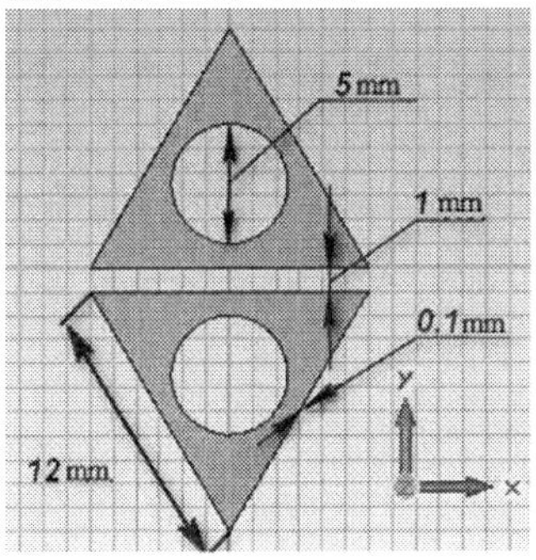

Fig. 1. Model of a triangular element.

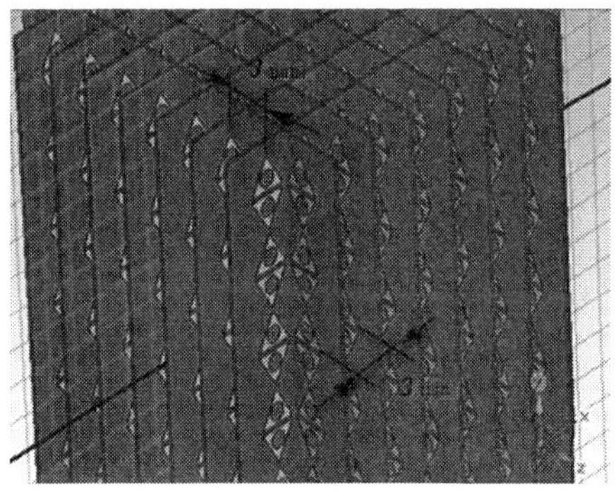

Fig. 2. Lattice model from triangular elements.

Along the black line (Figure 2), the structure passing through the center, necessary data, such as amplitude and phase of the field were taken off. Further for processing data were imported to a mathematical product of Mathcad 14 where necessary calculations were performed.

3 Simulation results

In Figure 3 the schedule of phase distribution of the field at a frequency of 4 GHz in the form of dependence of electric distance on distance of field of calculation along the line is submitted. To a layer it is possible to notice oscillation character which is bound to an interference of an incident and back wave, as well as in a case with amplitude distribution of the field. In a layer the negative phase attack which demonstrates decrease of an index of refraction is observed. Behind a layer there passes the free wave.

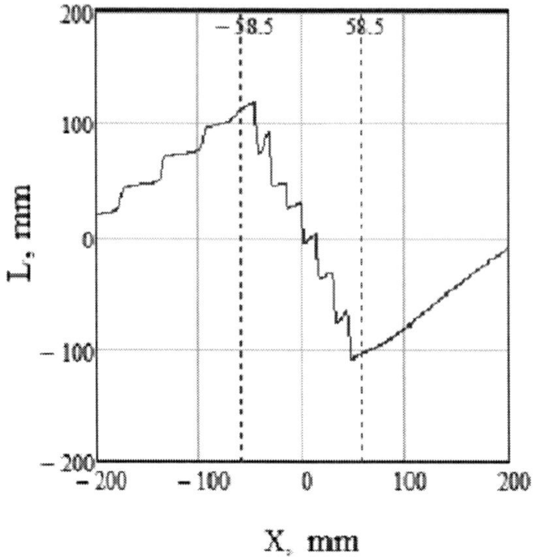

Fig. 3. Phase distribution of an electric field along a straight line at a frequency of 4 GHz.

For calculation of a real part of the equivalent index of refraction we use a formula:

$$n = \frac{\Delta L}{d} \quad (1)$$

where $\Delta L = l_2 - l_1$ –wave progression in a layer, d– thickness of structure from triangular elements. For calculation n, values l_1 and l_2 we take on limits of structure. The real part of the equivalent index of refraction for our structure at a frequency of 4 GHz is equal $n = -2.8$.

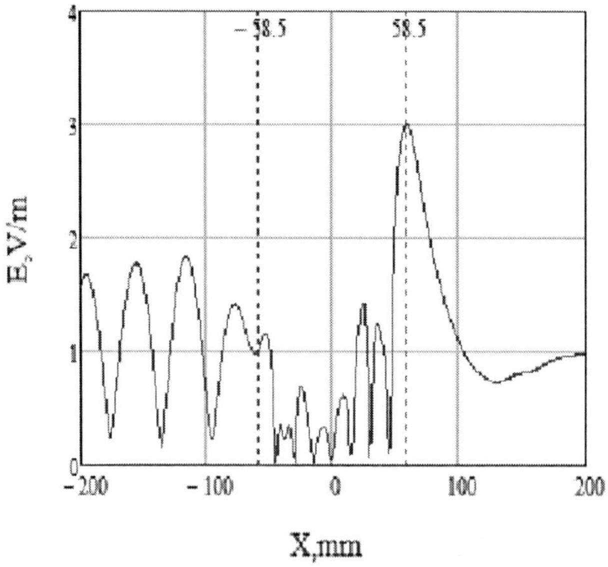

Fig. 4. Amplitude characteristic of a lattice of three layers at a frequency of 4 GHz.

Just pay attention to the distribution of the amplitude of the field E. Figure 4 shows the dependence of the amplitude on the distance, where 0 along the abscissa is the center of the lattice. Markers mark the edges of the model. It can be seen that an interference pattern is observed in front of the layer, which indicates the presence of a reflected wave. The maximum amplitude before the grating is 1.8 V. In the layer, the effect of multiple reflection of waves from the edges is observed, each minimum of the amplitude corresponds to the layer of the lattice. The amplitude above the amplitude of the incident wave whose maximum is equal to 3 V is observed behind the layer. This effect is due to focusing, which is one of the numerous properties of the material.

The negative value turned out not only on 4 GHz, in Figure 5 the schedule of a real part of the equivalent index of refraction for a frequency band 3÷7 of GHz is given below.

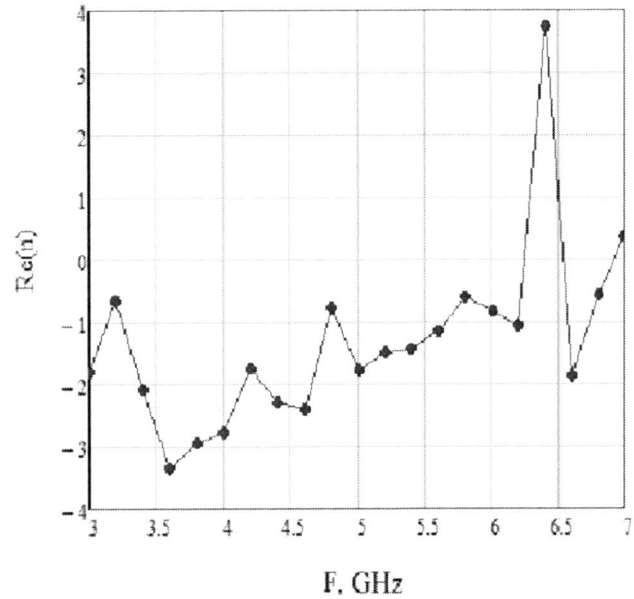

Fig. 5. Dependence of a real part of the equivalent index of refraction on frequency.

Points on graphics noted calculated values of the equivalent index of refraction. Calculation was carried out with a step to 0.2 GHz. The minimum value of the equivalent index of refraction is −3.4. At a frequency of 3.6 GHz. Apparently the schedule has the cut-up increasing character, but values are negative for a frequency band 3÷6.2 of GHz. At a frequency of 6.4 GHz there is a sharp increase, a real part of the equivalent index of refraction, to value $n = -3.7$ further we see just the same sharp decrease. Presumably it is bound to a resonance on which elements interact in such a way that the layer becomes a dielectric.

4 Conclusion

Analyzing the obtained data it can be stated that the structure of triangular elements with a cutout in the center, similar to the structure of ring conductors, according to the values of the real part of the equivalent refractive index, but works in a larger frequency range. This structure allows you to get rid of resonance effects with small changes in the value of the refractive index. From the amplitude distribution graph it can be concluded that a certain focusing zone forms behind the layer.

References

[1] V. Slyusar, Electronics **3,** 44(2010)
[2] W. Cai, U. Chettiar, A. Kildishev, and V. Shalaev, Nat Photon **1,** 224(2007)
[3] Alù and N. Engheta, Phys. Rev. **78**, 1(2008)
[4] A. Klokov, V. Yakubov, S. Shipilov, A. Gorst A. Mironchev, Electronic means and control systems: Proceedings of the X International Scientific and Practical Conference **1,** 146 (2014)
[5] V. Yakubov, A. Andretsov, I. Ponomareva, A. Mironchev, News of Higher Educational Institutions Physics **53,** 17(2010)
[6] J. Pendry, A. Holden, D. Robbins, W. Stewart, IEEE Trans. Microwave Theory Tech. **47,** 2075 (1999)

MATEC Web of Conferences 155, 01054 (2018) https://doi.org/10.1051/matecconf/201815501054
IME&T 2017

SOLAR POWER INSTALLATION FOR THE TERRITORY OF SIBERIA

Arman Boyakhchyan*[1], *Alyona* Okhorzina[1], *Aygul* Kshalova[2]

[1]Tomsk Polytechnic University, 634050, Tomsk, Russia
[2]Karaganda State Technical University, 100027, Karaganda

Abstract. This article describes combined solar systems with liquid cooling are considered in this article. Solar power systems have long been popular in Europe, but in the Siberian climate these systems were considered to be inappropriate. The paper argues that such installations may be used even in Siberia, and in some regions it will be economically profitable. In this article describe an installation that was assembled to carry out field experiments.

1 Introduction

The territory of Siberia occupies 70% of the territory of Russia, which is home to 25 million people. The most of the population lives in remote rural areas, where there is no centralized power supply. Because In the agriculture of Siberia, large-scale work on livestock farming is carried out, this requires fuel and energy costs. The solution of the problem of ensuring the energy needs of the population, agriculture and small industry in regions remote from centralized energy networks, provided they are environmentally friendly, can be solved by introducing autonomous alternative energy systems [1].

Solar panels use sunlight, not heat and not even sunlight. Therefore, in the winter they will work quite well, provided that the owner removes snow from them, although they will also work under snow. And sometimes, the snow thaws from the work of photocells.

In cloudy weather, the panels will produce less light than usual, but in general, there are rarely cases when the battery does not have enough time to charge during the day. But in a sunny, frosty weather, the batteries will be very effective [2, 3].

In winter, the tilt angle (Figure 1) of solar panels is very important, as the sun goes lower and this also affects their productivity. The vertical position also helps the batteries not to be covered with snow.

* Corresponding author: bojahchyan@yandex.ru

Fig.1. Solar panels in Siberia.

The creation of an economical power plant that converts solar energy into useful (thermal and electrical) energy will solve the problems of energy supply in the vast territory of Siberia, and will reduce losses in agriculture [4].

To increase efficiency and economy, including the advisability of using solar power plants on the territory of Siberia, it is necessary to use concentrators, cooling systems and orientation in the sun [5].

2 Experimental verification

For research, an installation was assembled, which is a hybrid system that converts solar energy into electrical and thermal energy. Sunlight, incident on the photoelectric module, is converted into electrical energy. The production of heat is due to the cooling of photovoltaic modules. Increase in energy production is achieved by introducing a low concentrating concentrator. The concentrator is a reflecting surface consisting of several segments, by means of which the concentration of solar radiation rises 2–3 times.

The incoming radiation flux falls on the concentrator and is reflected in the FM. The plane of the solar module is located perpendicular to the flux of reflected radiation. For work in the cold season, you can use an additional cover, which will also protect the concentrator and photovoltaic modules from atmospheric precipitation and reduce the outflow of heat to the environment.

To increase the power of the system used to track the sun.

To increase energy production without losses, it is necessary to introduce a cooling system that maintains the panel temperature below 40 ° C. At the same time, if possible, the system will circulate water in the system to ensure its heating.A laboratory experiment using a point heating source to test how the surface of the solar panel heats up and how heat dissipates from it.

The halogen spotlight NFL-FH1-150 W was used as a source of light and heat. During the heating process, the thermograms were recorded using the Testo 876 thermal imager. The experiment was carried out until the maximum heating of the solar panel surface was obtained (Figure 2).

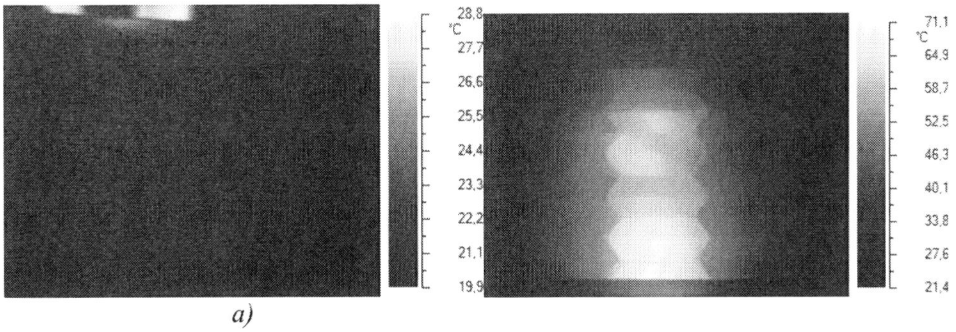

Fig.2. Thermograms: a) before heating; b) after the heating is turned on.

The real experiment was conducted in Tomsk. In the process of heat dissipation, tap water with a temperature of 150 ° C was applied to the heat exchange system. The system took about 20 minutes, so that the temperature at the maximum heating point decreased by 40 degrees.

A plant was assembled using a concentrating system. At the first stage, the heating time of the entire system was monitored on a sunny, clear day. At the second stage, the cooling system was connected, and observations were made for the cooling quality of the system Figure 3,4.

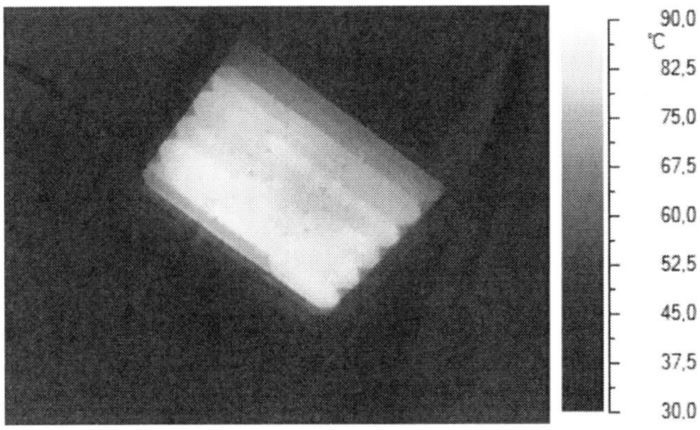

Fig.3. Thermogram during heating.

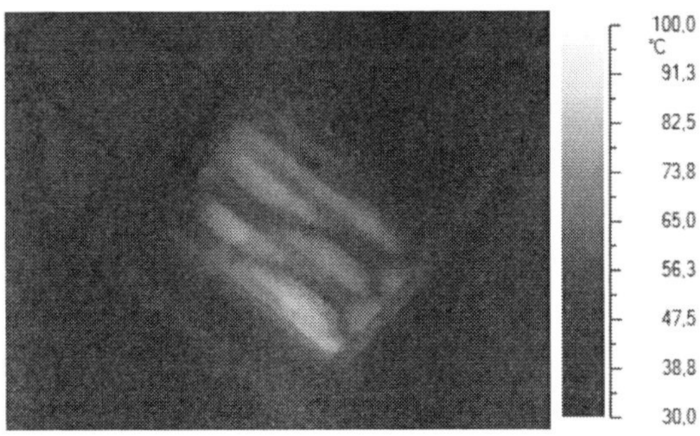

Fig.4. Thermogram with cooling.

3 Results of research

As a result of the full-scale experiment, it was revealed that this system works. Heating of the solar panel surface with concentrating panels occurs quickly. Reaching a dangerous temperature for solar panel at 90 ° C occurs within 10 minutes. The cooling system copes with its task.

Table 1. Experimental data

12-th of September	*V*=3.5 liters		*Sr*=560	*tamb*=26.5
time	*U*, B	*I*, A	*tpv*, ^{0}C	*tв*, ^{0}C
14:15	19.14	0.67	48	
14:21	18.15	1.23	44.9	
14:29			47	13
14:30			48.3	
14:48	17.97	1.87	45	40
14:55			48	15
15:01	18.17	0.52	43.3	
15:15	17.98	0.23	41.2	34
15-th of September	*V*=10 liters		*Sr*=550	*tamb*=17.5
time	*U*, B	*I*, A	*tpv*, ^{0}C	*tв*, ^{0}C
13:48	16.76	0.05	28.1	16
13:56	20	1.18	48.5	16
14:15	17.6	0.97	56.4	16
14:29	18.9	0.53	44.2	22
14:40	19.1	0.46	38.4	25.5
14:55	19.1	0.65	41.8	28
15:06	18.97	0.48	41.6	29
15:17	19	0.61	42.7	31
15:25	18.81	0.4	44	31

4 Conclusion

Experiments have shown that a combined solar installation allows the conversion of solar energy not only into electrical energy, but also to extract the useful thermal energy that is used to heat water. This installation should be used to increase the efficiency and profitability of using solar energy to generate useful energy.

References

[1] A.V. Yurchenko, F.V. Savrasov, V.I. Yurchenko, Izvestiya Tomsk Polytechnic University **314**, 43 (2009)

[2] N.N. Bakin, V.K. Kovalevsky, A.P. Plotnikov, A.V. Yurchenko, Optics of the atmosphere and ocean **11**,1337 (1998)

[3] A. Yurchenko, V. Syriamkin, A. Okhorzina, N. Kurkan, IOP Conf. Ser.: Mater. Sci. Eng. **81**, 012097 (2015)

[4] A. Yurchenko, M. Kitaeva, A. Okhorzina, Proceedings of 7th International Forum on strategic Technology, IFOST 2015, (2018)

[5] A.V. Yurchenko, M.V. Kitaeva, A.V. Skorokhodov, A.V. Okhorzina, Polzunovskii vestnik, 213 (2012)

SITUATIONS IN CONSTRUCTION OF 3D MAPPING FOR SLAM

Trang Nguyen Hoang Thuy[1,2*], *Stanislav* Shydlouski[2]

[1]National Research Tomsk State University, 634050, Tomsk, Russia
[2]Vietnam Maritime University, 484 Lach Tray, Le Chan, Hai Phong, Viet Nam

Abstract. Nowadays, the simultaneous localization and mapping (SLAM) approach has become one of the most advanced engineering methods used for mobile robots to build maps in unknown or inaccessible spaces. Update maps before a certain area while tracking current location and distance. The motivation behind writing this paper is mainly to help us better understand about SLAM and the study situation of SLAM in the world today. Through this, we find the optimal algorithm for moving robots in three dimensions.

1 Introduction

Along with the development of science and technology, in robotic mapping and navigation has been studied and more interested by scientists. For the purpose of fixing and improving positioning in short distances, simultaneous localization and mapping (SLAM) has marked significant advances in the development of science and technology. Previously, thanks to GPS technology (Global Positioning System) it is easy to pinpoint directions or give a route for the journey, now SLAM technology is also moving closer to human life with certain uses. It helps people to locate oneself and create a model of your surroundings, which is important in many applications. Thanks to the simultaneous localization and mapping (SLAM) technology, positioning approach commonly used indoors or underground where satellite navigation (GPS) is not viable because the enclosed space and the distance are too small under 6 meters. Research on SLAM has been started in the robotics community since 1986 in the papers of Smith and Cheeseman, sually with wheeled robots traversing a flat ground plane. Typically this was done by combining sensor readings (such as from a laser scanner) information about the control input (eg, steering angle) and the measured robot state (eg. counting wheel rotations). This could be a core difficulty of SLAM, such as creating a consistent and accurate map, making the best use of multiple unreliable sources of information [1]. Popular SLAM applications are home robots, unmanned aircraft, self-driving cars.

* Corresponding author: trangnht@vimaru.eud.vn

2 The concept of SLAM, localization and mapping

SLAM estimates the pose of a robot and the map of the environment at the same time. SLAM is hard by two main reasons. The first, a map is needed for localization (inferring location given a map), the second a good pose estimate is needed for mapping (inferring a map given locations).

SLAM: learning a map and locating the robot simultaneously [2].

SLAM technology was born in the robotics industry and is used by autonomous vehicles to concurrently map and navigate through an unknown environment. To do this, SLAM algorithms utilise information from sensors (often Lidar or imagery) to compute a best estimate of the device's location and a map of the environment around it.

SLAM algorithms are tailored to the available resources, hence not aimed at perfection, but at operational compliance. Published approaches are employed in self-driving cars, unmanned aerial vehicles, autonomous underwater vehicles, planetary rovers, newly emerging domestic robots and even inside the human body [3].

3 So what are the benefits of Geospatial SLAM?

The main advantage of Geospatial SLAM for 3D mapping applications is that scanning can be undertaken whilst mobile, and without the need for GPS. This facilitates rapid and accurate 3D acquisition of complex environments and is particularly effective for indoor mapping or surveys of enclosed environments that would not be possible using current GPS based mobile mapping solutions. When compared to traditional survey methods for measuring indoors, such as tape and Disto, or even static laser scanning, mobile indoor mapping can prove to be up to 10 times faster [4].

From 2014, Google has worked on mapping systems for in-house 3D mapping. This system includes both proprietary 3D hardware and software that processes the sensor data into the map. In 2016 Google released the Lidar SLAM real-time library. The SLAM algorithm uses LIDAR and IMU data to simultaneously identify the location of the sensor and create a tight map around it. But the solution will also have strengths and weaknesses so the company has plans to add new solutions to overcome the shortage in the existing SLAM algorithm that is the solution "Lifelong mapping '. By updating the SLAM algorithm so that an autonomous device can estimate the likelihood that a feature in its environment is still static or moving [5].

3D Mapping is a technology that allows you to change the visual characteristics of an object using a video projection and is able to turn almost any surface into a dynamic video screen.

SLAM is concerned with the problem of building a map of an unknown environment by a mobile robot while at the same time navigating the environment using the map.

SLAM consists of multiple parts; Landmark extraction, data association, state estimation, state update and landmark update.

SLAM is a key component of any truly autonomous robot. Much recent research has been done tackling the computational efficiency of SLAM and the data association and landmark extraction necessary for a robust SLAM method.

Environmental mapping involves creating a mathematical model of a real environment's spatial information. SLAM extends the requirements of this mathematical model; it must also jointly represent the robot's state and the position of extracted landmarks relative to the robot's location. The robot's state includes information on the robot's position and orientation. The basic SLAM framework involves odometry, landmark prediction, landmark extraction, data as sociation and matching, pose estimation, and map update. These processes are the backbone of every major SLAM method, and are performed in cyclic fashion.

SLAM algorithms must take account of a variety of parameters i.e. sensors, map representation, robot dynamics, environmental dynamics, and the integration of sensor measurements and the robot's control system over time.

The integration of these diverse parameters is most often accomplished using the two major SLAM-related algorithms. Kalman filters are routinely used, with the most popular being the Extended Kalman Filter (EKF) [6 –8].

Rao-Blackwellized Particle Filters (RBPF) [9, 10] have also been used in many SLAM methods.

Mapping the spatial information of an environment requires spatial sensors to which SLAM algorithms can be applied. The two most popular sensor modalities used in SLAM are raw range scan sensors and feature (landmark) based sensors (whether extracted from scans or images) [11, 12].

4 SLAM Problems

SLAM is a chicken-or-egg problem: i) a map is needed for localization. ii) a pose estimate is needed for mapping.

SLAM is currently one of the most widely studied, most interested in robotics. The problem of SLAM is clarified as follows. Consider a mobile robot consisting of two wheels connected to the engine, containing a control processor and a camera attached to it. Through the camera, the programmer can see objects around the robot, taking measurements of the distance from the robot to the surrounding objects, and processing the color in three dimensions when identifying the object. Know where there are objects, where is the way. By identifying that, the programmer will program a road map for the robot in three dimensions so that it can avoid the surrounding objects and follow the directions that have been driven so that the path is Robotic programming is optimal Figure 1. This is a concrete example of SLAM (Localization and Concurrent Planning).

Fig. 1. The robot moves in the gazebo environment.

The environment in which the robot moves is the indoor environment so it will not have the same structure as the pre-loaded robot environment. Robots will be tested in different locations to determine the best sensor (Figure 2). The robot recognizes the environment and the programmer creates the most accurate map so that it can move itself in the environment, avoiding the surrounding objects and that will be the optimal path for it.

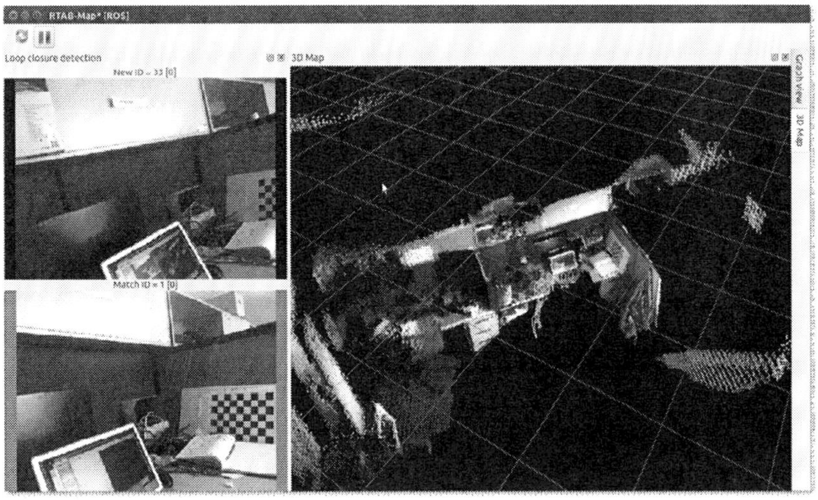

Fig. 2. Robot identifies the environment and creates maps in Rtabmap.

5 SLAM Applications

SLAM is central to a range of indoor, outdoor, in-air and underwater applications for both manned and autonomous vehicles (Figure 3). Examples: i) At home: vacuum cleaner, lawn mower; ii) Air: surveillance with unmanned air vehicles; iii) Underwater: reef monitoring; iv) Underground: exploration of mines; v) Space: terrain mapping for localization.

Fig. 3. SLAM technology in the industry.

Autonomous automobiles are now one of the applications of great interest and research. By applying a variety of techniques to detect ambient conditions such as radar, laser light, opacity and computer vision, especially localization and mapping algorithms (SLAM). Self-driving cars combine data from multiple sensors and off-line maps into current location estimates and map updates (Figure 4). It has simply moved itself with locating the system, real-time positioning (RTLS) and avoids obstructions.

Fig. 4. Self-driving cars and unmanned aircrafts.

This is the topic of simultaneous localization and mapping within the human body (body-SLAM) - Guanqun Bao - in 2014. A Body-SLAM technique that integrates motion information extracted from the video source and RF signal emitted by the capsule to enhance to the localization accuracy of the WCE and meanwhile reconstruct the path the capsule has traveled. The major contribution of this work is that we demonstrated the potential of using endoscopic images to aid the RF localization and possibility to map inside of human body (Figure 5) [13].

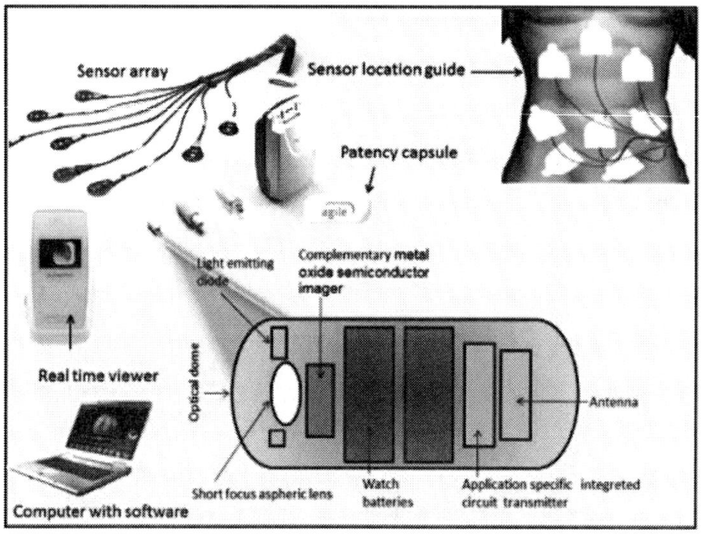

Fig. 5. Typical RF localization system.

Besides, now there are a lot of studies in SLAM:

- 2005 DARPA Grand Challenge winner STANLEY performed SLAM as part of its autonomous driving system.
- SLAM Simultaneous Localisation and Mapping at the Level of Objects [14].
- In 2009: Rafael Valencia, Ernesto H. Teniente, Eduard Trulls, and Juan Andrade-Cetto – 3D mapping for urban service robots [15].
- In 2012 Miguel Armando Migueis Pinto: SLAM for 3D Map Building to be used in a Matching Localization Algorithm [16].
- 3D Mapping using Real Earth Stencil (SLAM).
- 3D Mapping with Multimodal Plane SL.

6 Conclusion

The problem of simultaneous localization and mapping has seen great progress over the last 30 years. SLAM and related technologies are increasingly deployed in a variety of settings and practical applications. Slam technology will increasingly be relied upon to provide reliable metric positioning in situations where infrastructure-based solutions such as GPS are unavailable or fail to provide accurate, complete data. SLAM still constitutes an indispensable backbone for most robotics applications and despite the amazing progress over the past decades, existing SLAM systems are far from providing insightful, actionable, and compact models of the environment, comparable to the ones effortlessly created and used by humans.

References

[1] Osian Haines, *An Introduction to Simultaneous Localisation and Mapping*, URL https://www.kudan.eu/kudan-news/an-introduction-to-slam/ (2016).
[2] Wolfram Burgard, Cyrill Stachniss, Kai Arras, Maren Bennewitz, *Introduction to Mobile Robotics, SLAM: Simultaneous Localization and Mapping*, University of Freiburg (2012)
[3] P. Mountney, D. Stoyanov, A. Davison, A. Yang, *Simultaneous Stereoscope Localization and Soft-Tissue Mapping for Minimal Invasive Surgery*, MICCAI. Lecture Notes in Computer Science. ISBN 978-3-540-44707-8, **347-354** (2006)
[4] John Allan, Mark Reid, Ric Durrant, *Geospatial SLAM*, URL https://geoslam.com/slam (2012)
[5] Sean Higgins, *Google Releases LiDAR SLAM Algorithms, Teases Innovative Mapping Solution*, URL https://www.spar3d.com/news/software/google-releases-lidar-slam-algorithms-teases-innovative-mapping-solution/, Spar 3D editor, (2016)
[6] H. Wang, J. Wang, Le Yu, L. Zhen-ye, *A new SLAM method based on SVM-AEKF for AUV*. In OCEANS, **1-6**. IEEE (2011)
[7] J. Il-Kyun, S. Lacroix, *High resolution terrain mapping using low attitude aerial stereo imagery*, in Computer Vision, proceedings. 9th IEEE Conf., **946-951**, (2003)
[8] T. Suzuki, Y. Amano, T. Hashizume, *Development of a sift based monocular EKF-SLAM algorithm for a small unmanned aerial vehicle*, in SICE Annual Conf., Proceedings of IEEE, **1656-1659** (2011)
[9] M. Montemerlo, S. Thrun, D. Koller, B. Wegbreit, in AAAI/IAAI, 593 (2002)
[10] X. Li, W. Cui, J. S.Jia, EEE Conf., 779 (2010)
[11] H. Lategahn, A. Geiger, and B. Kitt, IEEE Conf., 1732 (2011)
[12] T.J. Chong, X.J. Tang, C.H. Leng, M. Yogeswaran, O.E. Ng, Y.Z. Chong, *Sensor Technologies and Simultaneous Localization and Mapping (SLAM)* (Universiti Tunku Abdul Rahman, Malaysia, 2015)
[13] Guanqun Bao, *On Simultaneous Localization and Mapping inside the Human Body (Body-SLAM)*, (2014)
[14] R.F. Salas-Moreno, R.A. Newcombe, H. Strasdat, P.H.J. Kelly, A.J. Davison, *SLAM++: Simultaneous Localisation and Mapping at the Level of Objects* (Imperial College London, University of Washington, 2013)
[15] R. Valencia, E.H. Teniente, E. Trulls, J.A. Cetto, *3D Mapping for Urban Service Robots*, (2009)
[16] M.A.M. Pinto, *SLAM for 3D Map Building to be used in a Matching Localization Algorithm*, (2012).

AUTHOR INDEX

AUTHOR INDEX

EDP Sciences – Web of Conferences
17, Avenue du Hoggar
Parc d'Activité de Courtabouf
BP 112
F-91944 Les Ulis Cedex A
France

ISBN 978-1-5108-5911-1

9 781510 859111

CURRAN ASSOCIATES INC.
proceedings
.com